NUNC COGNOSCO EX PARTE

TRENT UNIVERSITY
LIBRARY

WOOL

ITS CHEMISTRY AND PHYSICS

WOOL

ITS CHEMISTRY AND PHYSICS

by

PETER ALEXANDER *and* ROBERT F. HUDSON

SECOND EDITION

by

CHRISTOPHER EARLAND

PH.D., M.SC., A.R.C.S.

*Reader in the Department of Textile Industries,
Bradford Institute of Technology*

LONDON

CHAPMAN & HALL LTD

37 ESSEX STREET WC2

*© Peter Alexander, Robert F. Hudson
and Christopher Earland, 1963*

First published 1954
Second edition 1963

Catalogue No. 521/4

Printed in Great Britain by
J. W. Arrowsmith Ltd., Bristol

PREFACE TO THE SECOND EDITION

DURING the past ten years the considerable output of published work and the holding of three international conferences have clarified many aspects of the physics and chemistry of wool which were obscure when the first edition was published. Although revision was therefore necessary if this monograph is to retain its usefulness both for students and research workers, the original authors have not been working in the field of wool research for several years and consequently they did not feel able to carry out such a revision. I therefore undertook the task of preparing this second edition.

At the molecular level a helical configuration is currently accepted for the folded main protein chains of wool, and morphologically there is now little doubt that the controversial 'sub-cuticle' membrane is, in fact, a membrane surrounding the cortical cells. The advent of quantitative ion-exchange chromatography has resulted in the publication of many detailed amino-acid analyses for wool, although these in turn have raised the question of its constancy of composition. In addition, there is now a much clearer understanding of many of the chemical reactions undergone by wool, such as those involving its oxidation and the formation of lanthionine. These advances are reflected in the incorporation of considerable new material in Chapters 1, 8, 9, 10, 11 and 12 and at the same time the opportunity has been taken of carefully scrutinizing for errors and of bringing up to date the other chapters.

C. EARLAND

BRADFORD,
1963

AUTHORS' PREFACE TO FIRST EDITION

THE decision to write this book was taken after the authors had spent seven years of intensive research on a variety of scientific problems confronting the wool industry. In spite of the tremendous advances in our understanding of the physics and chemistry of wool fibres which have been made in the last two decades through the application of refined techniques, there exists no comprehensive monograph covering this subject. In view of the great diversity of the sources of information which include the journals of chemistry, physics and biology and the applied journals of the textile industry, there is a great need for a volume of these dimensions which we hope the present book will fill.

This book is intended for three classes of reader: firstly, workers directly engaged in research on protein fibres, particularly the younger scientists, who wish to gain a detailed knowledge of scientific researches on wool, to enable them to place their own specific problems in true perspective and possibly to draw ideas from an adjacent subject. Secondly, workers in all branches of protein chemistry may find this account valuable since wool has probably been studied more widely than any other protein. Many of the fundamental concepts of protein chemistry in general arose originally from researches on wool; thus the chemistry of the disulphide cross-link is based almost entirely on the reactions of keratin. The discovery that protein chains are folded in a regular way resulted from X-ray studies of wool fibres and human hair. Even physical theories and processes such as the formation of the salt link, the statistical theory of multimolecular sorption and diffusion in swelling media, were first studied and developed by scientists working on wool fibres before being extended to other systems. Much of this work is reported in highly specialized and technical journals which are not normally consulted by scientists working on general protein chemistry.

Lastly, we hope that this book will help the practical worker in the industry to utilize directly the enormous volume of data which has been accumulated and to assist him to develop new processes on a more theoretical basis than hitherto. To facilitate this use of the book, general explanations on an elementary level always precede the introduction of advanced ideas and developments. This method of treatment makes rather sharp changes in standard occasionally necessary and we hope the reader will not find this too disconcerting. No description of industrial processes has been given but an attempt has been made to indicate the practical significance of the various physical and chemical aspects.

At the present time many of the theories developed have not yet been fully established and some are the subject of considerable controversy. Although the authors have tried to present an impartial account of the most important opposing views they have not hesitated to indicate which, in their opinion, is the most acceptable at the present time. This has been done so that a sense of completion could be attained and does not imply a rigidity of viewpoint. The authors fully appreciate that in the course of time as new experimental data becomes available many of the theories which seem most plausible now will have to be modified or abandoned.

Diffusion processes and the related subject of molecular adsorption have been dealt with in considerable detail and the treatment given is general and applicable to many systems other than wool. This apparent digression is justified, in the view of the authors, by the fact that these subjects have advanced exceptionally rapidly in the last years and the application of this new knowledge to problems of the textile industry may bring about results of the greatest value.

The authors wish to stress that no attempt has been made to present a survey of the historical development of the wide range of subjects covered, or to provide a complete bibliography, which would have enlarged this book to quite an unreasonable size. Although an attempt has been made to provide accurate acknowledgements to the earlier works in each field, this may not always have been successful in view of the confusion prevailing in the literature. In conclusion the authors hope that by placing the emphasis on the more controversial subjects this book will act as a stimulus to research workers studying wool as well as to those interested in other fibres, particularly those of biological origin which must be of fundamental significance in the operation of living systems.

PETER ALEXANDER
ROBERT F. HUDSON

LONDON,
1952.

CONTENTS

	Page
Author's Preface to Second Edition	v
Authors' Preface to First Edition	vi

Chapter

1 The Morphological Structure — 1

The medulla. The cortex. The cuticle. The epicuticle. The cortical cell membranes.

2 Surface Properties — 15

Surface area. *Permeability method. Surface area by adsorption.* Electrical properties. Frictional properties. *Experimental methods. Results of frictional measurements. Theories of the directional friction effect.* The mechanism of felting. Polymer deposition.

3 Mechanical Properties — 55

The elasticity of wool fibres. The contribution of different bonds to the strength of the fibre. Entropy and free energy changes during elastic deformation. Permanent set. Supercontraction.

4 Sorption and Swelling — 86

Density and swelling. Sorption in pores. Sorption isotherms. Theories of sorption of water vapour. *Multimolecular isotherms. Solution theory.* Influence of swelling on absorption. Hydrate formation. Heat of sorption. Influence of swelling on electrical properties.

5 Rate Processes within the Fibre — 128

Penetration of neutral molecules into dry wool. Diffusion in a steady state. Diffusion in a non-steady state with constant diffusion coefficient. *Diffusion from a variable surface concentration.* Variable diffusion coefficient. *Measurement from distribution of solute. Measurement from rate of absorption.* The mechanism of the diffusion process. Diffusion in a swelling medium. *Intrinsic diffusion coefficients.* Rate processes involving the fully swollen fibre. *Diffusion across a liquid layer at the surface. Diffusion through the fibre. Diffusion from a constant surface concentration. Diffusion from a varying surface concentration. Diffusion accompanied by reaction within the fibre. Diffusion accompanied by absorption. The effect of temperature on the fibre diffusion process.* Rate of dyeing. The influence of the morphological structure.

CONTENTS

6 ACID-BASE CHARACTERISTICS — 180

Titration data. *The effect of temperature.* Difference in behaviour between soluble and insoluble proteins. *Influence of the electrical double layer.* Quantitative theories of acid combination. *Theory of Steinhardt and Harris. Theory of Gilbert and Rideal. The Donnan theory.* The absorption of weak acids. The titration of wool with alkali. Influence of the "salt link" on elasticity.

7 ION EXCHANGE AND DYEING EQUILIBRIA — 222

Evidence for anion affinity. Determination of anion affinity. Nature of anion affinity. Anion exchange and dying equilibria. Dyeing. The distribution of dye within the fibre. Neutral and basic dyeing.

8 THE DISULPHIDE BOND — 243

Reduction of the disulphide bond. Hydrolysis in neutral solutions. Reaction with alkalis. Reaction with peroxides. Photochemical oxidation. Oxidation with per-acids. Reaction with halogens. Relationship between disulphide-bond breakdown and the production of non-felting wool.

9 CHEMICAL REACTIVITY (excluding the disulphide bond) — 288

Acid hydrolysis. Alkaline hydrolysis. Enzymatic hydrolysis. The influence of heat. Ionizing radiations. The reactivity of tyrosine. Reactivity of the carboxyl groups. Reactivity of the amino groups. Reactivity of the thiol groups. Reaction with concentrated acids. Reactions with miscellaneous reagents.

10 FORMATION OF NEW CROSS-LINKS — 324

Formaldehyde. Mercury salts. Benzoquinone. Poly-functional alkylating agents. Dinitrofluorobenzene derivatives. Miscellaneous compounds. Rebuilding reduced disulphide bonds. The formation of polymers within the fibre.

11 CHEMICAL COMPOSITION — 346

Amino-acid composition. Terminal amino acids. The order of the amino acids along the peptide chain. Variation of the composition of morphological components. Fractionation into different polypeptides. The relationship between the different fractions.

12 STEREO-CHEMISTRY AND MACROMOLECULAR STRUCTURE — 376

The structure of β-keratin. The nature of the α-fold. The stability of the α fold in wool. Micellar structure. Chemical evidence for a three-phase composition. The physical properties of the micelle. Small-angle X-ray diffraction patterns.

AUTHOR INDEX — 399

SUBJECT INDEX — 411

ILLUSTRATIONS

NOTE

All figures are inserted in the text *except* the following, which are printed as separate plates:

Figs. 1.2–1.15 *between pages* 4 *and* 5

Figs. 11.2–11.3 *between pages* 370 *and* 371

Figs. 12.1–12.4 *between pages* 372 *and* 373

Figs. 12.9–12.14 *between pages* 388 *and* 389

CHAPTER 1

The Morphological Structure

ALL mammalian hairs belong to the same family of proteins, the keratins, and are closely related in chemical structure to all types of epithelial cell such as horn, skin and quills of feathers. Wool used in the textile industry ranges in diameter from 18 μ to 40 μ and Table 1.1 shows an approximate correlation between the diameter of

TABLE 1.1

Fibre diameter—top qualities

Quality	Average diameter in Microns (10^{-4} cm.)	Approximate maximum and minimum values of the diameter
80's	18·8	–19·25
70's	19·7	19·25–20·20
64's	20·7	20·20–22·00
60's	23·3	22·00–24·12
58's	24·9	24·12–25·65
56's	26·4	25·65–28·45
50's	30·5	28·45–31·55
48's	32·6	31·55–33·30
46's	34·0	33·30–35·10
44's	36·2	35·10–37·45
40's	38·7	37·45–39·20
36's	39·7	39·20–

the fibres and the quality as assessed by the industry. The staple length of fibres when removed from sheep varies considerably, and depends on a large number of factors, such as the breed of sheep and the position of the fibres on the skins.

The morphological structure of the wool fibre is extremely complex, and has been the subject of considerable controversy, but during the past few years the position has become much clearer. On the animal, the wool contains a large amount of grease, sometimes exceeding in weight that of the protein fibre itself. The grease is difficult to remove completely and is of complex composition which has not yet been fully determined. It is possible that a very thin film on the outside of the fibre may not be made of protein but since this is unlikely to exceed more than 0·1 per cent. of the fibre as a whole, no serious error is introduced if wool is considered as protein substance entirely. Each hair originates from a single follicle in the skin (see Fig. 1.1), and the detailed histology of the process

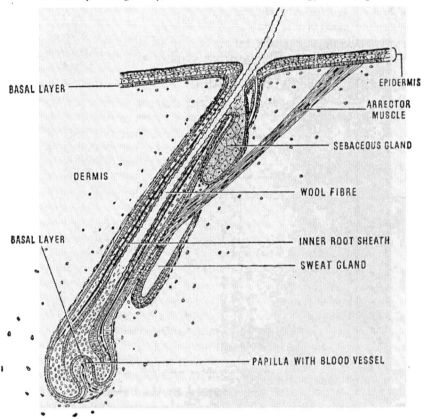

Fig. 1.1.—Cross-section through sheep skin showing development of wool fibre. (*Reproduced from 'Wool Science Review' by kind permission of the International Wool Secretariat.*)

has been investigated thoroughly, but falls outside the scope of this book. A considerable amount of information is available concerning the biochemistry of hair growth, and the influence of nutritional factors on the quality of the wool produced has been studied in

THE MORPHOLOGICAL STRUCTURE 3

detail and is of great practical importance.[1] Unlike silk threads, animal hairs are cellular structures, consisting of three distinct regions, the cuticle, the cortex and the medulla, all of which are made up of different kinds of cell. In a longitudinal section of a hair root, the development of the cortical cells can readily be followed. They arise as rounded close-packed elements at the base of the root and move along a shaft of the hair follicle where they elongate, become fibrous in appearance and keratinize, i.e. harden to produce the typical spindle-shaped cortical cells which can be isolated from the wool fibre.[2, 9] In theory, therefore, a wool fibre consists of two parts, that outside the skin and the part within the hair follicle undergoing the final synthetic processes. For all practical purposes, however, only the fully hardened part outside the skin is considered as wool and it is this material which forms the subject of this book.

The medulla

The relative proportions of the three components, i.e. cuticle or scales, cortex and medulla, vary widely from one type of hair to another and most coarse hairs contain a high proportion of medulla, so much so that in the early histological terminology the cortex was considered to form merely a tough film which surrounded the medulla and separated it from the scales. In the better types of wool it is completely absent and the cortex forms 90 per cent. of the fibre mass. The coarser types of wool and in particular the so-called kemp wool contain up to 15 per cent. of medulla cells which are many-sided in shape and quite distinct from the cortical and cortex cells in appearance. In general it can be said that the absence of a medulla distinguishes fine wools from all other hairs, and consequently it will not be considered further. Very little information is available concerning the chemical composition of this type of cell which is readily digested by enzymes which do not attack the cortex of unmodified wool.[3] At one time it was thought that it contained no cystine but this is now known not to be the case although in general the cystine content of the medulla is less than that of the fibre as a whole.[4] Figure 1.2 shows a cross-section of a non-medullated wool fibre and gives a good impression of the great complexity of structure, which is clearly revealed in the schematic drawing. Figure 1.3 is based on a most painstaking optical microscopic investigation by Reumuth.[5] It can be considered as a true representation which has not been modified significantly by a very large number of detailed electron-microscopic studies.

The cortex

As already mentioned the cortical cells comprise more than 90 per cent. of the total fibre mass of most wool fibres, and consist of longitudinal units which are of the order of 100 μ long and 4 μ wide (see Fig. 1.4). It has been assumed for a considerable time that the different morphological components are held together by a cement.[6] Later investigations, however, cast doubt on this assumption and it seems likely, though direct experimental evidence is lacking, that the fibres are held together as illustrated in Fig. 1.5 which shows connection of cortical cells by protruding fibrils.[7] Woods[8] isolated the cortical cells from wool and these could be made into a coherent film by orientating them. This was done by subjecting an aqueous suspension to an alternating current field in which they became aligned (see Chapter 2). A film of cortical cells showed many of the physical properties of the whole fibre, such as super-contraction, permanent set and reversible contraction giving rise to an α—β transformation in the X-ray diagrams. There can be little doubt, therefore, that all these properties reside within the cortical cells themselves and are not produced by rearrangements and reactions between the cells. The cortical cells are made up of fibrils which can be seen in the photo-micrograph, but more easily in the electron photo-micrographs shown in Fig. 1.6. The separation of the cortical cells into fibrils was first achieved by Hock, Ramsay and Harris[9, 10] by micro-manipulation under the optical microscope. These fibrils are in their turn, made up of still smaller units which have been variously termed sub-fibrils, micro-fibrils and proto-fibrils. Mercer and Rees[11] have shown that it is these sub-fibrils which are the smallest structural units, and suggest that they are held together within the actual fibril by a cement which can be digested, even if only slowly, by enzymes. As already mentioned, the idea of a cement between any part of the different morphological components is open to doubt, and it seems possible that an arrangement similar to that shown in Fig. 1.5 is also responsible for the cohesion of the micro-fibrils. In general, the division of the cortex into its component cortical cells and smaller units is achieved by enzymatic degradation. Although enzymes can only digest a very small portion of the fibre

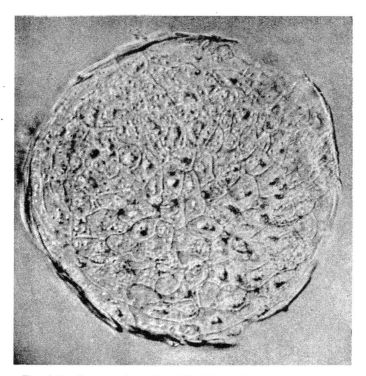

Fig. 1.2.—Cross-section of wool fibre[5] (magnification 750 times).

[Facing page 4

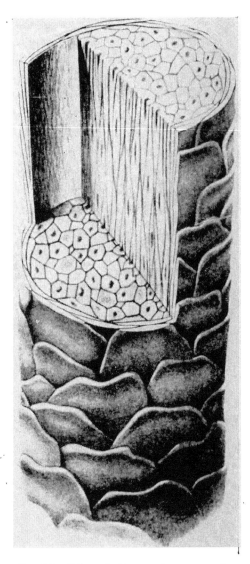

FIG. 1.3.—Schematic diagram of a wool fibre.[5]

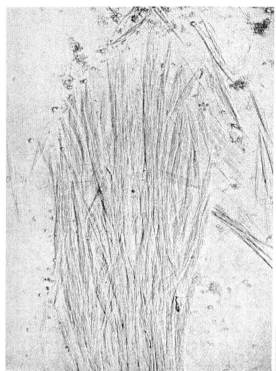

FIG. 1.4.—Spindle-shaped cortical cells showing subfibrils[5] (magnification 250 times).

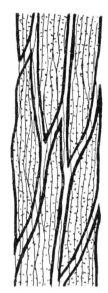

FIG. 1.5.—Diagram of fibrils interlocking cortical cells.[5]

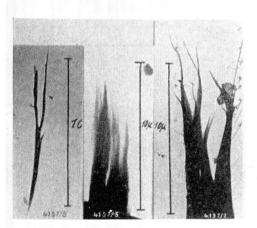

FIGS. 1.6 (a) and (b)—Electron-photomicrographs of cortical cell showing subfibrils. (*Kindly provided by Dr. H. Zahn.*)

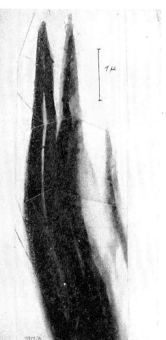

FIG. 1.6 (b).

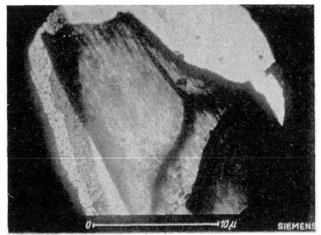

Fig. 1.7.—Aggregate of scales after enzygmatic separation. Electron-photomicrograph of similar cells. (*Kindly provided by Dr. H. Zahn.*)

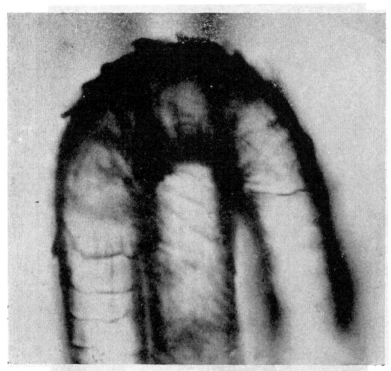

Fig. 1.8.—Knot tied in fibre showing ratchet-like separation of scales. (*Kindly provided by Mr. H. E. Millson.*)

FIG. 1.9a (*above*).—Electron-photomicrograph (3,000 times) of Exocuticle from wool (i.e. K_2 phase of scales), obtained by treating the fibres 96 hours with 1-N. NaOH at 5° C. according to a method of Zahn, and afterwards suspended in water by means of ultrasonics 5 min. 22,000 C/S (input 800 W magnetostrictive). Unshadowed. (*Kindly provided by Prof. Gralén.*)

FIG. 1.9b (*below*).—Electron-photomicrograph (9,000 times) of Endocuticle from wool (i.e. K_1 phase of scales), obtained from strongly over-chlorinated wool, violently shaken with water. Big triangular piece: endocuticle, background: epicuticle. Shadowed with Au 4:1. (*Kindly provided by Prof. Gralén.*)

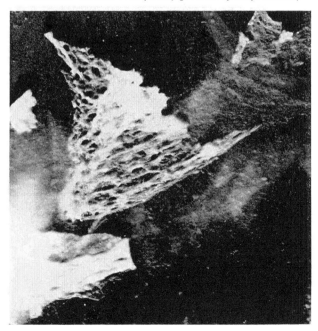

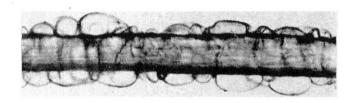

FIG. 1.10.—Wool fibre showing Allwörden sacs after treatment with chlorine. (*Kindly provided by Mr. H. E. Millson.*)

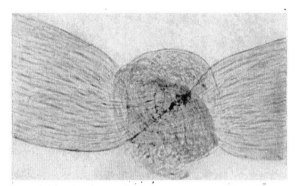

FIG. 1.12.—Cortical cell membranes isolated from oxidized wool by the method of Alexander and Earland.[45] The original fibre was knotted to demonstrate continuity of the membrane.

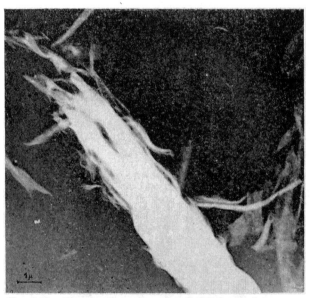

FIG. 1.13.—Electron-photomicrograph (6,500 times) of the material making up the resistant membrane shown in Fig. 1.12. The membrane was disintegrated by ultrasonics and shadowed with Pd. (*Kindly provided by Prof. Gralén.*)

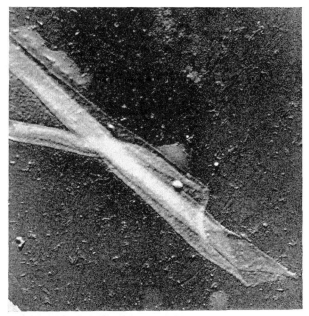

FIG. 1.11a.—Electron-photomicrograph (7,000 times) of Epicuticle from wool. The fibres have been chlorinated and the epicuticle loosened by shaking with water. Shadowed with MnAu. (*Kindly provided by Prof. Gralén.*)

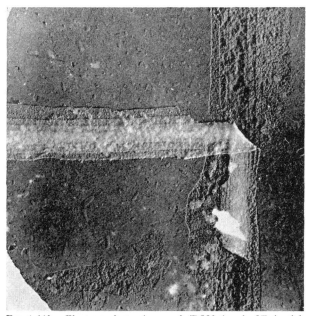

FIG. 1.11b.—Electron-photomicrograph (7,000 times) of Epicuticle from wool. Obtained as a residue when treating the fibres for several months with dilute Na_2S at room temperature. Collodion replica with a piece of the epicuticle stuck to the collodium Shadowed with Au. (*Kindly provided by Prof. Gralén.*)

Fig. 1.14.—Breaking up of the outer layers of human hair when these are stretched by more than 40%.[5]

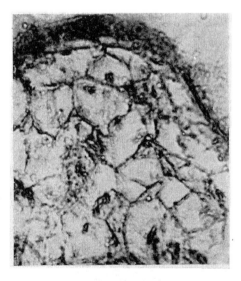

Fig. 1.15.—Photomicrograph of transverse section of peracetic acid treated Devon hog wool fibre after extraction with 2N sodium hydroxide solution, stained with methylene blue (×1000). (*Kindly provided by B. Manogue, M.S. Moss and R. L. Elliott.*)

(see Chapter 9) this limited attack is sufficient to enable the fibre to be split up into its morphological components. A similar separation can be achieved by the use of strong concentrated acids such as formic or sulphuric or by concentrated ammonia. Owing to the severe chemical degradation these methods have not been widely used. There is unfortunately no method available for breaking the wool into its component parts by purely mechanical degradation. Ultrasonic irradiation which is widely used for preparing specimens for the electron microscope is not effective in disintegrating wool unless this has received some prior enzymatic or chemical treatment. Farrant, Rees and Mercer[12] have obtained electron photomicrographs which indicate that the sub-fibrils are composed of aggregates of spherical or plate-like molecules with an average diameter of 100 Å. According to these workers, the fibrous keratins are made up like fibrin[13] or F-actin[14] by the lengthwise aggregation of globular protein molecules. As will be seen in Chapter 12, this concept can be used to develop a macromolecular structure for wool. It should be pointed out, however, that electron photomicrographs showing a corpuscular structure for wool have not been obtained by other workers, and the ideas of Farrant et al.[12] certainly cannot be accepted as established. It is quite possible that the globular appearance of the electron photomicrographs of Farrant et al. is due to faulty shadowing. At present the conventional idea of a structure consisting of long fibrous molecules is still generally accepted.

In 1953 an important observation on the fine structure of the wool fibre was made by Horio and Kondo,[15] who, by using a method of differential staining, showed that the wool fibre has a bilateral structure running the length of the fibre from root to tip. The structural asymmetry appears to be related to the natural crimp of the fibre and this and other aspects of the structure have been discussed by Mercer[16] who has pointed out that the apparent asymmetry of the cortex of crimped fibres was first implied in the work of Auber.[17] In the decade since its discovery, a very large amount of work has been performed on the dichotomous nature of the wool cortex and it is impossible to appraise the real position accurately. Thus Mercer et al.[18] found that the paracortex, the more stable segment, has a higher content of cystine than has the orthocortex; moreover the more reactive fraction of the cystine of keratin occurs predominately in the orthocortex. Fraser et al.[19] have indicated that there are differences in the contents of glutamic and aspartic acids that are related to the basophilic character of one segment as compared with the other. Mercer's explanation of the bilateral structure of wool fibres has been criticized by Barker[20] and Human and Speakman,[21] while Leveau[22] believes that the cystine contents of the two fractions are the same and that differences in reactivity may arise from steric factors.

The cuticle

The cuticle cells, generally known as scales, form the outermost layer of the individual hairs and vary in relative size and arrangement from one type of hair to the next. In all wool fibres, however, they are arranged in a way similar to that shown in the schematic diagram of Reumuth (see Fig. 1.3). The scales can be removed from the individual fibres mechanically by ball milling, grinding while frozen by liquid air, or scraping with a sharp edge such as a razor blade's, when products such as those shown in Fig. 1.7 are obtained. In general, enzymatic treatments provide the simplest method of separating the scales although in a degraded state.[9, 23]

The individual scales appear to be connected to one another but it is not known whether this results from an intercellular cement or the interlocking of fibrils which protrude from the individual cells.[7] The most characteristic feature of the scales is their overlapping arrangement and their general appearance is similar to that of tiles on a roof. In spite of the fact that when the scales are isolated mechanically they adhere together to some extent, there can be no doubt that they do not constitute a single membrane-like layer surrounding the whole of the fibre as has occasionally been suggested. Microscopical observation of the fibres which have been knotted[24] (see Fig. 1.8) clearly illustrates that the scales are separate and distinct cells.

The origin of the 10 per cent. of wool substance which goes into solution on treatment with enzymes is unknown, but there is some evidence to show that it comes in part at least from the scales, since these show a different electron-microscopic appearance when isolated by mechanical and by enzymatic methods.[25] Mercer and Rees[11] believe that the scales consist on the outside of an amorphous soft and enzymatically digested component, referred to as K1, and below this a more resistant layer, K2, which is grooved. On electron-microscopic examination the unaltered cell presents a smooth surface of uniform thickness. The scale remnants after enzymatic digestion possess an uneven and etched surface of parallel ridges (see Fig. 1.9). Moreover Zahn and Haselmann[7] were able to detect these ridges under the optical microscope from fibres which had been treated with formamide which removes the K1 phase, and they believe that these may be related to the so-called tono-fibrils which link the scales together. They further identify these ridges with fibrils similar to those occurring in the cortex. Mercer and Rees[11] have shown that after removal of the K1 phase the scales lose their sharpness of outline. The K1 phase becomes resistant to enzymes and other degrading agents by treatment with formaldehyde. It has been shown by Millson and Turl[16] that the scale tips or fringes of the scales are the first to be attacked by chemical reagents, which suggests that the actual edges are made up of the less-resistant K1 phase. It should, however, be borne in mind that the preferential attack at the edge may be due to physical factors. The K1 and K2 phases have also been given the names exo- and endo-cuticle respectively by Lindberg et al.,[26] but there seems to be little merit in introducing these new terms, since they give the impression of a continuity of these different layers over the whole of the fibre, which is not the case, and the original nomenclature introduced by Mercer and Rees[11] can usefully be retained. A very large number of electron-microscopic investigations of the scales of wool fibres have been carried out and it is difficult to decide priorities of discovery. It is satisfactory, however, that almost all the workers are in general agreement and that as a result the structures described can probably be accepted as essentially correct.

The epicuticle

As early as 1916 Allwörden[27] noticed that water bubbles are formed at the surface along the fibres when virgin wool is treated with solutions of chlorine and this reaction has subsequently been named after its discoverer (see Fig. 1.10). The rate of the appearance of the bubbles depends upon the time of treatment and concentration of the reagent[24] (see Table 1.2). At first this phenomenon was attributed to the swelling of a non-protein substance, called elasticum, somewhere within the wool fibre. This explanation, however, was effectively disproved by Müller,[28] who showed that the sacs which contained liquid were formed by a thin membrane situated on the surface of the scales. Each sac can be punctured with a microscope needle, thus allowing the liquid to flow out.[27] The actual bubble formation arises from the semi-permeable nature of the membrane which allows molecules of low molecular weight to penetrate readily but retains molecules of high molecular weight. Müller assumed that the chlorine reacted with protein in the scales below

TABLE 1.2

Time of appearance of Allwörden sacs after different chlorination treatments[24]

Sodium hypochlorite conc. g./l	pH	Time of treatment
1·0	4·5	30 to 45 seconds
2·0	4·2	15 seconds
0·2	4·1	12 minutes
2·0	8·3 to 8·8	12 minutes
0·2	8·5	sacs not observed

the membrane to give a soluble degradation product of high molecular weight which then caused the swelling under the semi-permeable membrane due to osmotic forces. The component of the scales which is solubilized by chlorine can probably be identified with the K1 phase. In general each sac or bubble can be seen to cover only one scale, but there are occasions when it appears as if a large sac covers several scales. Müller[28] in 1939 showed satisfactorily that a sac never covers more than one scale, but that a number of sacs may appear very close together, due to the overlapping nature of the scales, and become tightly compressed giving the appearance of covering more than one scale.

The presence of a very thin membrane on the surface of the fibres was indicated by an entirely different investigation in which Whewell and Woods[29] used an exceedingly sensitive staining technique for determining damage to wool fibres. They found that a virgin fibre did not stain with methylene blue, but that after the mildest mechanical damage, such as rubbing the fibre between fingers, it began to take up the dye. Most fibres which had undergone normal processing were stained with methylene blue indicating damage of the thin membrane. Ultimately, direct electron-microscopic evidence was produced by Lindberg, Philip and Gralén[30] who showed that fibres which had been dissolved in sodium sulphide left behind a very thin insoluble film in the solution which, because of its chemical inertness, is probably not made of protein. Direct evidence was obtained that this thin membrane was situated on the surface of the fibres by the fact that it could be obtained from chlorinated wool after washing and shaking in water. The chlorine attacks the K1 phase, loosens the top layer which then comes off and can be seen under the electron microscope (see Fig. 1.11). The Swedish workers called this membrane the epicuticle and believe that it is in general between 50–100 Å thick. Its chemical composition has not yet been determined since it is impossible to obtain specimens uncontaminated with other portions of the wool fibre in sufficiently large quantities for analysis.

Zahn[32] isolated the epicuticle from chlorinated wool by micromanipulation and the material he obtained appeared to consist mainly of protein, but the possibility that the preparation was contaminated with cuticular material or by protein degradation products has not been definitely excluded. The hydrolysate of a chemical extract from wool which was estimated to contain 30 per cent. of epicuticle was shown to contain sugars as well as amino acids.[33] This raises the possibility that the epicuticle may be a polysaccharide. Alexander[31] has suggested that by analogy with many other living systems, such as insects and fruit, all of which have an epicuticle of a fatty nature, that the epicuticle of wool is also made up of fats, possibly cross-linked in a three-dimensional network. On the other hand, Schuringa et al.[34] claim that the epicuticle is already in position on the outside of incompletely keratinized fibres close to the hair root; this they believe argues against its being fatty since the site of its formation makes it impossible to be a secretion from a sebaceous gland.

The Swedish workers believe that the epicuticle covers the whole of the fibre, but there is no evidence to support this suggestion, and it seems more probable that it forms the outermost layer of the individual scale cells. In no case has it been possible to isolate a fragment which is larger than an individual scale[32] and the work of Müller[28] on the Allwörden reaction does not support the suggestion that the thin membrane covers the whole of the fibre and suggests strongly that it covers one scale only.[9] Should this eventually be established the name epicuticle should be abandoned since this suggests a sheath surrounding the whole of the fibre.

Finally it should be emphasized that it is very difficult to remove the last traces of wool grease from a wool fibre and that small quantities of fat can be extracted continuously from a fibre even when this has already been Soxhlet extracted for many hours. It is well known that molecular films of grease are very powerfully held by surfaces and often cannot be removed with solvents in which they are freely soluble in bulk. If therefore the epicuticle should prove to be fatty in nature, it may be very difficult to decide whether it is the outermost layer of the scales or the final layer of wool grease.

Following on the work of Whewell and Woods[29] who found that the thin membrane on the surface of the scales greatly influenced the rate of diffusion of dye into the fibre, Millson and Turl[24] made a detailed study of dyeing from which they concluded that the epicuticle was capable of retarding the entry of dye into the fibre. Lindberg[35] reached a similar conclusion from studies of the rate of absorption of acid, and states that the extent of the damage to the epicuticle is important in determining the rate of diffusion into the fibre. This is discussed in greater detail in Chapter 5. Speakman[36] has pointed out that although the epicuticle may be important in controlling diffusion in completely undamaged wool fibres, there is no evidence to suggest that this is so in fibres which have undergone the normal technological processes such as spinning. The methylene-blue staining test reveals the extreme ease with which the epicuticle is removed or damaged, and shows that this membrane exerts only a small retardation into wool which has undergone even the mildest mechanical handling.

Similarly, Alexander, Gough and Hudson[37] found that the rate of diffusion of the hypochlorite ion into the fibre was not affected by treatment which removed the epicuticle and it was concluded that this did not influence the diffusion of these ions into the fibre. In this work the wool used had been handled by machinery in the spinning and knitting processes, and it is quite possible that the epicuticle was damaged extensively during these processes. Although the position is rather confused the epicuticle appears to be capable of preventing the diffusion of molecules into the fibre, although it is so easily damaged that it is unlikely to play any part in wool as normally handled.

Gralén[38] and his colleagues further believe that the epicuticle influences the frictional properties of wool fibres, but the evidence so far presented suggests that the epicuticle may be removed during the measurement and it seems unlikely that it plays a major part in the interesting frictional properties of wool which are of such great technological importance (see p. 25). One of the most elusive and yet most characteristic and important properties of wool is its feel or handle as assessed subjectively. So far it has not been possible to determine this property quantitatively and hence a scientific study has proved to be impossible. It has been proposed that the change in feel of a wool fibre in chemical reactions which are confined to the surface may result from the removal of the epicuticle and that the appearance of roughness or harshness may be brought about by uneven removal of the epicuticle along the fibre. In the absence of any direct evidence such a suggestion should be treated with reserve, especially since mechanical processes seldom bring about a marked deterioration in feel although they do disrupt the epicuticle.

The cortical cell membranes

Independently in 1937 and 1938 Reumuth[39] and Harrison[40] deduced from microscopic observations that a continuous membrane, which Reumuth termed the subcuticle, was accommodated between the scales and cortical cells of each wool fibre. In 1941 Lehmann[41] claimed to have isolated chemically such a membrane, which he renamed the epidermis membrane, by heating wool and hair fibres with formaldehyde at elevated temperatures, e.g. two hours at 140° C. The cortical and scale cells dissolved, leaving a thin film behind. Careful work by Mercer[42] and others with the electron microscope has since shown that the sheath left behind after such a reaction consists of epicuticle with adhering scales. Although appearing homogeneous in the optical microscope under ordinary illumination, phase contrast shows this preparation to be made up of scale cells, which confirms the electron-microscopic work.[42] Elöd and Zahn[43] also made a chemical preparation believing this to contain only the subcuticle which they termed the *Zwischenmembrane*, by supercontracting wool in phenol and then digesting it with trypsin to dissolve the disorientated keratin, leaving behind a thin membrane. This preparation was contaminated with scales and under the electron microscope was similar in appearance to the original preparation of Lehmann. Careful optical microscopic examination with phase contrast, showed the presence of a thin membrane in addition to the scales. It is possible that this membrane cannot be seen easily in the electron microscope because specimens for examination in this instrument have to be dispersed by ultrasonics before they transmit electrons, and that during this process the contaminating scales, but not the remaining membrane, may have been released. Zahn[44] has improved the enzymatic method of preparation by contracting the wool fibres in formamide, and by substituting pancreatin for trypsin as the proteolytic enzyme. In this way preparations of a membrane were made which did not contain any scales. This membrane, which is distinguished by its great chemical resistance, could also be isolated by treating wool with strong sodium hydroxide when the cortical cells and scale cells dissolve before the membrane. This method, however, is not suitable for studying the composition of the membrane in view of the extensive degradation which occurs. One of the simplest and most unambiguous methods of isolating this membrane consists of oxidizing all the disulphide bonds in the wool fibre with peracetic acid (see p. 266)[45] and then extracting with dilute alkali, e.g. 0·1N ammonia. During this process 90 per cent. of the wool fibre goes into solution and the material left behind (see p. 369) consists of a membrane virtually uncontaminated by scales, which was thought to be identical with the subcuticle predicted by Reumuth and Harrison and the *Zwischenmembrane* of Elöd and Zahn. Whatever the relationship between the different preparations may be, the mild peracetic method establishes unambiguously the existence of a relatively thick continuous membrane in the wool fibre. This membrane is of distinct chemical composition differing from that of the fibre as a whole.[45] Gralén *et al.*[46] have examined this membrane after disintegration by ultrasonics in the electron microscope and have seen the material shown in Fig. 1.13. On the basis of these pictures they claim that the membrane is made up of cells which are morphologically similar to the spindle cells. In the peracetic acid method of isolation the preparation is contaminated with portions of epicuticle which can be seen in the electron-microscopic photographs of the ultrasonically disintegrated membrane. As, however, the total amount of contamination due to the epicuticle is probably less than 2 per cent. of the whole of the preparation, it is unlikely that it influences the chemical and physical properties of the membrane isolated in this way. The fact that it can be isolated from wool, the epicuticle of which has been removed,[47] proves that the membrane obtained from oxidized wool bears no relation to the 'Lehmann' membrane which is now known to be an artifact.

All this work taken in conjunction leaves little doubt of the existence of a membrane, representing a substantial proportion, i.e.

8–10 per cent., of the whole of the fibre. Since it is made up in part of cells, Zahn and Haselmann[48] consider on morphological grounds that it forms part of the cortex and not the cuticle and have given it yet another name, namely cortex mantel or intermediate layer. From a chemical point of view, however, it is best to consider it as a separate component of wool and hair since (i) it can be isolated as a membrane by chemical treatments and retains its entity under conditions which disperse the rest of the fibre, (ii) its amino acid composition is different from that of the fibre as a whole, (iii) its crystalline structure, as isolated after oxidation with peracetic acid, is different from that of the remainder of the fibre (see p. 369).

Independently, Mercer[49] and Manogue and Moss[50] examined the membrane isolated by Alexander and Earland by treating cross-sections of wool and hair oxidized with peracetic acid with dilute alkali. Optical and electron photomicrographs showed quite clearly that the membrane consists of a honeycomb of membranes which originally surrounded each cortical cell (see Fig. 1.15). Although it has been established beyond doubt that the material contains cortical cell membranes, it has been pointed out by Johnson[51] that their thickness and weight (10 per cent. of the fibre) are too great to be exclusively cell membranes and they are probably thickened by inter- and intra-cellular material.

REFERENCES

[1] MARSTON, *Symp. Fibrous Proteins publ. Soc. Dyers & Col.* (Bradford) 1946, p. 207.
[2] MERCER, *J. Text. Inst.*, 1949, 40T, 640.
[3] ELÖD AND ZAHN, *Melliand Textilber.*, 1944, 25, 361.
[4] BEKKER AND KING, *Biochem. J.*, 1931, 25, 1077.
BLACKBURN, *ibid.*, 1948, 43, 114.
[5] REUMUTH, *Klepzig Textil-Z.*, 1942, 45, 288.
[6] STAKHEYEWA, *Bull. Soc. Chim. France*, 1936, 5, 647.
[7] SCHMIDT, *Z. f. Naturforschung*, 1946, 1, 1.
HASELMANN AND ZAHN, *Melliand Textilber.*, 1951, 32, 1.
[8] WOODS, *Proc. Roy. Soc.*, 1938, 166A, 76.
[9] HOCK, RAMSAY AND HARRIS, *J. Res. N.B.S.*, 1941, 27, 181.
[10] HOCK, RAMSAY AND HARRIS, *ibid.*, 1943, 31, 234.
[11] MERCER AND REES, *Aust. J. Exp. Biol. Med.*, 1946, 24, 147, 175.
[12] FARRANT, REES AND MERCER, *Nature*, 1947, 159, 535.
[13] PORTER AND HAWN, *J. Exp. Med.*, 1949, 90, 225.
[14] PERRY AND REED, *Biochim. Biophys. Acta*, 1947, 1, 39.
[15] HORIO AND KONDO, *Text. Res. J.*, 1953, 23, 373.
[16] MERCER, *ibid.*, 1953, 23, 388.
[17] AUBER, *Trans. Roy. Soc. Edinburgh*, 1950–51, 62, 191.
[18] MERCER, GOLDEN AND JEFFRIES, *Text. Res. J.*, 1954, 24, 615.
[19] FRASER, LINDLEY AND ROGERS, *Biochim. Biophys. Acta*, 1954, 13, 295.
[20] BARKER, *Text. J. Aust.*, 1954, 29, 586.
[21] HUMAN AND SPEAKMAN, *Text. Res. J.*, 1954, 24, 58.

[22] LEVEAU, *Bull. Inst. Text. France*, 1958, **74,** 75.
[23] BURGESS, *J. Text. Inst.*, 1934, **25T,** 289.
HOCK, RAMSAY AND HARRIS, *J. Res. N.B.S.*, 1942, **27,** 181.
ELÖD AND ZAHN, *Melliand Textilber.*, 1943, **24,** 157.
[24] MILLSON AND TURL, *Amer. Dyestuff Rep.*, 1950.
[25] HOCK AND MCMURDIE, *J. Res. N.B.S.*, 1943, **31,** 234.
ZAHN, *Melliand Textilber.*, 1943, **24,** 157.
[26] LINDBERG, MERCER, PHILIP AND GRALÉN, *Text. Res. J.*, 1949, **19,** 673.
[27] ALLWÖRDEN, *Z. angew. Chem.*, 1916, **29,** 77.
[28] MÜLLER, *Z. Zellforsch. u. Mikroskop. Anal.*, 1939, **A29,** 1.
[29] WHEWELL AND WOODS, *J. Soc. Dyers & Col.*, 1944, **60,** 148.
[30] LINDBERG, PHILIP AND GRALÉN, *Nature*, 1948, **162,** 458.
[31] ALEXANDER, *J. Soc. Dyers & Col.*, 1950, **66,** 349.
[32] ZAHN, *Text Rundschau*, 1952, **7,** 305.
[33] LAGERMALM AND GRALÉN, *Acta Chem. Scand.*, 1951, **5,** 1209.
[34] SCHURINGA, ISINGS AND ULTÉE, *Biochim. Biophys. Acta*, 1952, **9,** 457.
[35] LINDBERG, *Text. Res. J.*, 1950, **20,** 381.
[36] SPEAKMAN, *J. Soc. Dyers & Col.*, 1950, **66,** 469.
[37] ALEXANDER, GOUGH AND HUDSON, *Biochem. J.*, 1951, **48,** 20.
[38] GRALÉN, *J. Soc. Dyers & Col.*, 1950, **66,** 449.
[39] REUMUTH, Dissertation (Aachen), 1937.
[40] HARRISON, *Amer. Dyestuff Rep.*, 1938, **14,** 393.
[41] LEHMANN, *Melliand Textilber.*, 1941, **22,** 145.
[42] MERCER, LINDBERG AND PHILIP, *Text. Res. J.*, 1949, **19,** 678.
[43] ELÖD AND ZAHN, *Kolloid Z.*, 1944, **108,** 94.
[44] ZAHN, *Melliand Textilber.*, 1950, **31,** 695.
[45] ALEXANDER AND EARLAND, *Text. Res. J.*, 1950, **20,** 298.
Nature, 1950, **166,** 396.
[46] GRALÉN, LAGERMALM, PHILIP, ALEXANDER, ZAHN AND HASELMANN, *Text. Res. J.*, 1951, **21,** 236.
[47] BLACKBURN, *Biochem. J.*, 1951, **49,** 554.
[48] ZAHN AND HASELMANN, *Melliand Textilber.*, 1950, **31,** 225.
[49] MERCER, *Nature*, 1953, **172,** 164.
[50] MANOGUE AND MOSS, *ibid.*, 1953, **172,** 806.
[51] JOHNSON, Private communication.

CHAPTER 2
Surface Properties

Surface area

OWING to the small radius of an individual fibre, the surface area of a given mass of wool is extremely high, and this is reflected in some of its characteristic physical properties. The specific area may be measured by three independent methods (*a*) direct visual examination (*b*) permeability of a mass of fibres (*c*) physical adsorption. The first two methods lead to lower values in general, as only the broad geometrical surface is detected, small imperfections and cracks being neglected. The third method which is more sensitive, gives the total surface area available for molecular adsorption including surface irregularities and channels of molecular dimensions within the fibres. Thus the methods available lead to two distinct values for the surface area, each being significant under certain conditions and for certain processes. The first two methods will therefore be considered together.

Most textile fibres approximate to uniform cylinders of great length compared with the diameter, so that the surface area may be calculated from the average radius measured under the microscope. Actually the cross section is not perfectly circular, so that it is usual to determine the lengths of major and minor axes for each fibre and take the mean. The shape and size of all the fibres from a given source are only approximately the same so that at least several hundreds of individual fibres have to be examined to obtain a representative statistical value. As already mentioned in the first chapter, this method is widely used as the main criterion of wool quality owing to the parallelism between radius and quality (see Table 1.1).

For accurate work the projection microscope is used, with the fibres mounted in cedarwood oil to prevent any change in the regain of the conditioned wool.[1a] Although this particular method is extremely tedious, it has the great advantage of giving a statistical value of the mean deviation in fibre diameter at the same time.

Barker and King[1b] developed a method for determining the diameter by direct weighing of a fibre of known length on the microbalance, making use of the accurate values for the density of wool

(p. 88). The values obtained in this way and by the projection microscope method were found to agree closely.

The permeability method[2]

The specific surface area,* and consequently the mean fibre diameter, can be determined by an alternative method based on the simple laws of flow of a fluid in a cylinder. According to the Hagen-Poiseuille law, the rate of viscous flow V in a circular capillary of diameter d is given by

$$V = \frac{d^2 \Delta P}{32 \eta L}$$

where ΔP is the pressure difference along a length L, and η is the viscosity of the fluid.

Although a porous medium cannot be regarded as an assembly of uniformly aligned capillaries,[3] it is found experimentally[4] that a similar relation holds, with an unknown constant k, known as the shape factor, replacing the numerical constant of the Poiseuille equation, i.e.

$$V = \frac{m^2 \epsilon \Delta P}{k \eta L}$$

m is a characteristic dimension of the flow system known as the mean hydraulic radius and is equal to the ratio of the free volume to the surface area of the capillaries, and ϵ is the porosity of the medium, i.e. the fraction of the total cross-section free to the fluid. Thus,

$$m = \frac{\epsilon}{(1-\epsilon)S_0}$$

so that

$$V = \frac{\epsilon^3 \Delta P}{k \eta L (1-\epsilon)^2 S_0^2}$$

* The specific surface area S_0 is given by $S_0 = \dfrac{S_1 + S_2 + S_3 \dots S_n}{V_1 + V_2 + V_3 \dots V_n}$

where S_n and V_n are surface area and volume respectively of the nth fibre. This value is obtained from permeability and gas absorption methods, and must be distinguished from the average surface area S which is given by

$$S = \frac{1}{n}\left(\frac{S_1}{V_1} + \frac{S_2}{V_2} + \frac{S_3}{V_3} \dots \frac{S_n}{V_n}\right)$$

which is obtained microscopically.

This equation, first derived by Kozeny[5], has been verified experimentally mainly by the work of Carman.[4]

The constant k can be obtained by direct calculation only in a few of the simpler systems,[6] as it is a measure of the tortuous path of the fluid through the irregular channels. For uniform pores with an equilateral triangular cross-section $k = 1\cdot67$, for uniform circular pores $k = 2\cdot00$, and for rectangular pores of width ten times the length, $k = 2\cdot65$.[4] In general, however, k has to be determined empirically by using a sample of fibres of known diameter in the desired orientation. Cassie[7] has carried out a thorough investigation using a lock of fibres fitted into a small brass cylinder[8] so that the axes of all the fibres lie parallel to the axis of the cylinder; k was found to be of the order of 2·14, but was found to vary considerably with fibre diameter. Anderson and Warburton[9] found that k varies only slightly with porosity within the limits $\epsilon = 0\cdot4 - 0\cdot7$, although Sullivan[10] has found that at high porosities ($\epsilon > 0\cdot88$) the method breaks down completely.

TABLE 2.1

The effect of fibre arrangement on the shape factor k[9]

Fibre diameter microns	k Parallel	k Cut and randomized
35·8	1·64	5·54
28·3	2·06	5·08
23·6	2·08	4·75
19·5	2·10	4·78

Sullivan and Hertel[11] obtained the much higher value of 3·07 for k with the fibres parallel, compared with the value of 6·04 for the fibres perpendicular to the flow. Using these estimated values of k, the specific surface areas may be determined for other samples, and compared with the corresponding values obtained microscopically.[12] A comparison of the results of the two methods is given in Table 2.2 for a variety of fibres using a constant value for k for a given orientation.

Anderson and Warburton[9] have shown that the accuracy may be increased considerably by cutting the fibres into short lengths (c.

1·5 mm.) and packing them in a completely random manner in the cylinder. The pieces of fibre were randomized by shaking for a short time with petrol ether, which at the same time removes excess oil. This method of packing leads to very high values of k (see Table 2.1) and reduces the mean variation from 19 to 4 per cent.

TABLE 2.2

Comparison of specific surface area S_0 obtained by permeability and microscopic methods[12]

Material	Porosity	S_0 Permeability	S_0 Microscopic
Glass fibres ($\perp$ flow) ..	0·8075	$5·160 \times 10^3$ cm.$^{-1}$	$5·250 \times 10^3$ cm.$^{-1}$
Glass fibres ($\parallel$ flow) ..	0·8179	4·770	4·920
Cotton fibres ($\perp$ flow)..			
B.38, ⅞ in.	0·7660	3·720	3·690
B.38, 1 1/32 in.	0·7620	3·990	3·830
B.38, 1 3/16 in.	0·7560	4·210	4·410
Wool fibres ($\perp$ flow) ..			
Welsh wool	0·7090	1·205	1·167
Fine Noil wool ..	0·7160	1·916	2·068

Surface area by adsorption[15]

The determination of surface areas of solids by gas adsorption has been used extensively in recent years, and is now a standard method for this purpose. The method relies on the fact that the strength of binding in a monolayer is usually much greater than the forces between the molecules in the multilayers. Under these conditions the monolayer is largely complete before multimolecular condensation occurs, resulting in a sigmoid isotherm. In order to determine the surface area it is necessary to know (*a*) the volume of gas adsorbed in the monolayer and (*b*) the average area occupied by a single gas molecule. The point on the isotherm, corresponding to the completion of the monolayer was originally estimated empirically by selecting the most likely discontinuity,[14] but it is now obtained from theoretical considerations.[15] The average size of a sorbed molecule is usually obtained from the density of the liquified adsorbate, or from the diameter of the molecule determined by some other standard method.

The point at which a monolayer is complete was originally selected in a region just beyond the point of maximum inflexion (Point B), as this corresponds approximately to the limiting value of the Langmuir adsorption.[16] This point can however be determined more accurately by applying the equations developed by Brunauer, Emmett and Teller,[15] Harkins and Jura,[17] and others, which have proved highly successful. The B.E.T. method is based on an evaporation–condensation equilibrium between molecules in the monolayer held strongly on specific sites and those molecules in the other layers held by smaller forces. For simplification the heat of adsorption in the higher layers is assumed to be constant and equal to the heat of evaporation of the liquid (E_L). For adsorption on a free surface the general equation is of the form (p. 99),

$$v = \frac{v_m \cdot c \cdot p}{(p_0-p)[1+(c-1)p/p_0]}$$

where v is the volume of gas adsorbed at pressure p, p_0 is the saturation pressure at the particular temperature, and c is a constant representing the tendency of the molecules to enter the monolayer in preference to higher layers, and given approximately by $c \simeq e^{-(E_1-E_L)/RT}$. E_1 is the heat of adsorption of a molecule in the monolayer. Generally $c > 1$ so that the isotherm is composed of two distinct regions. The low-pressure region obeys the simple Langmuir adsorption, whereas the curve is convex to the pressure axis when v is large and p approaches p_0. The case which is of importance for condensation in pores of a given thickness to allow n layers to form is given by

$$v = \frac{v_m c x [1 - (n+1)x^n + nx^{n+1}]}{(1-x)[1+(c-1)x-cx^{n+1}]}$$

where $x = p/p_0$ and v_m is the volume of gas accommodated in the monolayer. The simple form may be rearranged as follows

$$\frac{p}{v(p_0-p)} = \frac{1}{v_m c} + \frac{c-1}{v_m c} \cdot \frac{p}{p_0}$$

which is suitable for application to the experimental data. From the linear plot of $p/v(p_0-p)$ against p/p_0, v_m and c may be derived.

Bull[18] has applied this isotherm satisfactorily to the adsorption of water on many proteins including silk and wool. By applying the isotherm to experimental data on a large variety of systems, Brunauer et al.[15] established the following important conditions.

(1) v_m is given approximately by Point B (Fig. 2.1), the rather sharp inflexion in the curve at relatively low pressures.

(2) $E_1 - E_L$ is a constant for a given gas, and is independent of the nature of the adsorbent.

(3) v_m, the volume of gas accommodated in the monolayer, leads to a value for the surface area, independent of the nature of the gas used, in the absence of swelling.

Bull obtained good linear plots for all the fibres studied (Fig. 2.2) although the calculated values of v_m using water lead to values of the

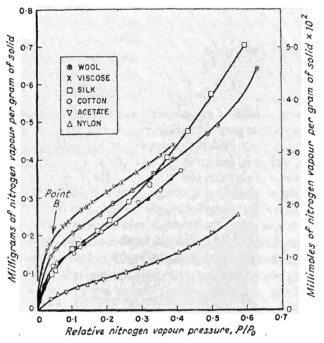

Fig. 2.1.—Adsorption isotherms of nitrogen on six textile fibres at $-195°$ C. (*Reproduced by permission*).[21]

surface area which bear no relation to the area determined by other methods, due to swelling (see p. 22). No correlation was found between c and the heat of wetting calculated thermodynamically (see p. 114), although the free energy of adsorption per molecule in the monolayer was the same for all the proteins investigated.

The chief theoretical objection to this treatment is that lateral interaction between molecules in individual layers is neglected,[19] although the heat of adsorption in such a layer is equated to the latent heat of the liquid E_L. From all points of view it would seem that molecular attraction within a layer is of greater importance in stabilizing the system than inter-layer attraction. This objection does not however invalidate the measurements of surface area or the composition of the primary layer since the fault disappears as the pressure decreases, and the general validity has been demonstrated by comparisons with the values obtained by several independent methods.[13] For example in the case of various samples of zinc oxide pigments,[20] direct microscopic counting, absorption of stearate, ultramicroscopic counting,

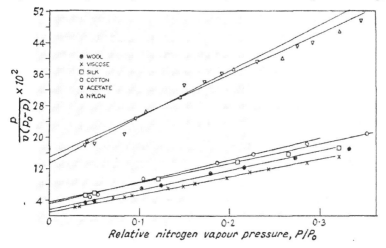

FIG. 2.2.—B.E.T. plots for obtaining surface areas accessible to nitrogen. (*Reproduced by permission*).[21]

permeability and adsorption led to values differing at the most by a factor of two.

From the values of the specific surface areas, given in Table 2.3, it is seen that the value for wool obtained by nitrogen adsorption is some 5 to 10 times greater than the geometrical area.[21] In addition, the area from water adsorption is approximately 200 times that obtained from the nitrogen adsorption.

The difference between the geometrical surface area and the specific surface obtained from nitrogen sorption measurements is probably not highly significant owing to the irregular surface of wool fibres. The agreement between the corresponding values for cotton is very good (compare Tables 2.2 and 2.3) and of the same order as found with spherical particles and powders. That the agreement is not so close in the case of wool may be attributed in part to the scale structure (see Chapter 1) but does not rule out completely the existence of transverse pores or fissures. The number of pores, however, must be limited and these data do not support a pore hypothesis such as that of Hedges, or the later theory of Hermans, to account for water adsorption (see Chapter 3). This conclusion is supported by diffusion studies with dry keratin fibres and horn (p. 92).

TABLE 2.3

Calculated surface areas accessible to water vapour and nitrogen vapour[21]

Material	Area I, available to H_2O mols. at 25° C. Sq. Metres per Gram.	Area II, available to N_2 mols at $-195°$ C.	Area I / Area II
Wool	206	0·96	215
Viscose rayon	204	0·98	208
Silk	140	0·76	184
Cotton	108	0·72	150
Acetate rayon	59	0·38	154
Nylon	45	0·31	145
Titanium dioxide	7	7·90	0·9

The internal surface of jute fibres has recently been measured by negative adsorption methods.[22] This method depends on the measurement of the small increase in concentration when a salt solution is shaken with the fibres, caused by the electrolyte distribution in the diffuse part of the Gouy double layer (p. 192).

In the case of jute, the internal surface area of the swollen fibres was found to be roughly a thousand times greater than the geometrical surface area, which is the same order as that obtained by the methods explained above for cotton (*c.* 500–1,000-fold) and wool (*c.* 1,000–2,000-fold). Although this method has not been applied to wool, it is interesting to note that the values of the pore size for 'red' jute (100 Å) and 'white' jute (100 Å) are of the same order as the value given by Speakman for the pore size (40 Å) of swollen wool fibres (see Chapter 4).

Electrical properties

The surface properties of wool fibres are in many ways unique, and it is not surprising therefore to find that they exhibit electrical properties not normally observed with other fibres and macromolecular systems. It has been known for many centuries that mammalian hair, in common with all insulators, exhibits the phenomenon of triboelectricity, i.e. the generation of static electricity on rubbing. Unfortunately no satisfactory theoretical explanation has been advanced to account for this behaviour, and the recorded experimental work is extremely meagre. Indeed it is still widely supposed that frictional electricity may be generated only by rubbing two unlike objects together, and that the sign of the charge imparted to a given substance depends only on its position in the triboelectric series. It has been observed from time to time, however, that electricity may be generated by rubbing together like bodies, and Faraday was one of the first to observe that two feathers behaved in this way. Subsequently many other solids have been examined, in particular by Shaw[23] who worked mainly with metals, and attributed the effect to surface strains on one of the solids which thus modifies the structure. This condition may be recognized by the gradual decrease in the charge on continued rubbing, so that in time it disappears or is even reversed. It is well known that in many cases triboelectric phenomena are highly irreproducible which can also be explained by surface strain, or alternatively in some cases by surface films of an impurity or uneven temperature increases. The effect of atmospheric moisture which is often advanced in explanation of these anomalies is by no means as pronounced as commonly believed.[24] In fact Medley[48] has recently found that the tribo charge on wool is independent of humidity and rubbing speed. The production of the electricity was attributed to corona discharge and even a single rub is sufficient to raise the potential to the breakdown value.

The triboelectric behaviour of wool is even more remarkable than that of other insulators. Martin[25] has recently made a series of important observations, the most interesting of which is probably that the effect is directional. Thus, if a fibre is pulled out of a bundle of fibres, all lying in the same direction, the charge is positive if the root end is removed first and negative if the tip is pulled. This effect is not observed if a descaled fibre is used, and the fibre develops a negative charge when pulled from the bundle in either direction. Further, as would be expected, no charge is developed if the fibre lies in the opposite direction to the others.

Martin showed the peculiar triboelectric effect of normal fibres to be due to the piezoelectric nature of the scale substance. He inserted a purified lock of merino wool tightly into a glass tube, taking care to keep all the fibres in the same direction, and filled the tube completely with molten shellac by vacuum suction. After cooling and breaking away the glass tube, a ¼-inch section was carefully cut and placed between two brass plates. It was found that the root end of the fibre became negative on compression, a force of 10 lb. producing a potential of the order of 1 volt. It was also found that a shearing force applied along the length of the stick produced a similar but smaller potential. This was taken to mean that the direction in which the charge separates is at a small angle to the fibre axis, and it was therefore identified with the length dimension of the scales.

In addition the fibres were shown to be pyro-electric.[26] The root end becomes negatively charged, and a potential of the order of 10 volts can be generated on immersion in liquid air. The pyro-electricity was attributed to the scales, as it was found that a suspension of scales in dry benzene rapidly aggregates causing flocculation on cooling in acetone–CO_2 or liquid air. Unfortunately a stable suspension of cortical cells could not be made, so it is not completely established that they are not piezoelectric to a limited degree.

Martin[25] has suggested that the piezoelectricity of the scales may be due to the unidirectional arrangement of the polypeptide chains, in contrast to the alternate arrangement in opposite directions suggested by Astbury[27] for the fibre as a whole. It is attractive to propose that the latter configuration is characteristic of the cortical cells, and provides the chief difference in structure between the cortex and cuticle.

The dielectric constant of keratin is also anisotropic as circular disc of cows' horn suspended in an alternating field orientate with the fibre axis along the field. As the properties of the fibres are determined largely by the cortical cells, the latter may also be assumed to have an anisotropic dielectric constant.

The piezoelectricity and directional triboelectricity are probably closely related to the anomalous frictional properties of the fibres in that all these phenomena arise from the peculiar molecular structure of the scale surface likened by Martin[28] to a 'molecular pile'. It is not to be assumed, however, that the electrical properties play an important part in the frictional behaviour and felting ability of wool, because the peculiar directional friction effect (p. 27) is observed under conditions where no surface electrical anomaly is observed (mainly in conducting solutions).[28, 29]

Frictional properties

As already indicated, the frictional properties of wool fibres and keratin proteins in general are anomalous, as the friction developed on moving a fibre against a second surface is directional, and is greatest when the fibre is moved in the root to tip direction. This is probably the most characteristic property of animal fibres, and can readily be demonstrated by rubbing a fibre between thumb and forefinger, when it always moves in the direction of the root.[30] As might be expected this behaviour is due to the scaley nature of the surface, and it can be shown readily that descaled wool will not migrate in any preferential direction.[31] This problem has received considerable attention, particularly through the many attempts to correlate friction and shrinkage. At present the whole subject is in a highly controversial state, and for this reason some attention must be paid to the experimental methods employed and the most important results obtained before the various theoretical considerations are discussed.

Experimental methods

For several years the only method available for the measurement of fibre friction was the 'violin bow' method introduced by Speakman and Stott.[32] In this method, fifty fibres, all pointing in one direction, are stretched between two bars, and the assembly placed on a piece of wool velvet. The angle is then measured when slip first occurs on tilting the cloth at right-angles to the fibre axis. The difference in the angles θ_1 and θ_2 for slip against and with the scales, expressed as a percentage of the smaller angle θ_2, is known as the scaliness.

$$\text{Coefficient of scaliness} = \frac{\tan\theta_1 - \tan\theta_2}{\tan\theta_2} \times 100 = \frac{\mu_2 - \mu_1}{\mu_1} \times 100$$

The violin bow method has the advantage of determining the mean of a large number of measurements on individual fibres, and also the same sample can be used in solutions of different pH. The construction of each bow is however tedious, so that various workers have developed alternative methods using single fibres.

Morrow[33] has compared the results of several different methods including the measurement of the force required to draw out a fibre which is pressed between two plates. Martin[29, 44] has used the similar procedure of subjecting a fibre placed between two optically ground plates to small oscillations at right angles to the fibre axis. This method was developed by Speakman, Chamberlain and Menkart[34] who obtained an absolute difference in scaliness by means of an instrument called the lepidometer which measures the tension developed as the fibre is rubbed at a standard rate between two opposed reciprocating surfaces. Mercer[35] used a 'slip-stick' method similar to that evolved by Bowden and Leden,[36] where the fibre is forced slowly against a piece of horn fixed on a piece of clock spring. The two surfaces stick together until the tension in the spring just exceeds the frictional force when slip occurs. The surfaces then stick together again and the process is repeated many times. The frictional force is measured by the deflexions of the spring corresponding to each slip position, which are recorded photographically. Mercer[35] expressed the results in terms of the coefficient of friction δ, defined as follows:

$$\delta = \frac{\mu_2 - \mu_1}{\mu_2 + \mu_1}$$

Modern methods pay attention primarily to the measurement of the individual coefficients of friction, and several variations of the method introduced by Lipson[37] for this purpose have been described. Lipson suspended a fibre round a cylindrical rod of polished horn and suspended weights on either end, such that slip occurs when the difference in tension $T_2 - T_1$ reaches a critical value. Under these conditions the frictional coefficient is given by the well-known expression,

$$\mu = \frac{1}{\theta} \log_e \frac{T_2}{T_1}$$

where θ is the angle of contact between fibre and drum. This method was used by Harris et al.[38] with wool felt wrapped round a metallic rod, by Martin and Mittlemann[39] using various materials for the cylinder, and by King[40] who has made a thorough study of the effect of the nature of the second surface.

As most of the frictional measurements have been made for the purpose of correlating them with felting, Lindberg and Gralén[41] consider that the friction between two fibres is more significant than the friction between a fibre and a second surface. They therefore constructed an apparatus to measure the inter-fibre friction. Originally a procedure based on the method of Bowden and Leden was employed[42] but this was found to give erratic results particularly in the anti-scale direction. This method was subsequently replaced by a procedure involving the twisting of two fibres to a given number, n, of turns in order to increase the contact area. The tension which had to be applied to one of the loose ends of the fibres to cause slip was thereby increased considerably, leading to more reproducible results. The coefficient of friction is given by an expression almost identical with that involving slip on a second surface,

$$\mu = \frac{1}{\pi n \theta} \log_e \frac{T_2}{T_1}$$

where θ is the angle between the two fibres in the twist.

Results of the measurements

The two frictional coefficients differ considerably with the type of wool used, and in general the difference in friction with and against the scales, now known as the directional friction effect, D.F.E., is closely related to the felting properties. Early results of Bohm[43] suggest that the D.F.E. decreases steadily with increasing coarseness of the fibres. The method used consisted of dragging sheets of fibres, mounted on a slide, across a glass surface which led to values difficult to reproduce. On the other hand, Harris et al.[38] found no correlation between the coefficients of friction and fibre diameter for forty-nine separate fibres.

The effect of the surface along which the fibre is drawn has been studied recently by King who found that little or no directional friction was exhibited on highly polished surfaces in support of the conclusions of Makinson.[44] It is interesting to find that on roughening the surfaces wherever possible, directional friction is produced in all cases by a lowering of the root-to-tip coefficient with little change in the opposite value (Table 2.4), contrary to everyday experience.[40] This may be due to contamination of the rough surface, although prolonged cleaning had no effect on the friction coefficients.

Various workers have observed that the coefficients of friction and the D.F.E. are much greater in water than in air.[38, 45, 46] According to some workers the anti-scale increases but the with-scale decreases,[35] although recent measurements by King[40] against horn and by Gralén[41] against a second fibre show that the latter also increases slightly (see Table 2.5).

TABLE 2.4

Friction between wool fibres (60 per cent. R.H.) and polished and rough surfaces[40]

Cylinder material	Polished		Rough	
	μ_1	μ_2	μ_1	μ_2
Glass	0·60	0·60	—	—
Casein	0·58	0·59	0·47	0·57
Ebonite	0·61	0·64	0·51	0·63
Sheep's horn	0·62	0·63	0·52	0·64
Cows' horn	0·48	0·54	0·44	0·54

TABLE 2.5

Static coefficients of friction by the fibre twist method[41]

	μ
Wool dry:	
With-scale	0·11
Anti-scale	0·14
Wool in water:	
With-scale	0·15
Anti-scale	0·32

Speakman, Chamberlain and Menkart,[34a] using the lepidometer, found that the effect is more pronounced at pH 1 and pH 11 than at pH 9, where the value was found to be a minimum. This was confirmed by Mercer[35] and also by Whewell, Rigelhaupt and Selim[47] who report the following values for the scaliness of normal wool: 23·5 per cent. in water, 29·4 per cent. in N/10 HCl and 21·5 per cent. in borax. Extensive work by Lindberg and Gralén[41] has confirmed that these differences are very small in the pH range 2–9, and are probably not responsible for the effect of pH on the degree of felting.

Martin[28] suggested that swelling agents increase the D.F.E. even in the presence of the vapour only, and this has been confirmed by

SURFACE PROPERTIES

King[40] (Table 2.6). Thus acetic acid has a much greater effect than ethyl alcohol in agreement with their respective swelling effects (p. 86). The importance of swelling was demonstrated by the increased D.F.E. obtained on swelling the ebonite cylinder with benzene. The effect is still greater when ebonite and wool are swollen; this is difficult to explain by the alternative explanation given by Makinson,[44] to be discussed shortly.

TABLE 2.6

Effect of swelling agents on frictional coefficient using wool fibres[40]

Swelling action	μ_1	μ_2
Horn and fibre swollen in ethanol	0·47	0·61
Horn and fibre swollen in acetic acid	0·62	0·80
Ebonite swollen in benzene, fibre unswollen	0·58	0·79
Ebonite unswollen, fibre swollen in water	0·62	0·72
Ebonite swollen in benzene, fibre swollen in water	0·65	0·88

The increase in the frictional coefficients and the D.F.E. with swelling has been explained semi-quantitatively by King[40] by the increase in contact area following the decrease in elastic moduli.

Several workers have pointed out that the frictional coefficients depend on the applied load, in a manner which is given closely by the relation

$$\mu = \mu_o + \frac{k_1 S}{N}$$

where N is the normal force per unit length of the fibre, S the contact area and μ_0 a small constant.[40, 41, 48, 49] This holds for all fibres including wool, although anomalies have been reported concerning the anti-scale friction. Mercer and Makinson[46] have found μ varies little with the applied load because they use the slip-stick technique, where N is almost independent of the applied load.[49] In the case of inter-fibre friction the area of contact S varies rapidly below a small limiting value of N, and then becomes constant; μ however increases regularly with decrease in N.

In all cases the frictional coefficients have been found to decrease with temperature, a change which has been shown to be reversible.[39,41] Considerable attention has been paid to the actual values of the two frictional coefficients and the differential friction. The frictional coefficients of a fibre against a second fibre are extremely low compared with the values for other textile fibres (0·22 to 0·5).[50] For example, the values in the dry state ($\mu_1 = 0\cdot11$; $\mu_2 = 0\cdot14$) are of the same order as well-lubricated metallic surfaces. These very low values have recently been attributed to the thin sheath known as the epicuticle which is thought to surround the whole of a fibre.[51] If this explanation is correct, the membrane would be expected to exert a considerable influence on felting and fibre migration, which are discussed in the following sections. At present, however, there is little convincing evidence to prove that the epicuticle forms a coherent sheath round the fibre, but merely forms the outer layer of each scale (see p. 8). Moreover, Speakman[52] has pointed out that the sensitive staining test of Whewell and Woods[53] reveals clearly that the epicuticle is very easily damaged by mechanical processing, and that whatever its influence on the physical properties of the natural fibre, it is unlikely to affect these properties after it has undergone ordinary processing such as carding, combing and spinning. At the present stage, therefore, no definite conclusions can be reached concerning the influence of the epicuticle on the frictional properties.

After mechanical abrasion, the coefficient of friction is considerably increased, but the difference is not removed until the scale stripping is complete.[54] Chemical treatments capable of rendering the wool unshrinkable tend to remove the directional friction effect[55] although this is not always the case[56] (p. 47) and the way in which the D.F.E. is reduced depends to some extent on the type of treatment.[57] The results of Mercer[35] show that halogens in aqueous solution reduce μ_1 and μ_2 to very low values, the difference varying little with pH. Alcoholic caustic potash and sulphuryl chloride in white spirit on the other hand increase both coefficients, in particular μ_1. In all cases the frictional coefficients are minimum in alkaline media, and in the case of the alcoholic potash treatment the D.F.E. in acid approaches that of normal wool (see Table 2.7).[57] This accounts for the greater tendency of these wools to felt in acid solution than in alkali[58] (see p. 43). The low coefficients of friction obtained in alkali have been explained by the formation of a gelatinous layer of degraded protein on the surface of the fibres, which hardens on immersion in acid.[59] This is reversible until the degraded layer is gradually removed by continued scraping when the friction in alkali increases.

An extensive series of measurements of the two frictional coefficients has been obtained by Lindberg and Gralén,[57] using the fibre twist method. They have also found that the result depends on the type of treatment employed, although in all cases increases in the coefficients were obtained. Thus mild chlorine treatments caused extensive increases in μ_1 and μ_2 in contrast to the results of Mercer using overtreated samples. In addition to the values shown in Table 2.7, treatments with papain according to the 'Chlorzyme' process[60] caused considerable decreases in the D.F.E., due to considerable increases in μ_1 and μ_2.[57]

TABLE 2.7

Frictional coefficients of treated fibres by stick-slip method[35]

Type of fibre	Rubbing surface	Conditions	μ_2	μ_1	δ
Untreated	Horn	Air dry	0·55	0·33	0·25
		Wet pH 4·0	0·66	0·32	0·35
		Wet pH 10·8	0·66	0·25	0·45
		Wet pH 1·3	0·72	0·32	0·38
Untreated	Wool	Wet pH 4·0	0·55	0·25	0·37
		Wet pH 10·8	0·54	0·20	0·46
*Chlorine-treated	Horn	Wet pH 4·0	0·11	0·11	0·00
		Wet pH 10·3	0·02	0·02	0·00
		Wet pH 1·3	0·12	0·11	0·04
Bromine-treated	Horn	Wet pH 4·0	0·25	0·21	0·09
		Wet pH 10·8	0·05	0·05	0·00
		Wet pH 1·3	0·17	0·15	0·06
Alcoholic caustic potash-treated	Horn	Wet pH 4·0	0·62	0·42	0·20
		Wet pH 10·8	0·53	0·35	0·20
		Wet pH 1·3	0·72	0·40	0·28
Sulphuryl chloride-treated	Horn	Wet pH 4·0	0·75	0·65	0·08
		Wet pH 10·8	0·45	0·42	0·03
		Wet pH 1·3	0·90	0·81	0·05

Coefficient of frictional difference $\delta = \dfrac{\mu_2 - \mu_1}{\mu_2 + \mu_1}$.

* Probably overtreated (see ref. 57).

TABLE 2.8

Fibre friction measured in sodium oleate after various fibre treatments[57]

Treatments	Time of treatment t min.	Distilled water μ_2	μ_1	s	Sodium oleate μ_2	μ_1	s
Untreated		0·30	0·13	4·4	0·27	0·10	6·3
0·05% Cl_2 pH 5	5	0·76	0·68	0·2	0·10	0·11	−0·9*
1 N $KMnO_4$ in 0·1 N H_2SO_4	10	0·68	0·37	1·2	0·68	0·45	0·8
1 N $KMnO_4$ in 0·1 N H_2SO_4 + 1% $NaHSO_3$	10	0·82	0·70	0·2	0·19	0·17	0·6*
0·1 N $KMnO_4$ in 0·1 N H_2SO_4 + 1% $NaHSO_3$	10	0·75	0·67	0·2	0·20	0·11	4·1*
0·01 N $KMnO_4$ in 0·1 N H_2SO_4 + 1% $NaHSO_3$	10	0·35	0·17	3·0	0·24	0·11	4·9*
2% KOH in alcohol (fibres air-dry)	10	0·81	0·67	0·3	0·64	0·36	1·2
2% SO_2Cl_2 in kerosene (fibres air-dry)		0·84	0·76	0·1	0·45	0·33	0·8
0·3 M Cl_2 in CCl_4 (fibres conditioned at 65% R.H.)	1	0·65	0·39	0·7	0·19	0·11	3·8*
0·3 M Cl_2 in CCl_4 (fibres conditioned at 65% R.H.)	10	0·83	0·57	0·5	0·06	0·06	0*
0·3 M Cl_2 in CCl_4 (fibres dried at 105° C. 1 hr.)	10	0·54	0·31	1·4	0·41	0·25	1·6
0·3 M Cl_2 in CCl_4 (fibres dried at 105° C. 1 hr.)	75	0·70	0·49	0·6	0·46	0·27	1·5
0·3 M Cl_2 in CCl_4 (fibres treated with NaOH)	10	0·65	0·45	0·7	0·10	0·10	0*
Scales removed by rubbing the fibres against emery cloth		0·68	0·61	0·2	0·56	0·47	0·3
Nylon staple fibre		0·45			0·41		

*When the apparatus is arranged to cover a wide range of frictional coefficients, the lower values of the coefficients have a higher relative error than the higher coefficients, thus making these values more uncertain.

$$s = \frac{1}{\mu_1} - \frac{1}{\mu_2}$$

SURFACE PROPERTIES 33

It was shown also that the coefficients of friction were greatly reduced in sodium oleate solution (Table 2.8), due partly to the layer of degraded protein and partly to the lubricating action of the anion active agent strongly absorbed on the surface (see p. 231).

Theories of the directional friction effect (D.F.E.)

Although the exact cause of the D.F.E. is still a matter of controversy, most workers now believe that it is due to the irregular structure of the scales. The development of a satisfactory explanation is extremely difficult to envisage at present as no general theory of friction has been established, but results of considerable importance have recently been accumulated by Bowden and Tabor[61] working mainly with metals. It has been established that contact between surfaces involves only a minute fraction of the total area, the resulting friction arising from the shearing of the junctions between such protuberances. From the magnitude of the frictional force it is concluded that in the case of metals the junctions are virtually welded together. Although the same mode of linkage is not possible in the case of non-metallic substances, Mercer[55] has advanced a similar type of explanation for wool fibres involving chemical bonding at the junctions.[62, 63] If this view is correct, some correlation between chemical reactivity and coefficient of friction is to be expected. Some support is provided by the observations of Shooter and Thomas[64] showing that very unreactive polymers, e.g. Teflon[68] and polythene,[64] have very low frictions, *viz.* 0·04 and 0·10 respectively. Polystyrene and polymethyl methacrylate (Perspex), which have a much higher proportion of polar side groups and aromatic groups, give high values (*c.* 0·5). To explain the directional friction effect, the formation of unsymmetrical bonds after the manner of Martin[28] has to be postulated.

It was thought for a long time that felting was caused by interlocking of the scales,[65] although no such interlocked fibres have actually been observed. A more likely explanation is given by the ratchet theory[66] which does not imply definite interlocking of fibres, but a mechanical ratchet effect when two fibres are rubbed together. This simple explanation immediately encounters several difficulties on a more detailed experimental examination of the peculiar frictional effect. The very different frictional coefficients in acid and alkaline media after various treatments (see Table 2.7) require an additional explanation such as a change in scale angle with changes in swelling,[67] and in some cases the formation of a degraded protein layer on the surface already mentioned (p. 30).

A simple ratchet effect can only operate on a rough surface or on such a soft surface that the scales can plough through the material when the fibre moves in the anti-scale direction. Many observations, summarized by Mercer[35] and Martin[28], have cast considerable doubt on the ratchet mechanism. Firstly, on close examination, little correlation is found between microscopic appearance and D.F.E. Thus, mohair and porcupine quill, which both have very smooth external surfaces, exhibit moderate and powerful effects respectively. In addition, the directional friction may be removed by several chemical treatments, in particular gaseous chlorine, alcoholic caustic soda and a mixture of permanganate and hypochlorite at pH 8·5, without affecting the scale structure visibly.[67] Polished porcupine quill and finger nail with no microscopically visible scales, although composed of overlapping layers of keratin, still possess D.F.E.

Secondly, Martin found that moistened fibres exhibit a small but finite directional friction effect against polished glass surfaces. Fibres readily migrate when placed between two such highly polished plates, and it is difficult to imagine the scale edges biting into the hard surfaces. Finally, Martin demonstrated that a ratchet surface similar to that of coarse wool may be prepared artificially, which exhibits no D.F.E. against rough or smooth surfaces. A lock of coarse fibres lying in the same direction was pressed into a resin-beeswax mixture to give negative casts, which were then filled with gelatin. On separating from the mould, the gelatin appeared as a large number of parallel fibres embedded in a flat surface. In addition to this observation, it is known that dry fibres with microscopically visible scales show no D.F.E. on a rough or smooth glass surface.

In the face of these objections, the simple ratchet theory had to be modified and emphasis has now been placed on the scale tips. To illustrate this proposed mechanism, Rudall[66] has constructed a model (Fig. 2.3) which consists of a piece of wood with serrated edges and small pieces of rubber glued to each tooth. These flexible scale tips cause a D.F.E. when the model is drawn across a plane glass surface as indicated in the diagram. This model has been accepted by Speakman[69] and in a modified form by Alexander[67] to explain the effects of the various anti-shrink processes on the frictional coefficients. In some processes where the scales remain on the fibres, e.g.

alcoholic caustic soda and the hypochlorite–permanganate mixture, the tips of the scales are considered to be removed, although in milder processes, e.g. acid permanganate, fluorine and chlorourea, the scale tips are only softened in such a way as to adhere to the neighbouring scales in alkaline media. This joint, however, may be broken by bending in acid media (in such an operation as caused by the milling

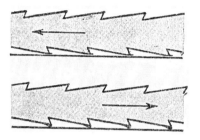

Fig. 2.3.—The Rudall Model, demonstrating the action of the scale tips resulting in a directional friction effect.[66] (*Reproduced from 'Wool Science Review' by permission of the International Wool Secretariat*).

stocks) when the scale substance is hardened, so that the fibres resemble the untreated ones. This explanation[67] also accounts for the observed differences in the D.F.E. values for wool treated with acid permanganate as measured by the violin bow and the Lipson technique.[37] (See Table 2.9.) In the first case the fibres remain straight,

TABLE 2.9

Directional friction effect and shrinkage[56]

Treatment	D.F.E.		Shrinkage per cent.	
	Violin bow	Lipson method	Washing machine	Milling stocks
Nil	31·0	0·23	24·0	25·1
4·0% Cl_2* applied at pH 2	1·0	0·06	1·1	1·8
0·5% $KMnO_4$†, pH 2 Immersion for 10 sec...	1·8	0·17	0·0	18·1
,, for 30 sec...	2·4	0·05	1·0	7·6

* on wt. of wool.   † % composition of solution.

while in the second method they are bent round a cylindrical rod (see p. 25).

Electron- and photo-micrographs, which have accentuated the scale tips, show that the free edges of the scales project from the surface in a similar manner to the artificial tips of the Rudall model (Fig. 2.3); Fig. 2.4 shows an electron-micrograph profile obtained by Mercer and Rees[70] and Fig. 1.8 shows the influence of bending the fibres[71] on the scale projection.

Fig. 2.4.—The scale profile under the electron microscope.[72] (*Reproduced from 'Wool Science Review' by permission of the International Wool Secretariat.*)

These views have been criticized by Martin, who maintains that no D.F.E. is demonstrated by the Rudall model if both rubber and glass are scrupulously cleaned. He attributes the normally observed positive effect to dust patches on one side of the rubber. According to Martin, the directional friction effect is a property of the whole of the scale surface, with the edges and tips playing a minor role. The friction is supposed to arise from unsymmetrical deformations of the molecules on the surface of the scales, based on the theory of Tomlinson.[72] According to this theory, the atoms of a body moving against a second body displace the atoms on the surface of the latter. At a certain critical displacement the atoms spring back to their original positions thus distributing their excess energy in the form of vibration to neighbouring atoms.

The presence of such an unsymmetrical molecular structure is assumed from the observations that wool scales and in fact all the keratins are piezoelectric (see p. 24), an effect which is always manifested in a given direction except in isolated cases. Martin stresses, however, that the electrical phenomena indicate only a peculiar type of surface structure, and are not primarily responsible for the D.F.E. Thus two parallel fibres lying in the same direction show no D.F.E.

but generate triboelectricity. Conversely, two fibres lying in the opposite direction do not develop triboelectric charges but show the D.F.E. In addition, a number of piezoelectric crystals, e.g. tourmaline, asparagine, show no directional friction.

Although this theory is attractive and capable of explaining several apparent difficulties of the ratchet theory, by virtue of the broad hypothesis involved, several objections have been raised, with the result that most workers now favour the scale tip mechanism. Thus Thomson and Speakman[73] have shown that the directional friction persists when the fibres are coated evenly with deposits of silver and gold of thickness $c.$ $0\cdot03$ μ (Table 2.10). Thicker films reduce the

TABLE 2.10

Effect of metal deposition on frictional properties[73]

Nature of fibre	Thickness of film μ	Scaliness	
		Violin bow (%)	Lepidometer (g)
Untreated	—	38·8	2·90
Silver coated	0·030	25·2	1·77
Untreated	—	39·2	—
Silver coated ..	0·056	5·3	—

effect considerably which is attributed to the sealing of the small gap (Fig. 2.4) between the scale edges, and increased rigidity of the scales thus preventing the bending required by the Rudall mechanism.

The ploughing action of the scale tips on a second surface has been examined more closely by Makinson,[44] who has developed simple relations for μ_1 and μ_2 in terms of the angle of friction, the average angle of contact of the two surfaces and the shearing force required to separate two asperities in contact. The expressions have been tested by obtaining a constant value of this shearing force for an assumed likely value for the contact angle, for both clean and lubricated surfaces. The scale tip theory can thus explain the action of lubricants in reducing the frictional coefficient[46] (see p. 28), which is difficult to explain on Martin's theory. Makinson shows that the directional friction effect is lowered considerably by setting the fibres in such a way as to flatten the scales on the surface of the fibre. The value is also extremely small (0·03 compared with 0·42) when the fibre is rubbed on a glass rod so that Martin's results are misleading in that his method of detecting a D.F.E. is hypersensitive. The fact that the D.F.E. on glass is of a different order from that obtained on polished horn, shows that the value is tending towards zero, but that no surface can be regarded as perfectly plane.

The mechanism of felting[74]

The extensive researches which have been made in the last two decades on the felting properties of wool before and after various treatments, leave little doubt that the D.F.E. of the fibres is the most important factor. As already indicated, considerable decreases are generally obtained when wool is rendered unshrinkable (e.g. see Tables 2.7 and 2.8). It was pointed out many years ago by Speakman, Stott and Chang[45] that the elasticity and ease of recovery from deformation also play a part in the felting mechanism. The early theories of the felting action were concerned primarily with explaining the strength with which fibres are held together in a felt. These considerations led to the theory of interlocking scales[65] and theories of chemical adhesion.[63] Both ideas do not attempt to explain the felting action, and have been rejected on many grounds. Berg[76] has pointed out that owing to the very large number of points of contact in a felt, friction is sufficient to confer considerable rigidity.

The process of felting involves relative movement of the fibres which may be caused either by mechanical rubbing or by a series of compression–extension operations. Compression or extension of a fibre during milling enables it to move in the direction of minimum friction, so that after recovery the process may be repeated so long as periodic forces operate.[28] The importance of elasticity has been demonstrated by Menkart and Speakman who showed that treatments with mercuric acetate and benzoquinone (see p. 332) reduce shrinkage due to the increase in rigidity via crosslinking, without decreasing the D.F.E. as measured by the lepidometer.[69] These results however are in contrast to the results of Bohm[43] who measured the D.F.E. against a glass surface. Although Speakman[69] has explained the discrepancies in terms of Rudall's model, further investigation is necessary.

In addition to the necessity for the fibres to migrate in a preferential direction during milling, they must do so in such a way as to cause a complex series of entanglements as found in a felt. This process was first considered by Shorter[78] who considered the mass of fibres to be fixed at certain points with other fibres creeping in such a way as to produce a self-tightening effect. This behaviour is represented diagrammatically in Fig. 2.5. The exact way in which the movement is produced is not made clear in this and similar theories, but according to Martin[28] the root ends of the fibre are primarily responsible. It was shown many years ago that single locks of wool felt at the root ends but not at the tip.[79] In addition, two locks fixed

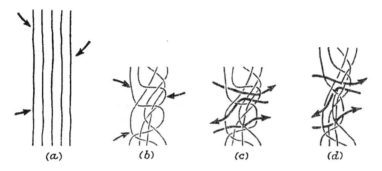

FIG. 2.5.—The Martin mechanism of felting. (*Reproduced from 'Wool Science Review' by permission of the International Wool Secretariat.*)

close together move apart on milling if the tips are facing but readily move together and felt if the root ends are adjacent. Martin[28] found that treating the tips of individual fibres with bromine followed by papain increased the shrinkage of a fabric composed entirely of these fibres, whereas a similar treatment of the root ends reduced the shrinkage considerably (Table 2.11).

In half the experiments the fabric was knitted from yarn composed of fibres all pointing in the same direction. As would be expected, such fabric felted less than the ordinary fabric, the difference being pronounced when the root ends were treated in order to remove the D.F.E.

The treatment of tips of fibres to increase felting has been known for a long time and is used in the process of carrotting fur.[80] This is usually attributed to the softening action enabling the tips to follow the more flexible root ends more readily in the formation of entanglements. Menkart and Speakman[81] have shown that felting is also improved by hardening the root end of the fibres with mercuric acetate,

TABLE 2.11

The effect on shrinkage of treating tips and root ends of fibres[28]

	Shrinkage on relaxed area after various times		
	1 hour	2 hours	3 hours
Untreated			
Orientated	24·5 (23·3)	34·0 (32·2)	—
Random	34·5 (32·1)	42·8 (39·4)	—
Tips treated			
Orientated	30·4 (27·0)	44·1 (44·2)	—
Random	34·4 (32·8)	42·0 (44·8)	—
Root treated			
Orientated	−6·8 (−2·9)	−2·1 (7·6)	9·1 (15·5)
Random	−1·3 (1·2)	20·6 (30·5)	33·4 (37·3)

Duplicate results are shown in parentheses

a considerable increase being obtained when this treatment is applied to fibres the tips of which have already been softened.

Martin[82] has explained these observations by drawing attention to the importance of compression during milling, which causes loops to form in some of the fibres. Other fibres migrate through these loops in a manner similar to a needle and thread (Fig. 2.5). This procedure is assisted by rigid root ends and soft flexible tips, and explains why a stronger felt is obtained by mixing a proportion of stronger non-felting material with the original fibres. It will be observed that this theory differs from the theory of Speakman in that the elasticity and ease of recovery play a very minor part in the felting operations.

The formation of entanglements has been attributed by Harris[83] to the curling tendency of fibres exhibited on compression, due to the supposed difference in elasticity of cuticle and cortex. Berg[76] has shown that carrotted fibres curl more readily than untreated fibres when heated in water, and Harris[83] demonstrated this effect by use of a model obtained by coating stretched rubber bands with cellulose acetate, and noting the curling on the removal of tension. Harris considers that this effect may be responsible for the superior felting qualities of fine wool compared with the coarser varieties, although Martin considers this difference to be small. It is unlikely that curliness can provide a complete explanation of the felting mechanism,

SURFACE PROPERTIES 41

but is probably a contributory factor, so that agents which prevent felting probable reduce both D.F.E. and the curling tendency.

As the physical properties which determine the degree of felting, namely D.F.E., rigidity and elasticity, vary with the external conditions, it is not surprising to find that the rate of felting is affected by such factors as temperature, pH of solution and type of felting machine.[84] In general, felting shrinkage is measured either by washing in soap solution or by milling, the latter usually being the more effective. Alexander[58] has pointed out that two types of felting action must be recognized, and these will in future be called 'Washing' and 'Milling'. The washing-machine test is usually performed in alkali which does not involve a high degree of compression, the fibre migration being produced in this case by the flow of water. In milling, however, the fibres are pounded by the milling stocks, causing alternate compression and relaxation. A similar movement is caused by hand-milling with soap which leads to similar results in general.

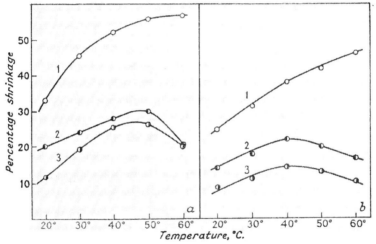

FIG. 2.6—Relation between percentage yarn shrinkage and temperature (1) normal wool; (2) alcoholic alkali treated; (3) sulphuryl chloride treated. (*a*) At pH 4·0. At pH 7·2.[84] (*Reproduced by permission.*)

The effect of temperature on shrinkage was first studied systematically by Speakman, Stott and Chang[45] who reported a maximum in felting rate at 42°–47° C. in soap solution. Speakman, Menkart and Liu[86], however, subsequently reported the optimum temperature to lie between 35° and 37° C. when fabric was milled in soap and borax in a specially designed hand-milling machine. This maximum is very close to the minimum swelling temperature of the fibres.[75] Mercer[84] has found no maximum on milling yarn at pHs of 4·0, 7·2 and 9·2 (Fig. 2.6), although in borax the rate of increase in shrinkage with temperature decreases sharply in the 30°–40° C. temperature range.

In acid solution there is no doubt that the shrinkage increases uniformly from room temperature to 98° C. as originally found by Schofield working with large batches. Mercer[84] has found, however, that after treatments with alcoholic potash and sulphuryl chloride to render the wool resistant to shrinkage, a maximum in felting rate is obtained on milling at various pH values from pH 4 to pH 9·2 (e.g. Fig. 2.6). These observations cast considerable doubt on the interpretation of shrinkage advanced by Speakman and based mainly on the effect of temperature involving a shrinkage maximum and on the effect of pH.[32, 45] A study of the extension and contraction of normal fibres shows a minimum in the hysteresis work loss for 30 per cent. extension, leading to a contraction around 40° C. A comparison with the reported shrinkage maximum led to the postulate that recovery from deformation by extension was an important factor in the mechanism of shrinkage. The detailed results of Mercer question this conclusion. Speakman[86] considers that the results obtained by Mercer cannot be compared directly with his own, owing to differences in the conditions of milling and estimating shrinkage when fabric and yarn are used. He considers that the maximum is only shown with fabrics where the fibres are held so firmly that fibre travel can occur only by alternate extension and contraction.

The effect of pH on shrinkage at a constant temperature was reported by Speakman, Stott and Chang[45] who found the value to remain constant from pH 4 to pH 10 increasing considerably on either side of this range (Fig. 2.7). It is known that the work required to stretch a fibre is constant in the isoelectric range (p. 214) but rises sharply at each end.[87] This similarity was also taken by Speakman *et al.* to prove the importance of elasticity in the shrinkage mechanism. Mercer[84], however, does not find that the shrinkage is constant in the isoelectric range, but decreases gradually from pH 3 to pH 9 where it is a minimum (Fig. 2.7).

The variation in shrinkage of treated wool with pH has also received limited attention and led to interesting and significant results. Mercer[84] noted that the degree of unshrinkability conferred by a certain shrink-resisting treatment may vary with pH. Thus his

SURFACE PROPERTIES

results show that wool treated with alcoholic potash becomes unshrinkable when milled in alkali, but in acid solution the wool shrinks to almost the same extent as the untreated samples. On the other hand, treatments with sulphuryl chloride render the fibres equally unshrinkable in acid and in alkali (Fig. 2.7). Mercer did not comment on this interesting observation, but Alexander[58] performed similar experiments and found the behaviour to be quite general. Treatments which remove the scales or loosen them so that they detach on milling, render the wool unshrinkable to both acid and alkali milling. Other treatments (Table 2.12) which do not remove the

TABLE 2.12

Shrinkage tests in acid and alkaline media after various treatments[58]

Treatment	Shrinkage in solution at pH 3	Shrinkage in solution at pH 10	Shrinkage in solution at pH 3, before solution at pH 10	Fibre damage after milling in solution at pH 10
Fluorine	21·0 (23·5)	5·2 (24·8)	28·8 (25·7)*	12
KMnO$_4$ (acid)	23·2 (27·0)	4·3 (21·2)	25·1 (27·1)*	16
Br$_2$ in CCl$_4$	21·5	0·5	—	25
Br$_2$ in H$_2$O	18·0 (26·7)	7·7 (19·8)	21·5 (26·7)*	23
NaOCl in H$_2$O	22·6 (30·2)	6·3 (30·0)	19·8 (28·0)*	32
Cl$_2$ in H$_2$O (pH 1·5)	4·0 (23·6)	3·3 (20·9)	4·2 (21·2)*	88
Cl$_2$ in CCl$_4$	3·2 (23·6)	0·0 (20·9)	1·7 (21·2)*	85
Cl$_2$ gas	3·8 (21·7)	1·4 (19·8)	—	83
SO$_2$Cl$_2$	7·2 (22·2)	4·2 (24·1)	11·4 (34·1)*	68
Alcoholic KOH	7·1 (22·2)	5·3 (24·1)	11·8 (34·1)*	59

Figures in parentheses give the shrinkage values of untreated control fabric.
* Control pieces had not been previously milled in alkaline solution.

scales, but merely modify them chemically, render the wool unshrinkable only to washing in alkali but not to acid and alkali milling. The alkaline test, which is usually carried out in the washing machine, does not involve a high degree of compression, as the fibre migration in this case is produced by the shearing action of the water. In milling, however, the fibres are pounded by the milling stocks, causing alternate compression and relaxation. A similar movement is caused by hand milling with soap, which leads to similar results in general. Phillips[88] considers that these shrinkage differences are due to uneven treatment of the fibres, as he has shown that a fabric made from yarn composed of 95 per cent. treated fibres shrinks to almost the same extent as an untreated sample. It is unlikely that this gives a complete explanation of the striking difference in behaviour of

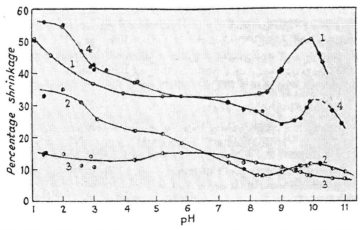

FIG. 2.7.—The relation between percentage shrinkage and pH at 25° C. (1) normal wool (Speakman, Stott and Chang)[45]; (2) alcoholic alkali treated wool; (3) sulphuryl chloride treated wool; (4) normal wool (Mercer)[35]. (*Reproduced by permission.*)

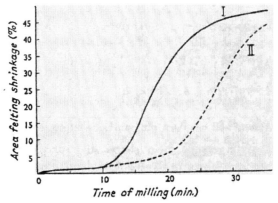

FIG. 2.8.—Representative rate of shrinkage curves[89] for chlorinated wool milled in acid solution (I) and soap solution (II).

samples treated uniformly with certain reagents towards acid and alkali milling, which is probably due to a more fundamental cause.

Carter[89] has pointed out that acid and alkali milling cannot strictly be compared for given times of treatment owing to the differences in the rates of shrinkage (Fig. 2.8). However, the results of Alexander cannot be due to this difference as the samples were compared with control pieces which shrunk to similar extents in acid and in alkali. In addition, on milling a fluorine-treated sample again, the shrinkage is less on a second milling in soap than in a second milling of a similar sample in acid.

Speakman[90] has explained the phenomenon by the greater swelling of degraded protein in alkali, and by assuming the swelling to be substantially reversible. Normal wool behaves in this way, however, so that according to this explanation it must be assumed that the degraded protein swells more in alkali, but not in acid solution, than does normal wool. For this to be the case, the alkali solution must cause further degradation which would probably not be reversible under the experimental conditions.

Alexander[56, 58] has suggested tentatively that these anti-shrink processes attack and soften the scale tips so that the softened scale substance acts as an adhesive in alkaline solution. The scales thus lie flat on the surface in alkaline solution and are not disturbed by the mild process of alkali washing. The milling action in acid bends the fibres, thus breaking the seals holding scales and surface together. In addition, the scale substance would become hard and brittle in acid solution so that the fibres migrate in the same way as the untreated fibres. This explanation is supported to some extent by the D.F.E. measurements discussed on p. 30.

As the exact mechanism of shrinkage of untreated wool is still obscure, a definite explanation of these observations cannot be given. Owing to the considerable theoretical and technical importance of the effect of various treatments on the shrinkage, further fundamental work on acid and alkali milling is highly desirable.

The chemical changes necessary for bringing about a change in the surface so as to remove the D.F.E. and felting are in general associated with attack on the disulphide bond. Alexander, Earland and Carter[91] showed that although this is a necessary reaction it does not by itself remove the D.F.E. and that an additional group in the wool macromolecule has to be attacked. It is considered that for the common non-felting reagents, this additional reaction occurs at the tyrosine residue (see p. 303). The difference between processes which bring about complete resistance to felting, and processes which only reduce the shrinkage in alkaline solution has not yet received an explanation in terms of chemical reactivity.[67]

Polymer deposition

The use of polymers to render wool unshrinkable has been developed comparatively recently because preliminary investigations indicated that a considerable weight increase was necessary. It was originally thought that the main effect of the polymer was in the resulting change in elastic properties as a result of cross-linking within the fibre. Thus Speakman and Barr[92] showed that treatment with mercury salts and quinones could produce a measure of unshrinkability in this way (see p. 332). Attention was then turned to the polymers and resins in common use in industry, but surprisingly unsatisfactory results were at first obtained. Speakman et al.[93] found that a 20 per cent. deposit of ethylene sulphide polymer, formed by exposing wool to a mixture of the vapours of water and the monomer at 50° C., affects the elasticity considerably and reduces the shrinkage. Similar results were obtained with vinyl and vinylidene compounds, using wool which had been impregnated with a suitable catalyst (persulphate) before being exposed to the mixture of water vapour and monomer.[94] The action was attributed to cross-linking, although it is known that large quantities of filler have a similar effect on the mechanical properties (see p. 325). On the other hand, the di-isocyanates[95] which can condense with amino, hydroxyl, carboxyl and thiol groups produce no change in tensile strength of the individual fibres when applied from a wide range of solvents. When a non-swelling solvent was used, however, the fibres were covered with a white deposit of polymer that reduced the shrinkage considerably. Again, however, very large quantities had to be used; for example, a deposit of 60–70 per cent. by weight can reduce the felting to approximately the same extent as a 3 per cent. chlorine treatment. Much better results were obtained by applying a mixture of *m*-phenylene di-isocyanate and hexamethylene diamine to give a weight increase of 14 per cent., followed by baking at 100° C. for one hour.

The reduction in felting by scale masking was clearly demonstrated by experiments using organosilicon polymers deposited from the vapour phase.[96] Samples of wool fabric were made unshrinkable with 3–5 per cent. of resin provided that the resin combined chemically with the wool surface. Previously Neish and Speakman[97] had produced unshrinkable wool by forming a deposit of silica on fabric by immersion in a solution of silicon tetrachloride in carbon tetrachloride. A high concentration is required however and the resulting handle is extremely harsh. It was found that effective resistance to shrinkage could only be obtained if hard solid polymers were produced.[96] Thus, only the trichlorides R–SiCl$_3$, where R is an aromatic or aliphatic group, or equivalent mixtures of silicon tetrachloride and the dichloride R$_2$SiCl$_2$ were effective. Extraction experiments showed that the effective polymers were held by Si–NH bonds to the wool fibre, but they could be extracted after hydrolysis. Thus, chemical bonding is necessary to obtain a uniform film-like deposit, and although the bonds are broken on subsequent milling, the wool remains unshrinkable. In support of the scale-making mechanism, it is found that the directional friction effect is considerably decreased, and in addition no inter-fibre adhesion could be detected.

Several recent studies have shown that scale masking is only a minor factor in the production of unshrinkable wool in many resin processes.[98] Measurements of the D.F.E. are in many cases highly irreproducible, due mainly to the very uneven deposit of polymer on the surface,[99] and also to possible changes in the fibres during the measurement. At present the silicones are the only resins that have been proved to reduce felting by scale masking only, and it has been suggested recently that the main action of polymers is the glueing of the fibres together within the yarn, thus preventing fibre movement. This mechanism, referred to by Royer as 'spot welding', can be recognized by the fact that the tensile strength of the yarn is greatly increased although the individual fibres are unaffected.[99]

It is well known that acid chlorination leads to an increased tensile strength of yarn, due to the adhesive properties of the degradation product, but the strength in water is invariably decreased. The 'spot-welding' mechanism is of considerable technical importance, as a high degree of unshrinkability can be obtained with a small amount of polymer. The handle of wool treated by any of the 'spot-welding' processes is quite good, although not quite as good as that given by some of the chemical methods. Speakman et al.[100] first produced evidence of this type of mechanism when they found that treatment of wool with anhydrocarboxyglycine in an inert solvent gave almost complete unshrinkability with only 4 per cent. weight increase. The polymer is formed by decarboxylation in the presence of water, the best shrinkage results being obtained with wool conditioned at pH 9.[101]

$$n \begin{array}{c} \text{NH} \text{---} \text{CO} \\ | \quad\quad | \\ \text{CH}_2 \quad \text{CO} \\ \diagdown \text{O} \diagup \end{array} \quad \xrightarrow{\text{H}_2\text{O or OH}^-} (\text{CH}_2 . \text{NH} . \text{CO})_n + n\text{CO}_2$$

Treatments using wool conditioned at a pH $<$ 5 did not reduce the shrinkage, because of the small size of the polymer (M.Wt. $c.$ 500). The polymer produced at pH 9 has a high molecular weight ($c.$ 2,000). Speakman thought that the shrinkage is reduced by scale masking and by the adhesive effect, although Alexander *et al.* have shown subsequently that the change in D.F.E. is highly variable. The maximum change is, however, relatively small compared with that caused by effective chemical treatments, and so it must be concluded that 'spot welding' is the more important of the two effects.

The polymer is probably held on the surface by physical forces and hydrogen bonding, as 100 per cent. lithium bromide solutions can remove it, to restore the original felting properties of the wool. Long chains only can bind with sufficient strength to maintain the connexion between adjacent fibres during milling, and the short chains produced by the acid catalysis are inadequate for this purpose. As anhydrocarboxyglycine is very sensitive to moisture, it requires careful handling which, together with the necessity for using organic solvents, is the chief disadvantage of the method. It was found that no other amino acids or related compounds were effective,[101] no doubt due to the low molecular weight polymers which are formed, and to the difficulty of alignment because of the bulky side-groups of the polymer. A promising polymer process for coating wool with nylon consists of depositing methylated nylon from alcoholic solution on the wool surface.[102] This polymer is then reconverted to nylon by subsequent hydrolysis. A recent variation[102a] of this process is that of interfacial polymerization of diamine and acyl chloride by a two-bath treatment of the fabric.

Most of the common resins have been applied to wool, although in most cases baking is required to give an unshrinkable finish. Several firms have brought out non-shrink processes based on melamine-formaldehyde resins (Lanaset). Their behaviour is in many ways similar to urea-formaldehyde resins, and the usual procedure is to pad unpolymerized methylolmelamine, a condensate of melamine and formaldehyde on to the wool, and to condense it to an insoluble resin by baking above 110° C. in the presence of an acid catalyst.[103]

$$
\underset{\text{Melamine}}{\begin{array}{c}\text{NH}_2\\|\\\text{C}\\/\,\,\backslash\\\text{N}\quad\text{N}\\\text{NH}_2-\text{C}\quad\text{C}-\text{NH}_2\\\backslash\,\,/\\\text{N}\end{array}} \quad +6\,.\,\text{H}_2\text{CO} \rightarrow \quad \underset{\text{Methylolmelamine}}{\begin{array}{c}\text{N}\,.\,(\text{CH}_2\,.\,\text{OH})_2\\|\\\text{C}\\/\,\,\backslash\\\text{N}\quad\text{N}\\(\text{HO}\,.\,\text{CH}_2)_2\text{N}-\text{C}\quad\text{C}-\text{N}(\text{CH}_2\,.\,\text{OH})_2\\\backslash\,\,/\\\text{N}\end{array}}
$$

Early workers concluded that the resin is formed within the fibre, because the cross-sections could not be stained with dyes.[104] Further work showed, however, that the deposit was confined to the surface, the lack of staining being due to the cross-linking action of excess formaldehyde, which is always present in the monomer.

McClearly and Royer showed that the torsional rigidity of the fibres is approximately double although the mechanical strength is completely unaffected.[105] They consequently attributed the main action of the resin to sealing of the scale junctions to form a stiff sheath round each fibre. At the same time it was realized that the spot welding of separate fibres played an important part in the action.[106] The latter effect was demonstrated by the low shrinkage resistance obtained if the fibres were treated in the raw or sliver state. The results of Stock and Salley[99] show that this resin treatment relies mainly on local adhesion between fibres as the tensile strength of the yarn is greatly increased.

Although the shrinkage figures are very good, the melamine process has not been applied on a large scale chiefly because of the poor handle, and also the lack of necessary baking equipment. The process has now been modified[107] by partly pre-condensing the resin and applying it in the colloidal form, which becomes fully cross-linked at temperatures below 100° C. In addition, preliminary modification of the wool surface with hydrogen peroxide enables a much smaller quantity of resin to produce the desired unshrinkability. This modified process has been shown to function by scale masking and by 'spot welding', although it is not known which of the two is the more important.[105, 106]

Urea-formaldehyde resins are found to have a similar effect on wool, and under some conditions can produce complete unshrinkability.[108] Direct application of methylolurea involving catalytic polymerization with acids and baking at 110°–140° C. has very little effect on the shrinkage unless more than 20 per cent. of the polymer is deposited. The handle of the wool is completely ruined and the polymer is removed on washing. Better results are obtained by padding the monomer at 110°–130° C. on the wool, and then polymerizing by further baking with a catalyst. In this way the monomer is probably anchored firmly on the wool surface before the polymer is formed. Photomicrographs show that 'spot welding' plays an important part in this process as in the melamine treatments.[56] It has been claimed that s-bisalkoxymethylureas produce less tendering than urea-formaldehyde condensates and are more readily polymerized on textile fibres.[109] Investigations have shown that the n-dibutyl ether is effective in producing unshrinkability with less than 10 per cent. of polymer.[110] The n-amyl compound is also effective and the sec. butyl ether shows some action, but the lower ethers are completely ineffective. The monomers are produced by the controlled reaction of urea with formaldehyde to give bishydroxymethylurea, $CO(NH.CH_2.OH)_2$. This is then alkylated by heating with excess of the corresponding alcohol.

$$CO.(NH.CH_2.OH)_2 + 2ROH \rightarrow CO(NH.CH_2.OR)_2 + 2H_2O$$

The subsequent polymerization probably occurs by a mechanism similar to that of bishydroxymethylurea, which first condenses to give a long chain structure, and subsequently undergoes cross-linking:

$$\begin{array}{c} \quad\quad\quad\;\; O \quad\quad\quad\quad\quad\quad\quad\quad\; O \\ \quad\quad\quad\;\; \| \quad\quad\quad\quad\quad\quad\quad\quad\; \| \\ -H.N-C-NH-CH_2-OR + H.N-C-NH-CH_2-OR \\ \quad\quad\quad\quad\quad\quad\quad\quad\quad\quad\quad\quad\quad\quad | \\ \quad\quad\quad\quad\quad\quad\quad\quad\quad\quad\quad\quad\quad\; CH_2-OR \\ \quad\quad\quad\quad\quad\quad\quad\quad\quad\quad\quad\quad\quad\quad\; \downarrow \\ \quad\quad O \quad\quad\quad\quad\quad\; O \\ \quad\quad \| \quad\quad\quad\quad\quad\; \| \\ -HN-C-NH-CH_2-N-C-NH-CH_2- + ROH \\ \quad\quad\quad\quad\quad\quad\quad\; | \\ \quad\quad\quad\quad\quad\quad\; CH_2-OR \end{array}$$

This polymerization is effected by heating above 80° C., so that the normal drying temperatures are suitable in contrast to the urea-formaldehyde polymerization. The resulting fabric has exceedingly high resistance to abrasion, equalled only by the A.C.G. process, and the tensile strength of the yarn is greatly increased. There is no doubt that this resin acts mainly by the 'spot-welding' mechanism, as microscopic examination showed abundant evidence of fibre bonding when the n-butyl ether was used, but not when the methyl ether was used.

Finally it will be mentioned that vinyl polymers produce unshrinkability by this mechanism. In all cases, the conditions are extremely critical[111] and direct polymerization on the surface does not in general lead to an unshrinkable finish. Thus polymers from butadiene and methyl methacrylate alone have no effect, whereas mixtures within a narrow composition range produce excellent shrinkage results without loss of handle.

In some cases, wool may be rendered unshrinkable by applying one polymer only in the presence of the correct quantity of plasticizer (see Fig. 2.9). Thus success was achieved with polymethyl metha-

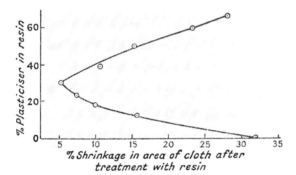

FIG. 2.9.—Influence of plasticizer (diethyl phthalate) content of resin (polymethyl methacrylate) on its ability to render wool resistant to felting. (Area shrinkage of untreated pattern 35%)[98]

crylate and polystyrene,[112] but not with vinyl chloride, acetate or cyanide. These resins are best applied in an emulsion form using an anionic detergent, and conditioning the wool with a cationic reagent In this way complete exhaustion from a dilute bath may be obtained.

The extensive experimental data which have been summarized

briefly above shows that the conditions are critical for a successful process leading to unshrinkability by the 'spot-welding' mechanism. Not only must the resin be firmly anchored to the fibre surface, but the mechanical properties of the polymer must be suitable to produce satisfactory bridge adhesion between adjacent fibres. The bridging material must be neither too rubbery nor too brittle, which no doubt explains the critical composition in the case of vinyl polymers and the critical molecular weight in the case of polyglycine.

REFERENCES

[1a] ANDERSON AND PALMER, International Wool Textile Organization, *Tech. Comm. Proc.*, 1948, 2, 5; [b] BARKER AND KING, *J. Text. Inst.*, 1926, 17, 153.
[2] SULLIVAN AND HERTEL, *Advances in Colloid Science*, N.Y., 1942, Vol. 1, 37.
[3] BAKHMETEFF AND FEODROFF, *J. Applied Mechanics*, 1937, 4, A-97.
[4] CARMAN, *Trans. Inst. Chem. Engrs.*, London, 1937, 15, 150; *J. Soc. Chem. Ind.* 1939, 58, 1.
[5] KOZENY, *Kulturtechniker*, 1932, 35, 478.
[6] EMERSLEBEN, *Physik. Z.*, 1925, 26, 601.
[7] CASSIE, *J. Text. Inst.*, 1942, 33, T185.
[8] BLOW AND MOXTON, *J. Soc. Chem. Ind.*, 1940, 59, 172.
[9] ANDERSON AND WARBURTON, *J. Text. Inst.*, 1949, 40, T749.
[10] SULLIVAN, *J. Applied Phys.*, 1941, 12, 503.
[11] SULLIVAN AND HERTEL, *ibid.*, 1940, 11, 761.
[12] SULLIVAN AND HERTEL, *Text. Res. J.*, 1940, 11, 30.
FOWLER AND HERTEL, *J. Applied Phys.*, 1940, 11, 496.
[13] BRUNAUER, *The adsorption of gases and vapours*, Oxford University Press, 1944; EMMETT, *Advances in colloid science*, New York, 1942, Vol. 1, 37.
[14] BRUNAUER AND EMMETT, *J. Amer. Chem. Soc.*, 1935, 57, 1754; EMMETT AND BRUNAUER, *ibid.*, 1937, 59, 1553.
[15] BRUNAUER, EMMETT AND TELLER, *ibid.*, 1938, 60, 309.
[16] LANGMUIR, *ibid.*, 1916, 38, 2221.
[17] HARKINS AND JURA, *ibid.*, 1944, 66, 1362; 1946, 68, 1941; see also FOSTER, *Faraday Society Discussion*, 1948, 3, 41.
[18] BULL, *J. Amer. Chem. Soc.*, 1944, 66, 1499.
[19] GILBERT, *Fibrous Proteins Symp.* (Soc. Dy. Col., Bradford), 1946, 96.
[20] Ref. 13(b), p. 23.
[21] ROWEN AND BLAINE, *Ind. Eng. Chem.*, 1947, 39, 1659.
[22] SCHOFIELD AND TALIBUDDIN, *Faraday Society Discussions*, 1948, 3, 51.
[23] SHAW, *Proc. Roy. Soc.*, 1928, 122A, 49.
[24] SHAW, *Phil. Mag.*, 1930, 9, 577.
[25] MARTIN, *Proc. Phys. Soc.*, 1941, 53, 186.
[26] MARTIN, *Min. Mag.*, 1931, 22, 519.
[27] ASTBURY, *Trans. Faraday Soc.*, 1933, 29, 193.
[28] MARTIN, *J. Soc. Dy. Col.*, 1944, 60, 325.
[29] *Wool Industries Research Association Bulletin*, 1941, 9, 4.
[30] MONGE, *Ann. de Chemie*, 1790, 6, 300.
[31] KING, *Biochem. J.*, 1927, 21, 434.
[32] SPEAKMAN AND STOTT, *J. Text. Inst.*, 1931, 22, T339.
[33] MORROW, *J. Text. Inst.*, 1931, T425.
[34a] SPEAKMAN, CHAMBERLAIN AND MENKART, *J. Text. Inst.*, 1945, 36, T91; [b]SPEAKMAN AND CHAMBERLAIN, *Nature*, 1942, 150, 546.
[35] MERCER, *Nature*, 1945, 155, 573; *J. Counc. Sci. Ind. Res. Australia*, 1945, 18, 188.
[36] BOWDEN AND LEDEN, *Proc. Roy. Soc.*, 1939, 169A, 371.

[37] LIPSON, *Nature*, 1945, **156**, 268; LIPSON AND HOWARD, *J. Soc. Dy. Col.*, 1946, **62**, 29.
[38] FRISHMAN, SMITH AND HARRIS, *Text Res. J.*, 1948, **43**, 475.
[39] MARTIN AND MITTELMANN, *J. Text. Inst.*, 1946, **37**, T269.
[40] KING, *J. Text. Inst.*, 1950, **41**, T135.
[41] LINDBERG AND GRALÉN, *Proc. Swedish Inst. Text. Res.*, 1948, **6**, 3; *Text. Res. J.*, 1948, **18**, 287.
[42] GRALÉN AND OLOFSSON, *Proc. Swedish Inst. Text. Res.*, 1947, **3**, 3; *Text. Res. J.*, 1947, **17**, 488.
[43] BOHM, *Nature*, 1945, **155**, 547; *J. Soc. Dy. Col.*, 1945, **61**, 278.
[44] MAKINSON, *Trans. Faraday Soc.*, 1948, **44**, 279.
[45] SPEAKMAN, STOTT AND CHANG, *J. Text. Inst.*, 1933, **24**, T273.
[46] MERCER AND MAKINSON, *ibid.*, 1947, **38**, T227.
[47] WHEWELL, RIGELHAUPT AND SELIM, *Nature*, 1945, **154**, 772.
[48] MEDLEY, *Nature*, 1950, **166**, 524.
[49] MAKINSON AND KING, *J. Text. Inst.*, 1950, **41**, 1.
[50] LINDBERG AND GRALÉN, *ibid.*, 1950, **41**, T331.
[51] GRALÉN, *J. Soc. Dy. Col.*, 1950, **66**, 465.
[52] SPEAKMAN, *ibid.*, p. 470.
[53] WHEWELL AND WOODS, *ibid.*, 1944, **60**, 148.
[54] LINDBERG, *Proc. Swedish Inst. Textile Res.*, 1948, **8**; *Test. Res. J.*, 1948, **18**, 470.
[55] See e.g. FRENEY, Fibrous Proteins Symp., Soc. Dy. Col., Bradford, 1946, 178.
[56] ALEXANDER, *J. Soc. Dy. Col.*, 1950, **66**, 349.
[57] LINDBERG AND GRALÉN, *Proc. Swedish Inst. Text. Res.*, 1949, **10**, *Textile Res. J.*, 1949, **19**, 183.
[58] ALEXANDER, Fibrous Proteins Symp., Soc. Dy. Col., Bradford, 1946, 199.
[59] SPEAKMAN AND GOODINGS, *J. Text. Inst.*, 1936, **17**, T607.
[60] W.I.R.A., PHILLIPS, MIDDLEBROOK AND HIGGINS, B.P. 546,915; U.S.P. 2,373,974.
[61] BOWDEN AND TABOR, *Proc. Roy. Soc.*, 1939, **169A**, 391; *Counc. Sci. Ind. Res. (Aust.) Bull.*, 1942, 145.
[62] ISHIDA, *Text. J. Aust.*, 1946, **21**, 7.
[63] JUSTIN-MUELLER, *Ref. Gen. Mat. Col.*, 1938, **42**, 378.
[64] SHOOTER AND THOMAS, *Research*, 1940, **2**, 533.
[65] WITT AND LEHMANN, *Chemische Technologie der Gespinstfasern*, Braunschweig, 1910, **1**, 86.
[66] RUDALL, quoted by SPEAKMAN, ref. [69].
[67] ALEXANDER, *Amer. Dyestuff Reporter*, 1950, **39**, 420.
[68] HANFORD AND ROYCE, *J. Amer. Chem. Soc.*, 1946, **68**, 2082.
[69] SPEAKMAN AND MENKART, *Nature*, 1945, **156**, 143.
[70] MERCER AND REES, *Australian J. Exp. Med. Sci.*, 1946, **24**, 147, 175; *Nature*, 1946, **157**, 589.
[71] MILLSON AND TURL, *American Dyestuff Reporter*, 1950, **39**, 647.
[72] TOMLINSON, *Phil. Mag.*, 1929, **7**, 905.
[73] THOMSON AND SPEAKMAN, *Nature*, 1946, **157**, 804.
[74] FRENEY, Fibrous Proteins Symposium, Soc. Dy. Col., Bradford, 1946, 178.
[75] BOXSER, *American Dyestuff Reporter*, 1938, **27**, 311.
[76] BERG, *Melliand Textilber.*, 1937, **18**, 438; *ibid.*, 1944, **25**, 110, 145, 183, 221.
[77] ARNOLD, *Textil-Forschung.*, 1929, **11**, 143.
[78] SHORTER, *J. Soc. Dy. Col.*, 1923, **39**, 270.
[79] C. D. H., *Deutsche Wollen-Gewerbe*, 1891, **23**, 1.
[80] See SPEAKMAN, Mather Lecture, *J. Text. Inst.*, 1941, **32**, No. 7.
[81] MENKART AND SPEAKMAN, *Nature*, 1947, **159**, 640.
[82] MARTIN, *J. Soc. Dy. Col.*, 1945, **61**, 173.
[83] HARRIS, *American Dyestuff Reporter*, 1945, **34**, 72.
[84] MERCER, *J. Counc. Sci. Ind. Res. Aust.*, 1942, **15**, No. 4.
[85] SCHOFIELD, *J. Text. Inst.*, 1938, **29**, T239.
[86] SPEAKMAN, MENKART AND LIU, *ibid.*, 1944, **35**, T41.
[87] SPEAKMAN AND HIRST, *Nature*, 1931, **127**, 665; *Trans. Faraday Soc.*, 1933, **29**, 148.

[88] PHILLIPS, *J. Soc. Dy. Col.*, 1944, 60, 330; *Fibrous Proteins Symp.*, *Soc. Dy. Col.*, Bradford, 1946, 190, 197, 201.
[89] CARTER, *Fibrous Proteins Symp.*, *Soc. Dy. Col.*, Bradford, 1946, 201.
[90] SPEAKMAN, *ibid.*, 202.
[91] ALEXANDER, CARTER AND EARLAND, *J. Soc. Dy. Col.*, 1951, 67, 23.
[92] BARR AND SPEAKMAN, *ibid.*, 1944, 60, 335.
[93] BARR AND SPEAKMAN, *ibid.*, p. 238.
[94] SPEAKMAN AND BARR, B.P. 559,787.
[95] BARR, CAPP AND SPEAKMAN, *J. Soc. Dy. Col.*, 1946, 62, 338.
[96] ALEXANDER, CARTER AND EARLAND, *ibid.*, 1949, 65, 107.
[97] NEISH AND SPEAKMAN, *Nature*, 1945, 156, 176.
[98] ALEXANDER, *J. Soc. Dy. Col.*, 1950, 66, 349.
[99] STOCK AND SALLEY, *Text. Res. J.*, 1949, 19, 41.
[100] BALDWIN, BARR AND SPEAKMAN, *J. Soc. Dy. Col.*, 1946, 62, 4.
[101] ALEXANDER, BAILEY AND CARTER, *Text. Res. J.*, 1950, 20, 385; B.P. 627,910.
[102] JACKSON AND LIPSON, *Text. Res. J.*, 1951, 21, 156; JACKSON, *ibid.*, 655.
[102a] B.P. 913,370 (1961).
[103] LANDOLT, *J. Soc. Dy. Col.*, 1948, 64, 93.
[104] MARSHALL AND AULABAUGH, *Text. Res. J.*, 1947, 17, 723.
[105] MCCLEARY AND ROYER, *ibid.*, 1949, 19, 457.
[106] MARESH AND ROYER, *ibid.*, 449.
[107] Anon., *American Dyestuff Reporter*, 1949, 38, 842.
[108] WOLSEY LTD., ALEXANDER AND BELL, B.P. 611,360.
[109] DUPONT, B.P. 537,971.
[110] ALEXANDER, CARTER AND EARLAND, *J. Soc. Dy. Col.*, 1950, 66, 579.
[111] RUST, U.S.P. 2,447,540; 2,447,772; 2,447,876; 2,447,877; and 2,447,878.
[112] WOLSEY LTD. AND ALEXANDER, B.P. 611,828 and 611,829.

CHAPTER 3

Mechanical Properties

The elasticity of wool fibres

THE mechanical properties of wool fibres have so far remained largely unexplained in terms of molecular structure and morphology. This is not surprising when it is realized that the elastic properties of homogeneous high polymers with well-established molecular structures have only been interpreted quantitatively with difficulty by the application of statistical mechanics. As this powerful mathematical tool can only be applied to systems of known structure, theoretical work on the mechanical properties of natural fibres is still mainly qualitative.

Following the pioneer work of Harrison,[1] Shorter,[2] and Karger and Schmid,[3] the most thorough investigation of the load–extension curve of wool under varying conditions has been made by Speakman.[4] In Fig. 3.1 the load–extension curves for wool at 25° C. and at humidities ranging from 0–100 per cent. are given. It will be seen, that in general the fibre may be stretched more readily and to greater extents as the moisture content of the wool is increased. Thus, although it is impossible to extend a dry fibre by more than 30 per cent., a wet fibre can be stretched to 60 per cent. and at very low rates of extension to 70 per cent. The curves shown in Fig. 3.1 were obtained at a fairly rapid rate of loading (1·8 grams per min.). With fibres conditioned at 100 per cent. humidity, the curves show an initial Hooke's law region of approximately 2 per cent. extension and then a rapid extension up to 25 per cent. Above this point, extension becomes increasingly difficult, except in the region of 50 per cent. where indications of an increased rate of extension similar to the region of very rapid extension, which immediately follows the Hooke's law region, are found. As the water content of the fibre increases from absolute dryness to saturation, the whole curve is foreshortened with respect to the abscissa, so that the limiting Hooke's law load for a wet fibre is only a quarter of that of a dry fibre.

Figure 3.2 shows how the form of the load–extension curve of fibres stretched in water changes with rising temperature. The limiting stress of the Hooke's law region is found to decrease continuously as the extensibility increases and the shoulder of the curves becomes less definite. Thus increasing the temperature has the same effect as increasing the humidity on the shape of the load extension curve, and

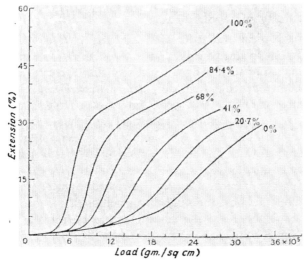

FIG. 3.1.—Load-extension curve of Cotswold wool for humidities varying from 0 to 100 per cent. at 25°C.[5] (*Reproduced by permission.*)

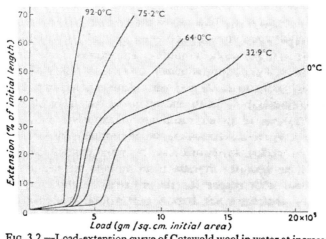

FIG. 3.2.—Load-extension curve of Cotswold wool in water at increasing temperatures.[4] (*Reproduced by permission.*)

makes the fibre more easy to stretch and enables it to be stretched to a greater extent. If a fibre is stretched quickly, and the tension released immediately after stretching, it will return to its original length but shows a marked hysteresis (Fig. 3.3.). Shorter[2] first pointed out that wool fibres invariably return to their original length even when stretched as much as 70 per cent. The actual breaking strength of the fibres under different conditions depends greatly on the rate of loading,[5, 6] and the elasticity is affected by the rate of applying stress even in the Hooke's law region, i.e. the value of Young's modulus depends on the rate of extension.[6a] Bull and Gutmann[8] consider the wool to behave as a gel beyond the Hooke region, which liquefies during elastic deformation, and consequently describe the stretching of a fibre in the region of 3–20 per cent. extension as a gel-sol transformation. Alternatively, Astbury and Woods[7] consider wool to consist of three phases of different elasticities. Direct chemical support for this elegant picture has recently been obtained and will be discussed in Chapter 12.

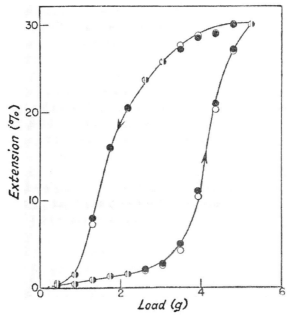

FIG. 3.3.—Stress-strain curve of wool fibre in water during two stress-strain cycles separated by 24 hours (O initial elongation; ● second elongation.[6] (*Reproduced by permission.*)

Speakman[9] showed that when a wool fibre is stretched rapidly in water at ordinary temperature the load–extension curve is reproducible if the extension does not exceed approximately 30 per cent. provided the fibre is released immediately after treatment and allowed to relax for several hours without tension before being stressed again (see Fig. 3.3). This discovery led to the development of the sensitive mechano-chemical method of analysis,[9] developed by Speakman, for detecting changes in fibres after physical and chemical treatments. Essentially, a single fibre is taken and stressed on a simple load–extension device while immersed in water, and from the area under the stress-strain curve the work required to stretch it by the arbitrarily chosen extent of 30 per cent. is determined. After a period of relaxation the fibre is given the proposed chemical or physical treatment, and after a suitable period it is restressed and the new value for the work required to stretch it by 30 per cent. determined.

Speakman, in his many researches, has always expressed the change in the work required to stretch 30 per cent. as a percentage of the original value, and called this the change in work. Harris on the other hand has usually expressed the two values as a ratio which he then terms the 30 per cent. index.[10] The decrease in work expressed by either method, provides valuable and sensitive information concerning the changes produced within the fibre by a chemical treatment. It is obvious that the change in the mechanical properties of such a complex substance as a wool fibre is likely to be influenced by many factors, and that great care must be taken in translating these changes into terms of molecular mechanisms involving bond-breaking or bond-forming processes. In spite of these uncertainties, however, the nature of many chemical reactions occurring within wool has been inferred almost entirely from observations of the changes in mechanical properties. Direct analytical methods for following changes within the fibre have only been applied within the last few years; these have in many cases confirmed the deductions made from the mechanical experiments.

If a fibre is stretched more than 30 per cent., it is irreversibly weakened and never regains its original strength even after long periods of resting. The same result may be achieved by stretching the fibre by 30 per cent. slowly in water, or by not allowing it to relax slowly immediately after application of the stress. All these treatments lead to a decrease in the work required to stretch the fibre to a given extension. The plastic flow which is responsible for this kind of damage, first described qualitatively by Astbury and Woods,[7] although it had long been known in the textile industry, has been the subject of a considerable number of investigations in different laboratories, notably those of Speakman[11] at Leeds and Eyring and Tobolsky[12, 13] at Princeton. The qualitative explanation first, advanced by Astbury and Woods, based on a three-phase system has probably not been improved upon by the detailed mathematical treatment of Eyring and Tobolsky, because of the complex and heterogeneous nature of the wool fibre. According to this explanation, rapid loading in water causes an extension which is confined largely to the amorphous regions of the fibre (i.e. the sol-gel transformation), whereas the molecular transformation or unfolding of the molecule (see p. 376) occurs only above 20 per cent. extension. The transformation from the one form to the other is far from complete at 30 per cent. extension, but when the fibre is maintained in a stretched condition this conversion continues as the tension relaxes.[14] The loss of tension may therefore be ascribed to the slow unfolding of the molecules, accompanied by rearrangement in the amorphous region. It is clear that if this latter process is prevented in some way from taking place to any appreciable extent, the fibre will return to its original state on release of the tension, because of the intrinsic contractile power of the macromolecule. If, however, irreversible changes have taken place in the amorphous region, the fibre will be weaker, although it may return to its original length. It will be shown later (p. 64) that hydrogen bonds between the molecules make important contributions to the strength of the fibre as a whole. It is possible that when a fibre is held in the extended form, changes in the hydrogen bonding occur while the main peptide fold is opening up, so that the arrangement of the grids with respect to one another is unlikely to be as perfect as in the original fibre. Alexander[15] showed from experiments in which the hydrogen bonding of the fibre was destroyed by strong solutions of lithium bromide, in which the fibre contracts and its strength is reduced to one-tenth of the original (see p. 79), that internal disarrangement of the fibre can lead to a weakening without a change in dimension. Yet if the fibre is removed from this solution, within half an hour it fully regains its original strength and shape. If, however, the fibre is left in the lithium bromide for several hours, on being removed from the solution it regains its original length but is weakened, e.g. its 30 per cent. index is 0·7 or there has been a reduction in work of 30 per cent. This loss in strength has been attributed to a molecular rearrangement within the fibre while the hydrogen bonds are broken, so that on removal of the lithium bromide a less perfect system of hydrogen bonds is rebuilt. By analogy with this behaviour, a similar type of rearrangement may account for the decay of tension in a stressed fibre, and the plastic flow of a fibre under constant load. Indirect evidence[16, 17] suggests that these phenomena are in part due to the breakdown of the disulphide bonds. Careful analytical work by Phillips[18] has, however, failed to detect any change in the cystine content after such treatments, so that the maximum number of broken disulphide bonds could only have been 3 per cent. of those initially present. A reduction of this magnitude is unlikely to influence markedly the mechanical properties of the fibres.

Wool fibres which have been held in a stretched condition in cold water so as to suffer loss of tension do not, as has already been mentioned, lose their ultimate power of recovery but only the speed at which they regain their initial length when the external stretching force is removed. For instance a fibre which was extended by 50 per cent. for almost one year required an immersion of 24 hours in water at room temperature before regaining its original length. Astbury and Woods[7] showed that this slow contraction of stressed fibres in water can be enormously accelerated by raising the temperature so that in steam the recovery is almost instantaneous. Although fibres treated in this way eventually regained their original length, they remained gravely weakened. In this connexion, it has been known for generations in the textile industry that unless wool fibres which have undergone severe stresses, such as those which occur in spinning, are allowed to rest at the highest possible humidity for several months, their mechanical properties are impaired. Jagger and Speakman[19] in a detailed series of experiments, determined the optimum conditions of relaxation necessary for maximum recovery of strength. These workers showed that a treatment at elevated temperatures in water repaired the damage most effectively, i.e. produced the greatest increase in strength. For example, a fibre stretched in water by 30 per cent. was kept under tension for 24 hours and then released in water for a further 24 hours, when it contracted to its original length but showed a reduction of work of almost 50 per cent. After treatment in water at pH 6 and 70° C., the same fibre was strengthened and only showed a reduction in work of 15 per cent. compared with its original value before being held under tension.

The kinetics of the slow stretching, or creep, of a fibre when held under constant tension has been studied by Speakman[20] who observed an exponential approach with time to the limiting extension of 70 per cent. at which the fibre breaks. When these results were plotted as log (length) against time, a linear relationship was obtained and the slope of the line is considered by Speakman to represent the rate of the unfolding of the molecules. This rate however varies greatly from fibre to fibre and according to Speakman and Ripa[17] is inversely proportional to the sulphur content of the wool. This deduction is based on analyses of a few fibres which are now known to be unreliable. The authors claim to have studied virgin fibres which contained less than 1 per cent. of sulphur, but this value is so extremely low, that the whole theory that rate of creep is inversely proportional to the number of disulphide bonds must be treated with reserve. A simple molecular interpretation of creep is not possible at the present time, especially as the cuticle is known to break down at extensions of more than 30–40 per cent. (see p. 9), which must influence the mechanical behaviour of the fibres. The temperature coefficient for this process was shown to correspond to an activation energy of 20 k. cal/mol. Meyer[28,29] believes the slow extension to be a measure of the internal viscosity of the fibre, and claims that it has a connexion with the $\alpha \to \beta$ transformation. From the data of Speakman he calculates an internal viscosity of 4×10^{13} c.g.s. units for wool fibres suspended in water.

The contribution of different bonds to the strength of the fibre

The disulphide bonds contribute largely to the wet strength of the fibre which decreases almost linearly with the cystine content. This was first shown by Harris and Brown[21] who reduced and methylated wool fibres to different extents and then determined their reduction in work or 30 per cent. index. More recently Alexander, Fox and Hudson[22] oxidized different proportions of the disulphide bonds with peracetic acid and then determined the strength recorded in Fig. 3.4. The strength of dry fibres, however, is not appreciably affected by the breakdown of disulphide bonds, and Fig. 3.5. shows the load extension curves of wool fibres in air in which different proportions of disulphide bonds have been broken. It can be clearly seen that a weakening only occurs after more than 60 per cent. of these have been reduced. The introduction of new cross-links (see p. 325) which often enhances the wet tensile strength, does not appreciably increase

62 WOOL: ITS CHEMISTRY AND PHYSICS

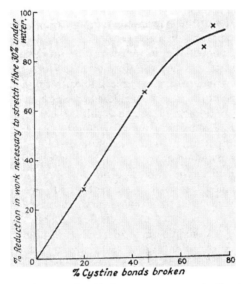

FIG. 3.4.—Relationship between wet strength of wool fibres and the number of disulphide bonds.[22]

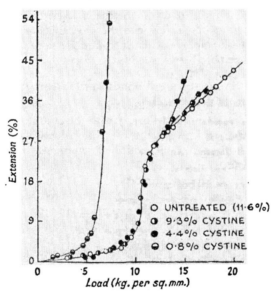

FIG. 3.5.—Relationship between dry strength of wool fibres and number of disulphide bonds present[21]; ○ untreated wool (11·6 per cent. cystine). ◐ 20 per cent. of the disulphide bonds reduced. ● 62 per cent. of the disulphide bonds reduced. ◉ 93 per cent. of the disulphide bonds reduced. (*Reproduced by permission.*)

the tensile strength of the dry fibre, and it can be concluded that the latter depends largely on the chain length and perhaps on inter-chain hydrogen bonding and not on covalent cross-links. In general the dry tensile strength of regenerated protein fibres is of the same order as that of wool but the former fibres are usually much weaker than wool when wet.

The dry strength of wool fibres is, however, greatly influenced by the number of peptide bonds (i.e. the length of the main chains), and decreases very much when 10–15 per cent. of the peptide bonds have been broken (see Fig. 9.1 p. 290). Peptide bond breakdown affects the wet tensile properties (see Table 9.1, p. 289) in a similar way, and from a practical point of view there can be no doubt that main chain hydrolysis has the greatest degradative effect. As the tensile strength of dry fibres is so much greater than that of wet fibres, it is clear that in its swelling action water breaks bonds which contribute to fibre strength. These bonds are probably weak hydrogen bonds formed between the peptide bonds. Besides these weak hydrogen bonds which can be broken by water, some hydrogen bonds can only be severed by more potent reagents such as phenol or lithium bromide. When a fibre is stressed in dilute aqueous phenol it swells considerably and becomes much weaker. The effect is fully reversible and on restretching the fibre in water alone it regains its original strength[23]

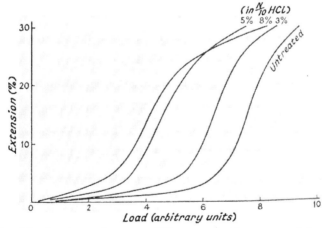

FIG. 3.6.—Load-extension curves of fibres stretched in aqueous solutions of phenol of different concentrations. The curve labelled 'Untreated' represents the load-extension curve in water (fibres allowed to swell in phenol solution for $\frac{1}{2}$ hr. before being stressed).[15]

(see Fig. 3.6). Strong solutions of lithium bromide (i.e. containing more than 50 g. of salt per 100 ml. liquid) cause a similar reversible weakening, although more dilute salt solutions increase the strength by dehydrating the fibre. With calcium and magnesium chloride this effect alone can operate, and the increase in strength is found to depend entirely on the concentration of the salt. Figure 3.7 shows the

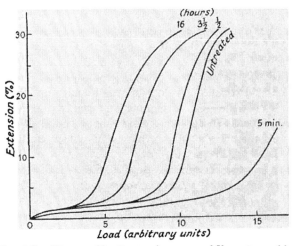

FIG. 3.7.—Change of load-extension curve of fibres stressed in 50 per cent. LiBr after different periods of immersion at room temperature. The curve labelled 'untreated' represents the load-extension curve in water.[15]

combined effect in lithium bromide solutions leading to an increase, and subsequently a decrease, in strength. The rate of breaking of hydrogen bonds depends on the diffusion of ions into the fibre and therefore follows the osmotic dehydration which is relatively rapid (Fig. 3.7). Fibres which have been weakened by reduction of the di-sulphide bonds, may regain their original strength on oxidation. Thus Harris demonstrated this to be the case if less than 50 per cent. of the disulphide bonds were broken by treatment with calcium thioglycol-late (see p. 246). Alternatively, the rebuilding can be carried out by incorporating new groups between the sulphur atoms by treatment with alkyl dihalides when again the fibre regains its original strength. In both cases, however, rebuilding of the links does not produce a fibre with the original tensile properties if more than 50 per cent. of the disulphide bonds were initially reduced. It is probable that after extensive disulphide breakdown, internal rearrangement takes place which prevents successful rebuilding of the cross-links. This rearrangement is prevented if the cross-linking takes place simultaneously with the reduction (see p. 341).

It is also possible to repair the strength of reduced fibres without cross-linking. Thus Harris[24] has shown that if the reduced disulphide bonds were treated with monohalides of high molecular weight, the fibres regained much of their strength although the disulphide bonds had not been rebuilt, and no new cross-links had been introduced. It is concluded that the interaction of secondary forces, arising from the high molecular weight residue introduced in the alkylation, increases the strength of the fibre, as treatment with alkyl halides of low molecular weight produces fibres which are permanently weakened. These observations clearly demonstrate that extensive chemical changes and breakdown of cross-links cannot always be detected by mechanical methods.

Entropy and free energy changes during elastic deformation

Elasticity which follows Hooke's law depends on inter-molecular attraction and on stretching the potential energy increases. Bodies which exhibit this normal elastic behaviour have a positive coefficient of thermal expansion and when held at a given extension the tension decreases on warming. This type of stretching can be said to increase the internal or free energy. Rubber-like materials on the other hand exhibit an entirely different behaviour since the chains are randomly folded, and on stressing assume a less random and therefore less probable configuration. The restoring force in this case is determined by entropy, i.e. the tendency to assume a completely random configuration. In such a system the tension increases on warming the material. Mathematically the position is summarized in the following equation:

$$K = \left(\frac{\delta U}{\delta l}\right)_T - T\left(\frac{\delta S}{\delta l}\right)_T = \left(\frac{\delta U}{\delta l}\right)_T + T\left(\frac{\delta K}{\delta T}\right)_l$$

K is restoring force at length l, U is the internal energy, T is the absolute temperature and S is the entropy. By carrying out extensions at different temperatures it is possible to calculate the contribution which the various factors make to the total tension. From Fig. 3.8 it can be seen that in the case of wool the entropy factor contributes very little to the total restoring force K, whereas in the case of rubber it contributes almost the entire energy. This condition was first established by Bull[25] for both wet and dry fibres. In an essentially similar investigation, Woods[26] reached the same conclusion. Elöd and Zahn[27] measured the change in tension with changes in temperature of a fibre which was held stressed at a constant extension and showed that this coefficient is positive for all types of keratin fibres even when almost all the disulphide bonds had been broken. These observations

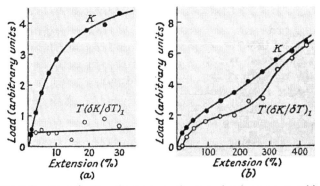

Fig. 3.8.—Contribution of entropy to the restoring force on stretching (a) supercontracted wool (b) rubber.[30] K represents the total restoring force and $T(\delta K/\delta T)_l$ represents the contribution of entropy to the restoring force.

confirmed that internal energy changes were responsible for the elastic behaviour of wool. The coefficient was found to be negative however when the fibres were stressed in phenol when hydrogen bonds were ruptured. Similarly disruption of disulphide bonds and rearrangements in micellar order by heating and similar treatments also produce negative temperature coefficients. Thus, so long as the fibre retains the α form, its elasticity is due to internal energy changes, whereas once its crystalline areas have been disorientated (this is facilitated by prior breakdown of disulphide bonds but it is not a direct consequence of this), it shows rubber-like behaviour. This is readily understood, as in the disorientated forms the internal molecular forces are reduced, the fibre no longer has its specific fold, and entropy changes alone contribute to its strength. It is important to realize that breakdown of the disulphide bonds without disorientation does not lead to rubber-like elasticity, and this supports the micellar picture of the wool molecule presented on p. 389.

The work quoted appears to provide powerful support for the theory of Astbury that the long-range elasticity of keratin fibres is a free-energy change resulting from attractive forces which tend to maintain the polypeptide chain in a folded configuration, but has been challenged by Meyer.[28] The latter maintains that the measurements of Woods, Bull, and Elöd and Zahn were all carried out in a non-equilibrium system, since the stress in the fibres studied by these workers diminished with time. Meyer stresses that wool fibres have an extremly high internal viscosity which masks the true elastic behaviour, and in the experiments of Bull and Woods the temperature coefficient of the complex process was determined and cannot be used for determining the nature of the elastic process. Valid determination can only be made by following changes in elasticity with temperature in fibres which no longer show plastic flow (e.g. which have been held under tension until no further change in length occurs or held at constant length until no further decay in tension occurs). In fibres swollen with water, the internal viscosity is so high that relaxation is too slow for experiments on the equilibrium elasticity to be made. Fibres swollen in glycollic acid, however, relax in a few days and Meyer et al.[29] measured their change in tension with temperature when maintained at constant length. These fibres showed complete rubber-like behaviour and no evidence was found for a change in free energy on stretching.

The theory of Astbury[30] that the long-range elasticity of wool fibres is a consequence of transformation of the polypeptide chain from a folded (i.e. shortened) α configuration to an extended β configuration (see p. 376) found its most powerful support in the thermodynamic data of Bull and Woods. If the views of Meyer have to be accepted that the long range of elasticity is essentially rubber-like, then the transformation from the α to β configuration does not contribute to the elasticity but is a direct consequence of extension. The change in the configuration in the crystalline areas of the fibre is then regarded as a recrystallization and on general thermodynamic grounds the more extended form is favoured by stretching. The long-range elasticity is considered to be confined entirely to the amorphous regions and should disappear if crystallization is complete. A reconsideration of the whole problem is required; on the present evidence neither point of view can be rejected and further experimental work is necessary.

Permanent set

Astbury and Woods[7] showed that when a moist wool fibre is stretched and held under tension for any length of time and allowed to dry, it remains elongated. This set however is not permanent, and on treatment with water the fibre usually returns to its original length in time (compare p. 60) and varying degrees of this so-called temporary set may be obtained. Thus a fibre which is set in hot water may not be released by cold water alone, but it can be released by water of the same temperature. Consequently a set imparted to fibres may for all practical purposes be termed permanent, although in actual fact the fibres can be relaxed under suitable conditions. It is generally acknowledged that intra-molecular hydrogen bonding is responsible for this behaviour, although the hydrogen bonds formed are not very stable and can be broken by hot water. If, however, fibres are stressed in steam for a period of at least 30 minutes (see Fig. 3.9), they acquire

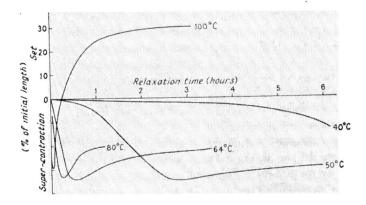

FIG. 3.9.—Setting and supercontraction of wool fibres after treatment with steam and water at different temperatures or various times. The fibre was stretched at 50 per cent. extension and held under tension in water or steam for the times given by the abscissae. The fibres were then steamed for one hour without tension. It is seen that the fibre obtains a set if it has been held under tension in steam for more than 30 minutes.[7] (*Reproduced by permission.*)

a set which cannot be undone by subsequent treatment with steam or boiling water. This condition is referred to as permanent set and usually an arbitary time of one hour in water at 100° C. has been used to determine the extent of permanent setting. It must, however, be realized that in many cases a set which remains after this period of boiling will decrease in longer periods in boiling water[31] (see Fig. 3.10). Speakman[32] concludes that the following reactions occur when wool fibres are set in steam: preliminary disulphide bond breakdown to the sulphenic acid or the aldehyde followed by cross-linking between the products of disulphide bond hydrolysis and the basic side chains, to form bonds of the following type:

(1) $R \cdot CH_2 \cdot SOH + NH_2 \cdot R' \rightarrow R \cdot CH_2 \cdot S \cdot NH-R'$
(2) $R \cdot CHO + NH_2 R' \rightarrow R \cdot CH = N-R'$

This chemical mechanism for permanent setting finds a measure of support from various observations, in particular the discovery that

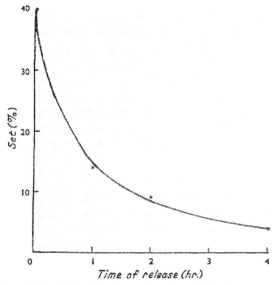

FIG. 3.10.—Effect of prolonged treatment in boiling water on fibre set at 40 per cent. extension by placing in a buffer of pH 5 at 100° C. for 2 hours.[31]

fibres of deaminated wool cannot be set,[32] and that the presence of acids also inhibits the setting of fibres.[33] Furthermore, blocking of the amino groups by reaction with dinitrofluorobenzene also greatly decreases the tendency of the fibres to acquire a permanent set.[34] After fibres have been set, the number of ϵ-amino groups available for reaction with dinitrofluorobenzene is reduced and this was interpreted as showing that these groups had taken part in the formation of new covalent links during setting.[34a] On the other hand, it must be pointed out that it has repeatedly been found that some of the amino groups in native proteins do not react with dinitrofluorobenzene because they are sterically inaccessible. It is therefore possible that during the setting process when the crystalline structure is changed from α to β (see p. 376) some of the amino groups become hidden without having been chemically changed. This suggestion is supported by the finding that in the β protein fibre, silk, many of the amino groups fail to react with dinitrofluorobenzene.

The influence of dyeing on the setting properties of fibres has not yet been satisfactorily decided. The evidence of Astbury and Dawson[35] (see p. 239) conflicts with the detailed work of Elliott and Speakman,[33] who have shown that fibres dyed with dyes of high affinity can no longer acquire a set. They conclude that the amino groups play an important role in the setting of fibres in steam or boiling water, since there can be little doubt that the dye molecules are associated in some way with the amino groups. It is doubtful whether this problem can be solved until more information is available concerning the actual position of the dye in relation to the crystallites and the amorphous areas within the fibre. Since charged amino groups prevent a fibre from acquiring a permanent set, it seems possible that it is the acidity of the fibre and not the actual deposition of dye which influences the setting of dyed fibres. If this view is correct then slight changes in experimental conditions of dyeing could account for the apparently conflicting results. Direct analyses do not support the chemical mechanism proposed by Speakman and although disulphide bonds can be hydrolysed by steam, Phillips[18, 36] failed to detect any appreciable reduction in the disulphide content of wool which had been steamed to give a permanent set. Moreover, no chemical evidence for the existence of S-N bonds could be found and it should be emphasized that their existence is postulated on the basis of mechanical experiments only (c.f., however, Speakman[36a]). In addition, Rudall[37] showed that fibres which had acquired a permanent set returned to their original length by treatments with hot concentrated solutions of urea, and Alexander[15] has shown that concentrated solutions of lithium bromide were also capable of reversing the set in fibres. Since neither of these reagents can break covalent bonds of the type envisaged by Speakman, it is difficult to reconcile these observations with the chemical theory of setting. All fibres set in steam are of course in the β configuration and the treatment with urea converts them back to the α form. A permanent set may be imparted to wool fibres by methods other than steaming, for example by holding them in an extended form for one hour in aqueous solution of pH 6–10.[31, 38] Fig. 3.11 shows the relationship between the pH of the setting medium and the set retained after release in boiling water, and Fig. 3.12 shows the rate at which set, permanent to an hour's boiling, is acquired in buffer solutions at 100° C. and pH 9 when the fibres are stretched by 30 per cent.

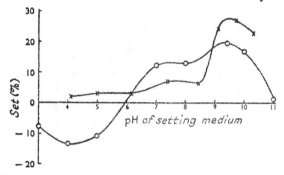

Fig. 3.11.—Permanent set of fibres after treatment in buffer solutions of different pH at 100° C. for 30 minutes while extended by 40 per cent.
O Data from Hind and Speakman.[38]
X Data from Blackburn and Lindley.[31]

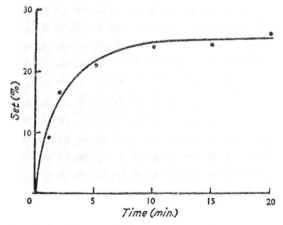

Fig. 3.12.—Rate at which set permanent to one hour's boiling is acquired at pH 9.[31] (*Reproduced by permission.*)

The fact that only solutions above pH 6 are effective in giving wool a permanent set strongly supports Speakman's mechanism, since the stability of the disulphide bond to hydrolysis greatly decreases with increasing pH. On the other hand, the set which is acquired in this way can be reversed by concentrated solutions of lithium bromide and urea. This, and similar observations, have brought various investigators to the conclusion that the set given to wool fibres by water at 100° C. and by steam, is not due to the formation of new covalent cross-links, but to the formation of strong hydrogen bonds which tend to retain the fibre in its new elongated form. According to this view, hydrogen bond breakers such as urea and lithium bromide release permanent set by breaking the new hydrogen bonds which are stable to boiling water. Thus, silk is readily dissolved by lithium bromide, but not of course by boiling water although the molecules are held together by hydrogen bonds only. It is clear, therefore, that hydrogen bonds stable to reagents such as boiling water and formic acid can exist. On this basis, there is no sharp dividing line between permanent and temporary set, and there is only a gradual transition in the strength of the hydrogen bonds formed during setting. This interpretation is in general agreement with the following view expressed by Astbury and Woods[7] who first discovered permanent and temporary set and who stated in their original publication: 'the phenomena associated with loss of tension and speed of recovery in water at ordinary temperatures are the first manifestations of those phenomena of temporary and permanent set which find their strongest expression under the action of steam.'

As the so-called permanent set obtained with steam or boiling water always results in the conversion of the fibre from the α to the β form, it seems probable that the powerful hydrogen bonds can exist only between the fully extended peptide chains and may be similar to those present in silk. It should be emphasized that the importance of the amino groups in the setting mechanism, which has been repeatedly demonstrated by Speakman, cannot be fitted into this picture but this is not altogether surprising since no information concerning the nature of the different hydrogen bonds present in wool is available.

In addition to setting by water and steam, fibres can be given a set at low temperatures by breaking the disulphide bonds, stretching them after disruption of the disulphide bonds and then rebuilding the latter, while the fibre remains in a stretched condition. If mild conditions are employed, the fibre retains its α form[15] although it may have received a set resulting in an increase in length of as much as 40 per cent. The principles of this low-temperature setting were discovered and developed by Speakman who reduced the disulphide bonds with sodium bisulphite or sulphite and then repaired the reduced bonds with metals (see p. 339).

As already mentioned (see p. 65), Harris and his co-workers developed Speakman's ideas, and their process of reducing stretched hair with calcium thioglycollate and treating the stressed and reduced fibre with an oxidizing agent, such as bromate or a persulphate, forms the basis of all modern chemical permanent waving. Harris also developed a process for setting wool fabrics in which a disulphide bond reducing agent and cross-linking compound are present simultaneously (see p. 339). Of increasing importance, however, are commercial processes in which the disulphide bonds are reduced by sodium or mono-ethanolamine bisulphite or by ammonium thioglycollate, followed by a steaming or blowing process. The Harris set, being caused by covalent linking, cannot be undone by powerful hydrogen-bond breakers such as lithium bromide solution, whilst the thioglycollate or bisulphite set is less stable owing to its dependence on hydrogen bonding (Farnworth[37a]).

Supercontraction

The phenomenon of supercontraction was first discovered by Astbury and Woods[7] in the course of experiments on the effect of steam on wool fibres. If the period of steaming of the stressed fibre is much shorter than that required to produce set (i.e. of the order of a few minutes only) the fibre is found to contract to a length less than the original value on being released from tension and placed in water at 100° C. or in steam. Appropriate data for a fibre which was stretched by 40 per cent., held in steam under tension for varying periods, and then released in steam are given in Fig. 3.9. Thus, for a period of steaming greater than 15 min., set is always produced, whereas for shorter periods supercontraction occurs. It is not necessary to use steam or boiling water on the stressed fibres to cause supercontraction if they are subsequently steamed without tension. From Fig. 3.9 it can be seen that if fibres under tension are treated for a sufficiently long time in water at lower temperatures (even as low as 20° C.), they will supercontract when subsequently steamed without tension.

Blackburn and Lindley[31] showed that it was necessary for the fibre to be stressed rapidly before being placed in steam under tension in order to obtain supercontraction. Thus, if the fibres are stretched by

40 per cent. over a period of 2–3 hours, then steamed for 2 minutes and released in steam, they will only supercontract by about 5 per cent. as opposed to 25 per cent. when the initial stretching takes less than 2 minutes. In addition, the steaming should follow rapidly upon the stretching operation for maximum supercontraction and if the fibre is stressed by 40 per cent. and left for 60 minutes at room temperature before being steamed for 2 minutes under tension, and then released in steam, supercontraction no longer takes place.

One of the simplest ways of causing supercontraction is to stretch the wool fibre in dilute caustic soda solution to a high extension and allow relaxation to take place whilst under tension in an alkaline solution.[7] On releasing the tension, the fibre contracts by about 10 per cent. It is important to note that the fibre after this contraction is still in the α form and gives a well-defined X-ray diagram. When the fibre which has been treated in this way is steamed it is found to contract still further (i.e. the total decrease in length may be as large as 50 per cent.) when the crystal structure of the fibre is completely altered and a partially disorientated β pattern appears. Zahn[55] found that treatment with caustic soda, even when the fibre is not under tension, changes it so that it is then capable of contracting in steam. One of the most important observations in connexion with supercontraction, due to Woods,[39] is that isolated cortical cells of wool keratin exhibit supercontraction in the same way as whole wool fibres, and that the mechanism of supercontraction is therefore intra-cellular and is not connected with the complex morphology of the fibre.

Although supercontraction originally referred to the contraction of a stretched fibre to a length less than its original, it is now also used to describe the contraction which occurs when unstretched fibres are subjected to certain chemical treatments. Supercontraction of unstrained fibres was first observed by Speakman[40] who found that after treatment with sodium sulphide, fibres contracted in length by 6·2 per cent., and after the reagent was washed out, contracted further on drying to give a total decrease in length of 20 per cent. It has since been found that many reagents which are capable of breaking the disulphide bond cause supercontraction to some extent, e.g. sodium bisulphite and sodium cyanide at 100° C.,[32, 41] solutions of silver and mercury salts, when applied at high temperatures for several hours[42] or at low temperatures for several days;[43] chloramides,[44] peracetic acid,[15] and chlorine peroxide[45] when applied at room temperature for a time sufficient to oxidize most of the disulphide bonds;

powerful reducing agents such as sulfoxylates and hydrosulphites.[46] It is surprising to find, however, that two reagents known to break disulphide bonds effectively, namely solutions of chlorine and solutions of sodium sulphite, do not bring about supercontraction.[47] It was concluded from these observations that the breaking of disulphide bonds is essential for supercontraction, and it was widely held formerly that contraction without breakdown of disulphide bonds is impossible. This view was strongly supported by the fact that when the disulphide bonds had been replaced by more stable bonds such as lanthionine, metal complexes[48] (e.g. S–Ba–S) or bonds of the type S–CH$_2$–S, the fibres no longer contracted in any of the media described above. Subsequent to this work, however, considerable evidence has been accumulated, mainly as the result of work carried out by Elöd and his colleagues, to suggest that another group of reagents which cannot possibly break disulphide bonds is capable of bringing about extensive supercontraction. The reagents investigated by Elöd were phenols[50] and formamide,[49] both of which were considered to act by breaking hydrogen bonds. More recently, strong aqueous solutions of lithium bromide have also been used. The influence of the concentration of formamide in water and the temperature of the solution on the degree of supercontraction can be seen from Fig. 3.13. No appreciable contraction takes place at temperatures below 100° C. Phenol[23] is a much more effective supercontracting

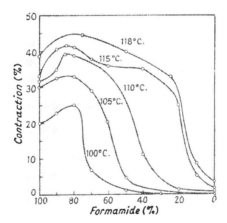

Fig. 3.13.—Supercontraction of horse hair after treatment for 3 hours in different mixtures of formamide and water at various temperatures.[49]

agent than formamide and causes supercontraction at lower temperatures (see Fig. 3.14). Figures 3.15 and 3.16 show that aqueous solutions containing 20 per cent. phenol and at a pH greater than 7 are most effective. In addition to phenol itself, the dihydroxy compounds catechol and resorcinol are also reactive. Hydroquinone, unlike its two isomers, does not bring about contraction and nitrated phenols also appear to be ineffective. Zahn[51] showed that the degree of

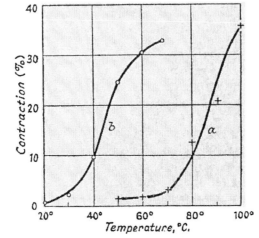

FIG. 3.14.—Influence of temperature on supercontraction of horse hair in 25 per cent. phenol.[23] (a) Untreated hair. (b) Reduced hair.

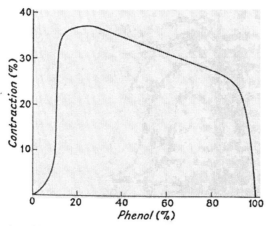

FIG. 3.15.—Supercontraction of horse hair in different phenol-water mixtures at 100° C.[23]

purity of the phenol was an important factor, and that certain commercial phenols contain impurities which prevent contraction from taking place. Zahn[52] believes that the temperature at which supercontraction in phenol commences represents the softening point of the keratin, and he compares this with the softening points of plastic materials. The temperature at which phenol brings about supercontraction varies with the disulphide content of the fibre used. Partially reduced keratin (see Fig. 3.14) contracts at much lower

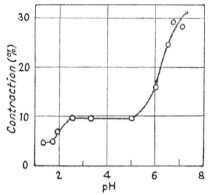

FIG. 3.16.—Influence of pH of phenol (20 per cent.) solution on supercontraction of horse hair at 100° C.[50]

temperatures than normal wool. When examining a whole series of different keratin fibres, Zahn was able to correlate the minimum temperature necessary for contraction with the cystine content of the fibres. The exact significance of a so-called softening point in a highly cross-linked system is, however, open to question and no completely satisfactory molecular interpretation for this interesting phenomenon has yet been advanced.

As already indicated, Elöd and Zahn[50] consider that phenols and formamide cause supercontraction by the breaking of hydrogen bonds, as no change in cystine content could be detected. Subsequent work with concentrated lithium bromide solutions, reported briefly on p. 64, has supported these general conclusions. Lithium bromide will bring about supercontraction, though slowly, even in the cold (Fig. 3.17) unlike the less powerful reagents phenol and formamide, which have to be applied at high temperatures causing intense swelling of the fibres. Lithium bromide is capable of dissolving silk and cellulose whereas phenol and formamide cannot disrupt the strong hydrogen bonds in these fibres. In the case of lithium bromide, only solutions of concentration greater than 50 per cent. cause super-contraction (see Fig. 3.18) which is completely reversible except for

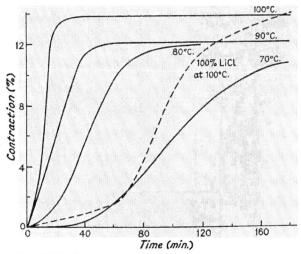

FIG. 3.17.—Rate of reversible contraction of wool fibres in 100 per cent. LiBr different temperatures. Also, rate of contraction in 100 per cent. lithium chloride at 100° C.[15]

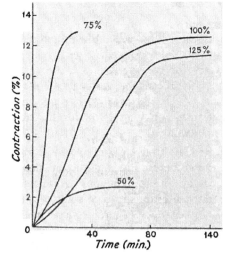

FIG. 3.18.—Reversible contraction of wool fibres in LiBr solutions of different concentrations at 85° C.[15]

prolonged treatments at elevated temperatures. Even after such treatments resulting in irreversible contraction, the cystine content of the wool is unaffected. If the lithium bromide solutions or the fibres are sufficiently acid to cause limited main chain breakdown, the supercontraction is always irreversible even after very short treatments. This indicates that molecular rearrangement, which occurs while all the stabilizing hydrogen bonds are broken, is the fundamental cause of this change in behaviour.

It is thought[15] that as lithium bromide diffuses slowly into the micelles the latter are distorted with consequent swelling to such an extent that the X-ray photograph is changed to that of disorientated β keratin. If the disorientation proceeds sufficiently, the fibre is unable to return to its original length when the lithium bromide is eventually removed by washing.

The hydrogen-bond breaking capacity of strong lithium bromide solutions is surprising since no other simple electrolyte solutions of corresponding concentrations have the same effect. Lithium, however, being the smallest cation except for the proton, has a very high polarizing power and is highly hydrated in solution. It is probable, therefore, that the lithium ion when incompletely hydrated combines with polarizable groups in the same way as it combines with water of hydration. Thus, in a protein it may co-ordinate with the imide groups and prevent these from forming hydrogen bonds. This hypothesis[15] receives some support from the fact that lithium bromide only begins to exert its effect in solutions which contain more than 50 g. per 100 c.c. of water, when the amount of water available is only sufficient to provide 10 molecules for each lithium ion. The exact number of molecules of water which are required to hydrate a lithium ion is not known, but is thought to be of this order of magnitude. Thus, according to this explanation solutions containing more than 50g. of lithium bromide per 100c.c. contain incompletely hydrated lithium ions which then co-ordinate with groups in the protein. When such a fibre is placed in the water the lithium ions are immediately hydrated and hydrogen bonding within the fibre can again take place, resulting in no change in steric arrangement unless relative movement of the chains has occurred due to prolonged immersion in the solution.

Reversible contraction was first obtained by Whewell and Woods[47] who used solutions of cuprammonium hydroxide. The results of a typical experiment are given in Fig. 3.19. If the fibres are washed in hydrochloric acid they return to their original length, unless the treatment in cuprammonium has been carried out for too long a period. When in the contracted state the fibre gives no X-ray pattern of any kind (i.e. there is one central black spot only and no arcs or spots) but after the supercontracted fibre is washed with acid the original α photograph is obtained. It seems highly probable in this case also that the reagent acts primarily by breaking hydrogen bonds,

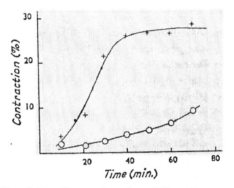

Fig. 3.19.—Contraction of wool fibres in cuprammonium hydroxide (-+-+-); reversing supercontraction by washing with acid (-O-O-).[47] (*Reproduced by permission.*)

since this reagent dissolves cellulose and silk, but as it is very alkaline, attack on the disulphide bonds also occurs. It is therefore impossible to determine the contribution which hydrogen and disulphide bonds respectively play in supercontraction. Although the fibre returns to its original length after a treatment with cuprammonium followed by acid, it is nevertheless weakened, presumably because of the disulphide bond breakdown due to the alkalinity of the solution. Except for such minor differences, therefore, there is little doubt that the effects of lithium bromide and cuprammonium are very similar.

Supercontraction produced by steaming a stretched fibre may also be due to breakdown of hydrogen bonds for, as already mentioned, no reaction of cystine can be detected. Since steam is only able to exert its hydrogen-bond breaking power on stretched fibres, i.e. fibres in the β configuration, it would appear that hydrogen bonds of the stronger type can be broken more easily in stretched fibres. However, the whole mechanism of contraction in steam requires further study and it would be dangerous at this stage to draw definite conclusions.

Unstretched fibres also contract when heated in water at 135° C. in a sealed tube (see p. 300). Under these conditions, breakdown of disulphide bonds occurs and it is difficult to determine whether breakdown of the latter or breakdown of hydrogen bonds is responsible for the supercontraction. Dry heat above 200° C. produces supercontraction (see p. 301) but here again destruction of cystine occurs.

As a result of the work[15] with peracetic acid, it has been possible to determine the requisite number of disulphide bonds which have to be broken before supercontraction occurs as a result of disulphide breakdown only. From Table 3.1 it can be seen that no contraction

TABLE 3.1[15]

Contraction of wool oxidized with peracetic acid

Time of treatment in 1·6 per cent peracetic acid at 22° C.	Percentage of cystine oxidized to cysteic acid	Contraction %	
		after thorough rinsing in cold water	followed by 5 min. in water at 95° C.
0·5 hr.	46	0·0	1·0
1·0	71	0·0	22·0
25·0	92	20·0	41·0*

* Fibre gives a disorientated β X-ray diagram; in all the other cases the fibres were in the α configuration.

takes place until more than 90 per cent. of the disulphide bonds have been oxidized. If the oxidation treatment is followed by immersion in water at 95° C. it is possible to obtain supercontraction after 70 per cent. of the disulphide bonds have been broken. This suggests that in treatments with disulphide-bond breaking reagents, such as bisulphite at 100° C., more than 70 per cent. of the disulphide bonds have to be broken before supercontraction occurs. Unfortunately no direct chemical evidence is available to substantiate this deduction. From a theoretical point of view it is significant that after the peracetic acid treatment, which leads to a contraction of approximately 20 per cent., the α structure of the fibre is retained and an X-ray photograph very similar to that of an untreated fibre obtained. However, if the oxidized fibre is placed in water at 95° C., in which it contracts further, the crystalline structure is completly altered and an X-ray photo of disorientated keratin is obtained. It would appear from this observation that disorientation of the crystalline structure of the keratin fibre is greatly facilitated by rupture of the disulphide bonds. This conclusion is in agreement with the observations made by Elöd and Zahn[50] that contraction in phenol occurs at lower temperatures after some of the disulphide bonds have been reduced, and that fully reduced wool contracts severely on being placed in boiling water. It should be emphasized, however, that no contraction occurs even in water at 100° C. if less than 50 per cent. of the disulphide bonds have been broken, and this disposes of the suggestion made by Speakman that the breakdown of a very small number of disulphide bonds may be responsible for the supercontraction of fibres after treatment such as steaming.

The supercontraction behaviour of wool can be interpreted on the basis of the model[15, 53] (see p. 390), which proposes that the morphological sub-units are made up of two phases, polypeptide chains of high molecular weight in the form of micelles held together by hydrogen bonds set in a three-dimensionally cross-linked cement. Breakdown of disulphide bonds leads to a contraction of the cement which then becomes rubber-like and results in a relative movement of the micelles without destroying the crystalline structure. The micelles, however, are disrupted by the breakdown of hydrogen bonds and the crystalline structure of the fibre is then lost. According to this model the breakdown of some inter-micellar disulphide bonds facilitates the disruption of the micelles, and less powerful reagents can then break the hydrogen bonds which hold each micelle together. For example, water at 95° C. can disrupt the micelles which contain no disulphide bonds, whereas water at 140° C. or a strong hydrogen-bond breaking solution, is required to break up these when the disulphide bonds have not been disturbed.

It will be mentioned in Chapter 4 that wool shows a marked swelling anisotropy and optical birefringence properties which are associated with an orientated structure. After contraction in formamide, phenol or lithium bromide, however, the fibre loses both these properties.[53] This observation agrees with the suggestion advanced above that these reagents disrupt the crystallites. Disulphide-bond breakdown alone does not produce disorientation of the crystallites, and after oxidation with peracetic acid the swelling anisotropy of the fibre is increased.[15]

Various workers have investigated the effect of reactions with the acidic and basic side chains on the extent of supercontraction. Thus, Speakman has shown that deaminated fibres contract more than untreated fibres in bisulphite, and concluded that the amino groups prevent contraction. When the amino groups are blocked by acetylation,[53] however, the contraction in bisulphite is unchanged, and it seems probable that the greater contraction of deaminated fibres is due to the general degradation (see p. 316) which occurs in the reaction. Blackburn and Lindley[31] found that the contraction in bisulphite is decreased considerably by esterification of the carboxyl groups although the contraction in hydrogen-bond breaking solutions is unaffected (see Table 3.2).

TABLE 3.2[53]

Influence of chemical modifications on the contraction of wool fibres in solutions of lithium bromide and bisulphite

Treatment of wool fibre	Contraction (per cent.) in solution of	
	lithium bromide	bisulphite
None	14	28
0·1% HCHO; 60° C.; 30 min.	14	10
0·3% ,, ,, ,, ,,	13	3
Esterified with epichlorhydrin	16	12
Esterified with methyl alcohol	15	18
Acetylated with acetic anhydride	12	30

The influence of new cross-links introduced by chemical treatment on the supercontraction of wool fibres is extremely complicated. The cross-links may be formed in only one of the phases and thus influence the supercontraction in one type of medium and not in another. Thus, after a fibre has been treated with formaldehyde (see p. 327), it contracts in the hydrogen-bond breaking solutions of lithium bromide but not in bisulphite (see Table 3.2). Contraction in phenol and similarly in bisulphite seems to be very greatly influenced by the introduction of new cross-links. If the disulphide bond is replaced by a more stable link, e.g. lanthionine by treatment with alkali or cyanide, fibres no longer contract in solutions which are capable of breaking disulphide bonds, whereas they still contract in lithium bromide (see Table 3.3) in agreement with the proposed two-phase

TABLE 3.3[15]

Influence of an alkali pretreatment which introduces —$CH_2 \cdot S \cdot CH_2$— *cross-links on supercontraction in different media*

Treatment	Alkali pretreatment	Percentage contraction
5% Sodium metabisulphite for 1 hr. at 100° C.	None	20·3
	pH 9 buffer for 3 hr. at 100° C.	1·0
100% LiBr for 3 hr. at 100° C.	None	13·5
	pH 9 buffer for 3 hr. at 100° C.	19·6
2% Peracetic acid for 24 hr. at 20° C. followed by water at 95° C.	None	41·0
	pH 9 buffer for 3 hr. at 100° C.	0·0

model. The contraction in phenol is difficult to explain since it is decreased on the introduction of new cross-links but not influenced by an alkali pretreatment which introduces lanthionine links. Phenol, unlike lithium bromide, swells the wool fibre very considerably and is adsorbed by the fibre by replacing bound water. The swelling is no doubt due to the increased size of the phenol molecule compared with water. There are, however, a number of organic acids (e.g. formic acid) which swell the fibre without bringing about supercontraction and the two effects are therefore not directly related. The only obvious difference in the reaction between phenol and formic acid lies in the pH of the system, and extensive swelling may be unable to cause supercontraction in acidic solutions. This view is supported by experiments of Elöd and Zahn[50] that the contraction in phenol increases with increase in pH (see Fig. 3.16). Strong aqueous solutions of sodium xylene sulphonate[54] bring about pronounced swelling and at higher temperatures supercontraction in a similar way to phenol.

REFERENCES

[1] HARRISON, *Proc. Roy. Soc.*, 1918, **94A**, 460.
[2] SHORTER, *Trans. Faraday Soc.*, 1924, **20**, 228.
[3] KARGER AND SCHMID., *Z. tech. Phys.*, 1925, **6**, 124.
[4] SPEAKMAN, *Proc. Roy. Soc.*, 1928, **103B**, 377.
[5] ASTBURY AND STREET, *Phil. Trans. Proc. Roy. Soc.*, 1931, **230A**, 75.

[6] HARRIS, MIZELL AND FOURT, *J. Res. N.B.S.*, 1942, **29**, 73.
[6a] SIKORSKI AND WOODS, *Proc. Leeds Phil. Soc.*, 1950, **5**, 313.
[7] ASTBURY AND WOODS, *Phil. Trans. Proc. Roy. Soc.*, 1933, **232A**, 336.
[8] BULL AND GUTMANN, *J. Amer. Chem. Soc.*, 1944, **66**, 1253.
[9] SPEAKMAN, *J. Text. Inst.*, 1947, **37T**, 102.
[10] SOOKNE AND HARRIS, *J. Res. N.B.S.*, 1937, **19**, 535.
[11] SPEAKMAN AND SHAH, *J. Soc. Dy. Col.*, 1941, **57**, 108.
[12] HALSEY AND EYRING, *Text Res. J.*, 1946, **16**, 329.
[13] KATZ AND TOBOLSKY, *ibid.*, 1950, **20**, 87.
[14] MERCER, *J. Text. Inst.*, 1948, **39T**, 246.
[15] ALEXANDER, *Ann. N.Y. Acad. Sci.*, 1951, **53**, 653.
[16] SPEAKMAN, *J. Text. Inst.*, 1936, **27T**, 231; *Nature*, 1941, **148**, 141.
[17] SPEAKMAN AND RIPA, *Nature*, 1950, **166**, 570; *Text Res. J.*, 1951, **21**, 215.
[18] LINDLEY AND PHILLIPS, *Biochem. J.*, 1945, **39**, 17.
CARTER, MIDDLEBROOK AND PHILLIPS, *J. Soc. Dy. Col.*, 1946, **52**, 203.
[19] JAGGER AND SPEAKMAN, *J. Text. Inst.*, 1951, **41P**, *Nature*, 1949, **164**, 190.
[20] SPEAKMAN, *Nature*, 1947, **159**, 338.
[21] HARRIS AND BROWN, *Symp. Fibrous Proteins*, 1946, p. 203 (Publ. Soc. Dy. Col., Bradford.)
[22] ALEXANDER, FOX AND HUDSON, *Biochem. J.*, 1951, **49**, 129.
[23] ZAHN, *Z. f. Naturforschung*, 1947, **2b**, 286.
[24] PATTERSON, GEIGER, MIZELL AND HARRIS, *J. Res. N.B.S.*, 1940, **25**, 451.
[25] BULL, *J. Amer. Chem. Soc.*, 1945, **67**, 533.
[26] WOODS, *Nature*, 1946, **157**, 229.
[27] ELÖD AND ZAHN, *Kolloid Z.*, 1949, **113**, 10.
[28] MEYER AND HASELBACH, *Nature*, 1949, **164**, 33.
[29] MEYER, WYK, GONON AND HASELBACH, *Trans. Faraday Soc.*, 1952, **48**, 669.
[30] ASTBURY, *Proc. Roy. Soc.*, 1947, **134B**, 303.
[31] BLACKBURN AND LINDLEY, *J. Soc. Dy. Col.*, 1948, **64**, 305.
[32] SPEAKMAN, *J. Soc. Dy. Col.*, 1936, **52**, 335.
[33] SPEAKMAN AND ELLIOTT, *Symp. Fibrous Proteins*, 1946, p. 116 (publ. Soc. Dy. Col., Bradford).
[34] FARNWORTH AND SPEAKMAN, *Nature*, 1948, **161**, 890.
[34a] SPEAKMAN AND ASQUITH, *Nature*, 1952, **170**, 798.
[35] ASTBURY AND DAWSON, *J. Soc. Dy. Col.*, 1938, **54**, 6.
[36] PHILLIPS, *Symp. Fibrous Proteins*, 1946, p. 39 (publ. Soc. Dy. Col., Bradford).
[36a] SPEAKMAN AND ASQUITH, *Proc. Inter. Wool Text. Res Conf. Aust.*, 1955, C.302.
[37] RUDALL, *Symp. Fibrous Proteins*, 1946, p.15 (publ. Soc. Dy. Col., Bradford).
[37a] FARNWORTH. *Text. Res. J.*, 1957, **27**, 632.
[38] HIND AND SPEAKMAN, *J. Text. Inst.*, 1945, **36T**, 19.
[39] WOODS, *Proc. Roy. Soc.*, 1938, **166A**, 76.
[40] SPEAKMAN, *J. Soc. Chem. Ind.*, 1931, **50T**, 1.
[41] SPEAKMAN, *Nature*, 1933, **132**, 930.
[42] ELÖD, NOWOTNY AND ZAHN, *Kolloid Z.*, 1942, **100**, 293.
[43] SPEAKMAN, *Nature*, 1941, **148**, 141.
[44] ALEXANDER, CARTER AND EARLAND, *J. Soc. Dy. Col.*, 1951, **67**, 17.
[45] DAS AND SPEAKMAN, *ibid.*, 1950, **66**, 583.
[46] BROWN, PENDERGRASS AND HARRIS, *Text. Res. J.*, 1950, **20**, 51.
[47] WHEWELL AND WOODS, *Symp. Fibrous Prot.*, 1946, p. 50 (publ. Soc. Dy. Col., Bradford).
[48] SPEAKMAN AND WHEWELL, *J. Soc. Dy. Col.*, 1936, **52**, 380.
[49] ELÖD AND ZAHN, *Kolloid Z.*, 1944, **108**, 94.
[50] ELÖD AND ZAHN, *Melliand Textilber.*, 1949, **30**, 17.
[51] ZAHN, *ibid.*, 1949, **30**, 517.
[52] ZAHN, *Kolloid Z.*, 1949, **113**, 157.
[53] ALEXANDER, *ibid.*, 1951, **122**, 8.
[54] ALEXANDER AND EARLAND (unpublished).
[55] ZAHN, *Kolloid Z.*, 1950, **117**, 102.

CHAPTER 4

Sorption and Swelling

Introduction

THE considerable affinity of wool fibres for water is a well-known property, which at first sight appears to be difficult to reconcile with their water-repellent nature.[1] The latter property depends essentially on the nature of the surface which is known to differ considerably in structure from the remainder of the fibre, whereas the absorption of a swelling liquid proceeds mainly internally. In addition to water, most liquids containing hydroxyl groups are found to swell the fibres to some extent, although the higher alcohols experience considerable difficulty in entering the fibres.[2] In spite of the extensive research into processes of this kind with wool and similar polymolecular systems, the exact mechanism is still in some doubt, and no one theory has been found to satisfy the experimental data completely.[3] As neutral molecules penetrate a fibre, adsorption, hydrate formation, swelling and probably solid solution occur simultaneously, causing continual changes in the structure of the fibres as the sorption proceeds, although apparently not in the crystalline regions (see p. 388). In addition, considerable uncertainty as to the part played by the various morphological components still exists.

The adsorption of nitrogen[4] at low temperatures has no effect on the fibre dimensions, and consequently has been used to estimate the available surface area (see p. 18). Polar molecules, however, are absorbed much more strongly and penetrate into the interior of the fibre with accompanying changes in dimensions. At one time it was thought that the liquid gradually filled a series of relatively large micro-pores randomly distributed throughout the dry fibre.[5] The pore theory was supported by heat of wetting data (see p. 115), and by the results of Speakman[6] on the effect of size of molecules on the penetration into the fibre. This theory also explains immediately the fact that the surface area, as determined by nitrogen absorption, is greater than that obtained microscopically. Subsequent work has, however, largely discredited the pore theory, which has been superseded by modern theories based on multimolecular sorption. Before discussing these theories in detail, however, it is proposed to review briefly the most important experimental work which has paved the way for the more recent developments.

Density and swelling

As with all fibres, and many other macromolecular substances, the swelling produced by the absorption of water and the lower alcohols by wool is highly anisotropic. The increase in diameter is approximately 17·5 per cent. whereas the change in length is only 1·2 to 1·8 per cent. when saturated with water. Table 4.1 gives values of the change in diameter with regain.

As the dimensions of the fibres change, corresponding changes in the density are observed. It was found by King[7] that the apparent

TABLE 4.1

Change in fibre diameter with moisture content (regain*)

Regain percentage	Diameter swelling percentage	Regain percentage	Diameter swelling percentage
0	0·00	15	7·0
2	0·64	20	10·0
5	2·00	25	13·0
7	2·90	30	16·0
10	4·40	33	18·0

*Regain is the weight of water sorbed divided by the dry weight of fibre.

density of dry wool measured by the standard specific-gravity bottle method, varies considerably with the buoyancy medium (Table 4.2). As in the case of cotton, it is observed that swelling liquids lead to very high values of the density, whereas most organic liquids lead to a constant value of c. 1·306.

This was originally explained by supposing the water or alcohol to be highly compressed in the pores within the fibre, whereas the organic liquids are unable to penetrate the pore system completely, leading to low densities. Helium, on the other hand, gives a true measure of the density; filling the free volume completely without causing any swelling.

These views have, however, been critized by Hermans, Hermans and Vermas,[8] who have pointed out that density and pore structure

TABLE 4.2

Apparent density in various liquids[7]

Medium	Sp. gr. of liquid	Apparent sp. gr. of wool (dry)
Methyl alcohol	0·79149	1·408
Water	1·00000	1·396
Ethyl alcohol	0·79684	1·388
Naphtha	0·75173	1·330
Paraffin B.P.	0·88746	1·324
Amyl alcohol	0·80890	1·318
Carbon tetrachloride	1·59000	1·309
Oleic acid	0·89164	1·306
Nitrobenzene	1·20200	1·303
Olive oil	0·91700	1·306
Toluene	0·86150	1·306
Benzene	0·87540	1·304

are macroscopic concepts, whereas it is now known that any free space within the fibres is of molecular dimensions only, arising from the inefficient packing of the bulky side chains. Thus, organic liquids such as carbon tetrachloride and benzene cannot penetrate the fibre, but merely envelop it completely. In these media therefore, the measured density corresponds to the fibre as a whole, including any free volume within the fibre. When a swelling liquid is used, an increase in density results owing to the filling of such free space by a rearrangement involving more efficient packing of molecules and side chains in the swollen fibre. According to this view, the water is not under any compression as visualized by the earlier explanations. The argument was illustrated by Hermans[9] by contrasting the packing of large equal spheres (representing amino acid or hexose units), and of the same large spheres with smaller ones (representing water molecules). It is readily seen that the latter arrangement leads to the greater density.

Hermans introduced this concept to explain the change in the density (measured with benzene as the buoyancy liquid) with water content. The following results (Table 4.3) obtained by King[7] show an initial increase, explained above by the filling of free space with water molecules, followed by a continual decrease to the value of 1·267 at saturation. Similar changes are observed with cotton and other fibres.

TABLE 4.3

The specific gravity determined in benzene, and the apparent density in water at various regains[7]

Regain percentage	Spr. gr. of wool at 25° C.	Apparent density in water at 25° C.
0	1·304	1·3965
2·00	1·307	1·3860
3·75	1·314	1·3770
5·90	1·314	1·3660
7·50	1·312	1·3590
12·00	1·313	1·3390
17·00	1·312	1·3205
17·30	1·308	1·3193
20·00	1·304	1·3100
22·85	1·297	1·3000
30·20	1·277	1·2780
33·00	1·267	1·2710

Hermans and his colleagues[9] think that these results are significant if the relation between the specific volume v_a and the weight of water, a, absorbed per gram. of dry fibre is considered. Then the specific volume is related to the density d_a as follows:

$$v_a = (1+a)/d_a$$

The relation between v_a and a, represented graphically in Fig. 4.1, shows that at low regains the volume of the water introduced appears to be less than the corresponding volume of liquid water as explained above.

As the water content increases, however, the slope of the graph approaches unity, that is the volume increase of the fibre equals the volume of water sorbed. If this linear portion is extrapolated, it meets the ordinate at v_c which gives a value of the specific volume of the dry fibre excluding the internal free volume. As water is capable of occupying this volume, this value may be identified with the specific volume of the dry fibre measured in water, i.e. the reciprocal of the density measured in water. The volume $v_0 - v_c$ thus represents the volume of free space within the fibre, which decreases as the absorption proceeds (Fig. 4.1).

In spite of the considerable evidence for this explanation quoted by Hermans in the case of cellulose, the contraction in volume would follow directly from the chemisorption of the water to form definite hydrates in the initial stages of the absorption (see p. 111).

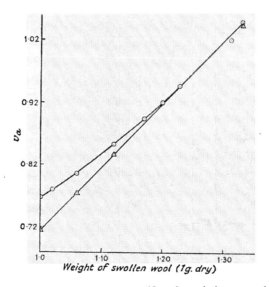

Fig. 4.1.—The change in specific volume (v_a) measured in inert solvent ⊙ and in water △, with swelling.[7]

Sorption in pores

The highly anisotropic swelling of wool fibres and similar polymer systems is explained by the association of a limited number of polymer chains into definite units or micelles with their axes orientated almost uniformly parallel to the fibre axis.[6] According to this general arrangement, molecules of liquid can penetrate between the micelles forming a gel of relatively low water content. The extent of penetration of liquid into the micelle may be investigated by X-ray analysis. In the case of wool, most of the swelling is known to be inter-micellar, as the X-ray spacings are only affected slightly by water absorption[10] (see p. 388). Absorption of formic acid, however, which causes considerable swelling (see p. 208) completely destroys the X-ray picture, showing considerable penetration into the micelle. The swelling of gelatin and collagen in water is similarly found to be mainly intra-micellar.

Estimations of the dimensions of the micelles and the inter-micellar distance in dry and fully swollen fibres were made by Speakman from observations on the penetration of molecules of different sizes into the fibre. Thus Speakman[6] originally observed that the strength of fibres when immersed in water, methanol or ethanol was decreased by almost the same amount, whereas in alcohols of molecular weight greater than propanol, no decrease was observed. Thus, it may be concluded that molecules of radius greater than that of propyl alcohol are incapable of penetrating the fibre rapidly. It is found, however, that pyridine and heptyl alcohol can enter the fibres freely in the presence of water. Using the change in mechanical properties as the criterion of penetration by neutral molecules, the data of Table 4.4 show that few molecules can penetrate the fibre rapidly.

TABLE 4.4.
The effect of sorption on the mechanical properties of a fibre[6]

Solute	P.E. for 30% extension (g./cm./c.c.) $\times 10^5$	% Extension at break	Breaking load g./cm.$^2 \times 10^5$ (initial area)
Water	1·43	50·4	14·9
Glycol	1·6	45·2	15·9
Methyl alcohol	1·72	40·9	14·2
Ethyl alcohol	2·44	43·2	17·9
Propyl alcohol	4·43	37·1	23·0
Butyl alcohol	5·01	31·3	22·7
Amyl alcohol	5·02	33·8	24·1
Glycerol	5·35	32·8	25·1
Acetone	4·17	30·1	18·2
Pyridine	4·70	31·2	21·1

The assumption of a definite pore size which can admit or exclude solvent molecules depending on their diameter is now known to be untenable, as King[11] later showed that dry wool can absorb propyl alcohol and acetone to an extent of at least 16 per cent. (see p. 128). Although in these cases the rates are very small, King has advanced arguments to show that the rate-controlling process is the same as that in the case of water and the lower alcohols.

This criticism casts considerable doubt on the calculations of the pore size in the unswollen fibre made by Speakman,[6] but the other estimations of micellar dimensions are probably valid. Thus the average pore size of the swollen fibre was determined by estimating primarily the micelle thickness, d, from the fractional increase in diameter on swelling, S, given by,

$$S = z/d$$

where z is the increase in pore size due to the swelling. By studying the elastic properties in mixtures of methyl and octyl alcohols, the concentration at which octyl alcohol molecules are just admitted can be determined. The increase in pore size is then given approximately by the difference in length of the two alcohols, assuming that the alkyl chains lie at right angles to the surface of the micelles. This procedure led to a value of $c.$ 200 Å for the width of each micelle. Similarly the increase in pore size when the fibre is completely saturated with water is given by $200 \times 0 \cdot 175 = 35$ Å, so that the actual size of the pores in the fully swollen fibres is approximately 40 Å, assuming a value of 5 Å for the original pores. The average separation distance of the structural units in the swollen fibre may be referred to as a pore size, and it is in this sense that it is used here.

It is difficult to estimate the other dimensions and the actual shape of the micelles with certainty, although the anisotropic swelling indicates that the length is greater than the width. The elastic properties vary in a similar way to the fibre dimensions.[6] Thus the breaking load is not greatly affected (Table 4.4), whereas the rigidity is much more sensitive to water absorption, and is reduced in the ratio of $14 \cdot 7 : 1$ by changing from dryness to saturation. This close parallelism indicates that the number of divisions in the solid structure is much greater in the cross-section than along the fibre. Assuming that the micelles are ten times longer than they are wide, the internal surface area was found to be approximately 10^6 cm.^{2}/g. This compares favourably with the value of 2×10^6 cm.^{2}/g. determined by Rowen and Blaine[4] by application of the B.E.T. theory to the absorption isotherm. In addition, the value of 40 Å for the average thickness of the swollen pores is in agreement with the value which may be assessed from the value of v_m calculated from the B.E.T. theory (see p. 18), and the total volume v absorbed at saturation. If v_m is accommodated on the total surface area of the micelles, the pore diameter is given approximately by $2 \cdot \dfrac{v}{v_m} \cdot (10 \cdot 6)^{0 \cdot 5}$ Å, where $10 \cdot 6$ Å^2 is the surface area occupied by a single water molecule. This leads to a value of approximately 35 Å, in close agreement with the value obtained by Speakman. A pore size of this order is also indicated by an examination of the size of dye which can readily penetrate the swollen fibre, although in this case, aggregation is liable to complicate the interpretation.

Sorption isotherms

All macromolecular systems are found to give isotherms which represent a combination of two of the three following processes, (*a*) solution sorption given quantitatively by Henry's law, (*b*) Van der Waals' adsorption, given quantitatively by the Langmuir isotherm, and (*c*) capillary condensation or multimolecular adsorption, which leads to considerable increases in the rate of sorption with respect to the quantity sorbed.[12] The isotherms of some of the simpler polymers, e.g. the vinyl polymers,[13] may be expressed in terms of (*a*)

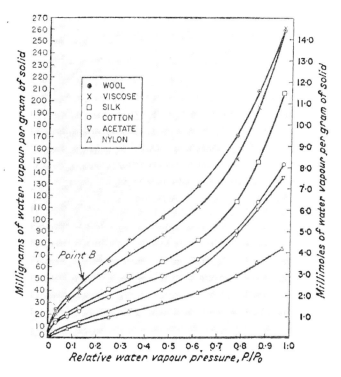

FIG. 4.2.—Adsorption isotherms of water on six textile fibres at 25° C.[4] (*Reproduced by permission.*)

and (c), and casein[14] gives a linear isotherm initially, due to the sorption of water on the amino groups.

Most porous solids, however, including the globular and fibrous proteins, give the general sigmoid type of isotherm[12] shown in Fig. 4.2, which is usually interpreted in terms of processes (b) and (c). Katz[15] has observed that swelling crystals, e.g. haemoglobin and the vitrellins, behave similarly. The question thus arises as to whether the shape of the isotherm is due to the swelling, or whether absorption may be considered to occur on sites on the 'internal surface' within the fibre. As sigmoid isotherms are also observed in cases of purely physical adsorption on rigid systems,[12] the latter alternative is the more widely supported, although many workers have objected to the use of the term internal surface, and prefer to regard the changes as chemical compound formation followed by either Raoult[16] or Flory-Huggins mixing.[17] Some authors consider that the mechanical restraint caused by swelling also modifies the form of the isotherm, and have obtained theoretical 'stress-free' isotherms of a completely different form.[18] These alternative explanations will be considered in some detail in the following section.

Two further experimental points are of considerable practical importance. In order to determine the regain at various relative humidities it is necessary to obtain samples of dry wool. Complete dehydration is almost impossible as the last traces of water are held very strongly. It is highly likely that a small amount of water is necessary for stabilization of the molecular network, as there are indications that complete dehydration causes irreversible changes in the fibres.[19] This is difficult to establish as slow degradation occurs at the temperatures which are required (see p. 300). It is therefore common practice in research on textile fibres to dry them under suitable arbitrary conditions. Thus, as the weight was found to decrease gradually on drying in a vacuum oven at increasing temperatures, Bull[20] chose the temperature of 105° C., which has been widely adopted. The same isotherm for egg albumin has been obtained recently by heating over phosphorus pentoxide at 80°–90° C. in vacuum,[21] and drying over sulphuric acid seems to be a satisfactory alternative.

A further feature of the adsorption is that the quantity of water adsorbed at a given partial pressure is variable, and dependent on the previous history of the fibre[20, 23] (Fig. 4.3). Thus, starting with a fully swollen fibre and gradually reducing the relative humidity, the desorption curve is followed. On reversing the procedure, starting with the dried fibre, the characteristic absorption curve is followed. A further desorption leads to an isotherm lying between these two and gives a complete cycle which may be repeated indefinitely. According to Bull,[20] the hysteresis is only observed between 15 per cent. and 65 per cent. saturation, whereas some workers[24] have found that the difference between the desorption and absorption curves disappears after the cycle has been repeated several times.

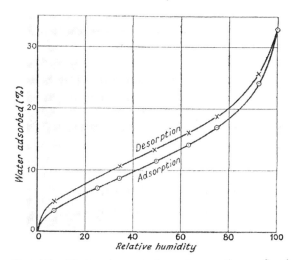

FIG. 4.3.—Hysteresis curves for water sorption on Southdown wool.[23] (*Reproduced by permission.*)

A similar hysteresis cycle is observed in many similar polymer-water systems, although apparently it is not exhibited by vinyl polymers containing polar groups.[25] Cotton,[9] wood[26] and other cellulosic materials show considerable hysteresis, which has been the subject of much experimental and theoretical study recently. Casein and benzoylated caseins, and in particular tobacco mosaic virus,[28] show pronounced hysteresis,[25] and this phenomenon is probably general for all proteins. It is interesting to record that the crystalline peptide alanylglycine which absorbs one water molecule at low humidities and a second water molecule as saturation is approached, retains both water molecules on desorption to a very low regain value.[29]

The hysteresis phenomenon in textile fibres is of considerable commercial importance, for example in the buying of 'conditioned wool', since the relative humidity of storage does not determine the water content. The theoretical significance is not entirely clear, although various suggestions have been advanced.[27, 30, 31] It is possible that swelling may cause structural changes in the gel which modify slightly the strength of binding of the water molecules already sorbed.[26, 32, 33] Then, as the polymer returns to its original configuration, the equilibrium is displaced until the fibre approaches the dry state. Barkas[27] has developed a detailed treatment for the absorption of water by wood, based on the compression of the water molecules by the elastic restraint. As swelling proceeds, shear resistance forces cause inelastic alterations in the structure of the solid, which thus modify the isotherm to different extents when approached from opposite sides. The possibility must not be overlooked, however, that the hysteresis is merely a time phenomenon, due to the relatively slow return of the molecular chains to their original configuration after distortion although at present the available evidence tends to discount this view.

Theories of the sorption of water vapour

It has already been mentioned in the introduction that the earliest theory advanced to account for the main experimental results considered the majority of the water to be accommodated in relatively large micropores.[5, 30] This was concluded originally from the thermal data of Hedges (see p. 115), which were taken to show that the adsorption takes place in three loosely defined stages, (a) chemical bonding up to 5 per cent. regain, (b) the filling of pores with liquid water up to 25 per cent., and (c) an osmotic solution process characterized by little heat change between 25 per cent. and 33 per cent. regain. The effect of water on elasticity[34] (Fig. 3.1 p. 56), was found to support this theory as the first limited absorption has little effect on the elongation, but causes considerable evolution of heat. The second stage affects the rigidity considerably and is accompanied by a smaller heat change (see Table 4.7). The final absorption affects the elasticity very little and almost no heat is evolved.

The existence of micropores of definite and limited size within the dry fibres, as for example in the case of silica gel, has been disproved by various considerations. Thus the regular lateral swelling and the high energy of activation required for diffusion of water and alcohols into the fibre (p. 144) show that the originally compact structure must be forced open to accommodate the adsorpate. Capillary condensation is also unlikely, as pointed out by Cassie,[18] as the water molecules are initially held firmly on the most hydrophilic groups (see p. 112), thus leaving the remaining free surface extremely hydrophobic.[1] In support of this deduction, it is found[35] that the advancing contact angle at the surface is slightly greater than 90°. This causes a slightly convex surface so that a reduction in equilibrium vapour pressure below the saturated vapour pressure of the vapour cannot occur. On the other hand, liquid air readily wets wool fibres[18], and as the molecules are extremely small, considerable condensation in capillaries would be expected in the capillaries if present. The magnitude of the adsorption at $-183°$ C., however, does not permit any such condensation.

The absorption of water by cotton was explained initially by Peirce[30] by considering two kinds of water, namely that chemically bound (α) and free water (β). Using a simple evaporation-condensation process, he derived the following relation between the amount of water absorbed in the two different ways

$$C_a = 1 - e^c$$

where c is the total number of water molecules absorbed per sorption (group hexose for cotton, amino acid residue for wool), and C_a the number of α molecules absorbed per group. Speakman[36] has constructed the isotherm for wool relating the amount of β water sorbed

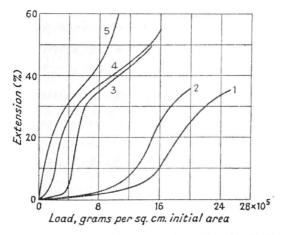

FIG. 4.4.—Load-extension curves in various liquids.[34] (1) n-Butyric acid. (2) Propionic acid. (3) Water. (4) Acetic acid. (5) Formic acid. (*Reproduced by permission.*)

to the rigidity, obtaining a linear relation except at high regains, assuming that the α water has no effect on the rigidity. The α water was considered to be held on hydrophilic side chains (see p. 112) as the heat of wetting is of the same order as the heat of hydration of OH and COO^- groups. From the β isotherm, it is found that one water molecule is absorbed for each residue present. The β water which is responsible for the considerable changes in rigidity was thus considered to associate with each peptide link. Gilbert[37] has, however, criticized drastically the derivation of this isotherm as the assumptions made are incompatible with accepted thermodynamic principles, although it has been applied widely to textile fibres. Hydroxylic liquids other than water cause fibre swelling by a similar mechanism, as indicated by the similar elasticity changes (Fig. 4.4).

Multimolecular isotherms

The most satisfactory quantitative treatment so far given is based primarily on purely physical processes, and does not require a detailed picture of the mechanism.[12] The general theory considers a monolayer to be held strongly on specific sites, which provides secondary sites for the condensation of more molecules in the form of multilayers. The original postulates of Brunauer, Emmett and Teller[38] have been severely critized[37] in recent years in spite of the considerable success of the resulting equations.[12] The chief objection is that lateral interaction between molecules in individual layers is neglected although the absorption energy in such a layer is equated to the heat of evaporation. From all points of view, it would seem that molecular attraction within a layer is of greater importance in stabilizing the system than inter-layer attraction. It is also questionable whether this model can be adopted for a compact elastic structure such as wool, as it requires an essentially open rigid structure, with pores of sufficient dimensions to allow multilayers to form without mechanical restraint. As already discussed in some detail (p. 92), wool does not contain pores, as wool fibres and horn keratin membranes transmit diffusing molecules by activated diffusion and not by the laws of Knudsen and Poiseuille (p. 16) referring to flow in capillaries.

Subsequent treatments of multimolecular adsorption have used the principles of statistical mechanics.[32, 39] Cassie[32] considers that as condensation occurs below the saturation vapour pressure of the liquid, the free energy of the absorbed liquid must be less than that of the bulk liquid, attributed not to a decrease in heat content, but to a decrease in entropy (see p. 116). This is attributed to the entropy of mixing of water molecules in different aggregates within the fibre. The mixture was assumed to be ideal as the interchange involves no change in energy as the molecules are of the same size. The entropy of mixing concept has been widely critized, as in contrast to the extrapolated data of Hedges, the calculations of ΔS from the effect of temperature on the isotherm (p. 117) do not indicate a positive change in entropy at high regains. In addition, as pointed out by Gilbert,[37] an entropy of mixing due to the exchange of like molecules is not supported by general thermodynamics,[40] and it seems unlikely that a small entropy increase is to be interpreted in this way, but as an entropy of dilution.

An alternative statistical approach was subsequently suggested by Hill[39], which leads to a final equation identical with the general equation of B.E.T. theory (p. 19). If A is the number of molecules of vapour sorbed per unit mass at a particular vapour pressure x, equal to p/p_0, a certain number X is considered to be absorbed very strongly on specific or localized sites which are assumed to be of equal energy. Then if B is the total number of localized sites, the number of distinguishable ways X identical molecules may be distributed among these sites is given by[40] $B!/(B-X)!X!$. Similarly the number of distinguishable ways $(A-X)$ identical molecules may be distributed on top of X molecules of the first layer is given by $(A-1)!/(A-X)!(X-1)!$ The complete partition functions* for the absorbed monolayer Q_S and the multilayer liquid Q_L are then given by:

$$Q_s = \frac{B!}{(B-X)!X!}[j_s e^{E_1/kT}]^X$$

$$Q_L = \frac{(A-1)!}{(A-X)!(X-1)!}[j_L e^{E_L/kT}]^{(A-X)}$$

where $j_s \cdot \exp(E_1/kT)$ and $j_L \cdot \exp(E_L/kT)$ are the partition functions

* The partition function is a statistical concept giving a quantitative measure of the distribution of the energy of a system in the permissible quantized levels. It is related to the energy of these various levels, $e_1, e_2 \ldots$ as follows:

$$Q = \sum^i e^{-e_i/kT}$$

As it is directly related to the free energy as follows:

$$-F = RT \log_e Q$$

the equilibrium between the various components at constant temperature can be represented simply in terms of their complete partition functions.

for single molecules in the first and higher layers respectively. E_1 and E_L are the potential energies of the two molecules, k is the Boltzmann constant and T the absolute temperature.

The distribution of molecules at equilibrium can be found from the condition that the free energy of the system as a whole must be minimal. Thus, as the complete partition function is related to the free energy F by,

$$F = -RT . \log_e Q$$

at equilibrium

$$\frac{\delta \log_e Q_s . Q_L}{\delta X} = 0.$$

From the combined expression for $\log_e Q_s Q_L$ the following general relation results,

$$(A-X)(B-X) = \beta X^2$$

where

$$\beta = (j_L/j_s) \exp(E_L - E_1)/kT$$

From the definition of x, it follows that $x \equiv p/p_0 = (A-X)/A$ so that $A/B = cx/(1-x)(1-x+cx)$, where $c = 1/\beta$.

This equation can be transformed into the following familiar forms of the B.E.T. isotherm,

$$\frac{x}{A.(1-x)} = \frac{1}{Bc} + \frac{c-1}{Bc}x$$

and

$$\frac{x}{n.(1-x)} = \frac{1}{n_1 c} + \frac{c-1}{n_1 c}x$$

where n is the number of moles sorbed, and n_1 is the number of moles of sorption sites per unit weight of protein.

The statistical treatment leads to a fundamental interpretation of the evaporation-condensation term c, which has no definite significance on B.E.T. theory. The reference zero of energy is taken as that of the liquid state, so that $E_1 - E_L$ is the heat evolved when one mole of liquid is absorbed on the localized sites. It may be shown that the coefficient β is equal to $p_\frac{1}{2}/p_0$ where $p_\frac{1}{2}$ is the vapour pressure required to fill half the localized sites.[32] Although B has the same significance as v_m of the B.E.T. method, the sites are not restricted to the surface, and may be distributed throughout a gel.

Bull[20] has applied this isotherm satisfactorily to many proteins, including silk and wool, to obtain the data given in Table 4.5.

TABLE 4.5

B.E.T. constant, c, and weight of water in the monolayer, n_1, for the sorption of water vapour by proteins[20]

	c	n_1 g./100 g. dry protein	Protein area m.²/mg.
Nylon	4·40	1·92	0·068
Silk	12·78	4·07	0·144
Wool	11·13	6·58	0·233
Collagen	17·80	9·52	0·337
Gelatin	17·40	8·73	0·309
†Egg albumin	11·58	5·65	0·200
†β-lactoglobulin	8·57	6·67	0·236

† Lyophillized.

Good linear plots of $p/n(p_0-p)$ against x were found for all the proteins studied (Fig. 4.5) for values of $x < 0.6$–0.5. The value of n_1 thus obtained is much less than the value obtained by studying the films of soluble proteins on water surfaces,[41, 42] but has been interpreted in terms of hydrate formation on the more reactive groups (see p. 111). The failure of the simple isotherm to explain the experimental data at high humidities[20, 43] has led to various modifications of the original postulates. The assumption that the potential energies of the molecules in the higher layers are identical, and equal to E_L is particularly unlikely. Dole[16] has extended the above treatment to include variable heats of absorption in the different layers. His general isotherm cannot be applied directly to the experimental data owing to its complex nature, and the number of unknown quantities involved. Under the appropriate conditions however, it reduces to the Langmuir and B.E.T. equations. If all the molecules, including those in the monolayer, are absorbed with energies equal to the heat of vaporization of water, the following equation, which is a type of Raoult's law for a system in which the sites are not restricted to the surface, is obtained[44]:

$$x/n = 1/n_0 - x/n_0$$

where n is the number of molecules sorbed on n_0 possible sites. This theory has been found to explain the data satisfactorily in the case of insoluble collagen and polyvinyl alcohol[25], although some differences are found in the case of soluble plakalbumin.

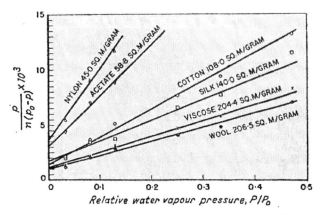

Fig. 4.5.—B.E.T. plots for obtaining surface areas accessible to water vapour.[4] (*Reproduced by permission.*)

Recently, the isotherm derived by Bradley[45] has been applied by Mellon and Hoover[46] to experimental data for ovalbumin, silk and wool over the entire range of x (0·05 — 0·95) and over a wide temperature range (35° — 100° C.) with considerable success, as shown in Fig. 4.6. This theory is based on the original theory of De Boer and Zwicker[47], who attributed the condensation of non-polar molecules on ionic solids to induced dipole forces. The induced dipole is thus transmitted from the molecules in the first layer to the molecules in the nth layer, so that the molecules are held by decreasing forces as n increases. Bradley[45] derived a similar isotherm for the absorption of polar molecules with a permanent moment μ which may be expressed as follows,

$$\log 1/x = K_2 . K_1{}^n$$

where K_2 depends on the nature of the sorptive polar groups, and K_1 is a function of μ. K_1 also depends specifically on the nature of the sorbed molecule. This isotherm automatically takes into account the change in energy of the molecules with the number on the absorbed layer, and as may be seen by the plot of log log(1/x) against n, it is very successful (Fig. 4.6).

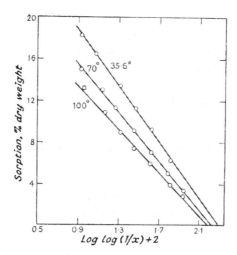

FIG. 4.6.—Absorption isotherms for the Bradley theory.[45]

Solution theory[15]

The application of absorption isotherms to polymer-water systems has been critized by many workers on the grounds that such theories can only be applied to systems with a definite rigid surface, and that the internal surface of a polymer or protein is meaningless. Thus, in many cases an alternative approach has been adopted, based on the solution of polymer in the solvent, and Hermans[9] accepts the interpretation of Katz[15] that after the initial hydrate formation, the water absorbed at higher regains forms a solid solution.[48] It is pointed out that the behaviour of fibre-water systems is formally analogous to the addition of water to sulphuric acid and phosphoric acid, which initially form hydrates, and yield sigmoid absorption curves. At saturation regain however, these homogeneous solutions, like soluble proteins, can absorb an infinite amount of water. The solution theory has been applied satisfactorily to the sorption of organic solvents by rubber[49] and similar substances.[43, 50] These cases are simpler than the wool-water system as no absorption is expected on specific sites, and the polymer chains are known to be much more flexible than the highly cross-linked chains in wool keratin.

The solution theory has several advantages over multimolecular theories, overcoming in particular the difficulty of deriving an isotherm on the basis of two-dimensional absorption, and treating the sorbate as liquid water. Barrer[51] has proposed a method of combining a molecular sorption process with a subsequent solution process. He suggests that an initial energetic absorption occurs on localized sites (CO and NH) according to the Langmuir isotherm, followed by an independent process of sorption by mixing with polypeptide chains in the same way as the sorption of liquids by rubber and plastics. This leads to a sigmoid isotherm, with an evolution of heat and decrease in entropy initially, followed by a final endothermic heat change and small entropy increase (see p. 116). In this way, the sorption may reach a limit as $p \to p_0$, which may be related to the heat of mixing.

Rowen and Simha[43] have applied the quantitative theory developed from statistical considerations by Flory and Huggins[17] to the absorption by cellulose and several proteins. The partial molal free energy ΔF for the solution process is given approximately by,

$$RT . \ln . x = -\overline{\Delta F}_{sol.} = RT . (ln . v_1 + v_2 + \mu v_2{}^2)$$

where v_1 and v_2 are the volume fractions of the condensed vapour and polymer respectively; μ is a semi-empirical parameter determined by the heat and entropy of mixing. This expression is found to predict the experimental results for many systems including cellulose and wool in the region of high regains ($x > 0.6$), that is over the exact range where the multilayer theory breaks down completely. This change-over is well illustrated by Fig. 4.7, which compares theoretical predictions and experimental results for both solution and multimolecular absorption theories. The parameter μ may be obtained from activity data when this is available. Conversely, the activity may be calculated for various regains from the derived values of μ. From such calculations, it is found for the salmin-water[20] and cellulose nitrate-acetone[50] systems that the activity increases regularly from 0 to 1·0. In the case of wool and cellulose, however, the activity values reach unity at definite polymer-water compositions, above which activities greater than unity are obtained. This indicates that in such cases phase separations of pure solvent and solid-liquid mixture occur at these critical concentrations.

Hailwood and Horrabin[52] adopt a solution approach, but consider that an ideal solid solution is formed between polymer hydrates and water. They assume that α water forms a hydrate with a definite unit in the fibre (cf. Peirce[30]), and that the β water is in simple solution with polymer and polymer hydrate. This solution is supposed to be ideal, but this assumption has been criticized severely by Cassie,[18]

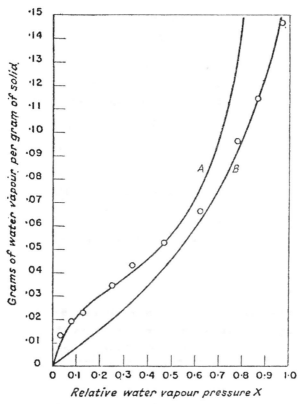

Fig. 4.7.—Comparison of multimolecular adsorption (A) and solution (B) theories.[43] Experimental points shown O. (Reproduced by permission.)

Gee,[53] Barrer,[54] and Ubbelohde.[55] A series of equilibria are set up for the various hydrates:

$$K_i = w_i/w_{i-1} \cdot a', \quad \text{where} \quad i = 1, 2, 3, \ldots n, \text{ and}$$

where w_i represents the activity of the ith hydrate in the polymer, and a' the activity of dissolved water in the solid phase. It follows that $w_i = w_0(a')^i \cdot K_1 \cdot K_2 \ldots K_i$, where w_0 is the activity of the unhydrated protein. If only one hydrate is considered to be formed for the sake of simplicity, and if activities are replaced by mole fractions, the following equation may be derived,

$$\frac{M.r}{1800} = \frac{K.h}{1-K.h} + \frac{K.K_1 h}{1+K.K_1 h}$$

where $K = a'/a$, M is the residue molecular weight, r is the regain, h is the relative humidity expressed as a percentage and a is the activity of water in the vapour phase. In the vapour phase the activity may be equated to the fractional humidity i.e. $a = h/100$.

The success of the isotherm was based on the following correlation with experimental data for wool, silk, hair, nylon and cotton.

(a) If suitable values of K_1, K and M are chosen, the theoretical isotherm agrees well with the experimental curve up to 100 per cent. relative humidity.

(b) Values of the heat of absorption, assuming zero heat of mixing, as the solution is supposed to be ideal, agree satisfactorily with the experimental values of Hedges[5] except at very low humidities.

(c) The partial molar volumes of each constituent of the solution were shown to be independent of composition assuming the calculated values of M, and the free energy of solution $\Delta G°$ and King's data for the specific volume of wool at various regains.[7]

This theory was used to assess the relative proportion of crystalline and amorphous regions in the fibre from the calculated values of the residue molecular weight. If the molecular weight corresponds to a fundamental unit in the protein, e.g. the mean amino acid residue, the difference between calculated and actual value should give a measure of the extent of the inaccessible regions, i.e. the crystalline phase (p. 392). In all cases, M is found to be greater than the actual mean value for the amino acids, and from these values the following crystalline to amorphous ratios were derived: Wool,[10] 0·80; Silk,[10] 3·76, Nylon, 5·67; Cotton,[9] 2·08. In view of the work of Mellon et al.[56] who have found that reprecipitated wool keratin gives an isotherm identical with the original, such calculations of crystalline to amorphous ratio are of doubtful validity.

This treatment has been criticized for several reasons but mainly because of the assumption that the three constituents form a perfect solution. Usually maximum deviations from ideality are observed when small and large molecules are mixed. If considerable deviations from ideality do occur, the calculated value of M can bear no relation to the structure of the fibre. In addition, the isotherm contains three parameters which have to be evaluated to give the most suitable values. If these are suitably chosen, a sigmoid isotherm must be obtained.

Cassie[18] points out that as the β water is identified with liquid water, the regain approaches infinity at 100 per cent. relative humidity so that the agreement at high humidities is no better than that for a generalized multimolecular theory. Their calculations of $\Delta G°$, however, involve the assumption that the lack of agreement at high humidities is due to compression.

Hermans[9] agrees largely with the Hailwood and Horrabin picture, and has developed a similar theory for the somewhat simpler case of cellulose. In this case, the calculation of the crystalline to amorphous ratio agrees well with values obtained from direct X-ray studies of the hydrates associated with each glucose anhydride unit, density determinations and heat of wetting data.

Influence of swelling on absorption

In the light of the preceding discussion, it will be apparent that wool falls between two well-defined classes of absorbents, those which lack rigidity and can form a simple homogeneous solution, e.g. sulphuric acid, and those which are completely rigid, e.g. silica gel and titania, but which possess large micropores and a constant internal surface area (see Table 2.3), on which multisorption by physical forces may occur. Gels differ from solutions in that the latter can withstand hydrostatic pressures only, whereas the former can withstand shearing stresses.[15] A fibrous gel is normally distorted to an extent depending on the rigidity modulus, and may swell laterally on absorption until the swelling is overcome by the hydrostatic pressure developed. The compression of the water has been shown to affect the absorption curve markedly in the swelling of wood.[26] In this case, the swelling is restrained by the well-defined cell walls, so that the concept of swelling pressure is more easily visualized than in the case of fibres, and Barkas related the hydrostatic pressure directly to the mechanical properties of the system. Although similar theoretical treatments of the effect of stress in wool fibres on the extent of sorption have been developed, very few direct measurements have been made. However, Treloar[57] has found that a small increase in water content at a given humidity is produced by increasing the stress along the axis of the fibre. This observation is in complete disagreement with the results of White and Stam,[58] but is consistent with the general thermodynamics of the system. The differential swelling, S_x, along any coordinate x, due to the compressive force X, is defined as follows,

$$S_x = \frac{V}{x}\left(\frac{\delta x}{\delta n}\right)_X$$

where V is the volume of the swollen gel under consideration, and the partial differential $(\delta x/\delta n)_x$ is the increase in the length x per unit mass of liquid sorbed, under constant force X. If this force alone is acting and is now increased by dX causing an increase in vapour pressure p to maintain a constant liquid content,

$$S_x \cdot dX = -v \cdot dp = \frac{V}{x}\left(\frac{\delta x}{\delta n}\right)_X \cdot dX$$

where v is the volume occupied by 1 g. of vapour at pressure p. By using the identity

$$\left(\frac{\delta n}{\delta X}\right)_p = -\left(\frac{\delta n}{\delta p}\right)_X \left(\frac{\delta p}{\delta X}\right)_n$$

it follows that

$$\left(\frac{\delta n}{\delta X}\right)_p = \frac{V}{v} \cdot \frac{1}{x}\left(\frac{\delta x}{\delta p}\right)_X$$

It follows, therefore, as the length of a fibre increases with the relative humidity, that the water content increases with an increase in stress.

The effect of lateral stress, which would have a much more marked effect than longitudinal stress on the sorption, is very difficult to determine experimentally, but may be estimated by the above general type of treatment. Cassie[59] was the first to modify the experimental sorption-isotherm to take into account the lateral stress produced by swelling, and preferred to calculate the change in partial free energy caused by the elastic constraint rather than the swelling pressure. The total free energy change is given by,

$$\Delta G = \overline{\Delta G_1} + \overline{\Delta G_2}$$

where $\overline{\Delta G_1}$ is the partial free energy change due to the absorption process under no constraint, and $\overline{\Delta G_2}$ is the elastic contribution to the free energy. From the general equation for the free energy of the process, dividing the intrinsic energy into two parts, one referring to the elastic changes, and the other referring to the stress-free absorption process,

$$\Delta G = \Delta H - T\Delta S = \Delta E_1 + \Delta E_2 + \Delta(pv) - T\Delta S$$

If μ is the chemical potential for n moles of water sorbed on the gel and μ' the chemical potential for an idealized stress-free gel,

$$\mu - \mu' = RT \log p - RT \log p'$$

where p and p' are the vapour pressures of water in equilibrium with the real gel and the stress-free gel. It may be shown that,

$$RT \ln \frac{p}{p'} = \left(\frac{\delta E_1}{\delta n}\right) + p\left(\frac{\delta V}{\delta n}\right) = \frac{\delta H_E}{\delta n}$$

where H_E is the heat sorbed in swelling the gel, and $\delta H_E/\delta n$ is the differential heat of swelling (see p. 114).

Warburton[60] has derived an expression for $\delta H_E/\delta n$ in terms of the mechanical properties of the gel, assuming that swelling causes the same stresses as those caused by external means:

$$\left(\frac{\delta H_E}{\delta n}\right) = 2TVE_E \cdot \frac{a_x S_x + \sigma_{xz}(a_x S_z + a_z S_x)}{(1 - \sigma_{xx} - 2\sigma_{zx}\sigma_{xz})}$$

where E_E is the transverse Young's modulus, σ_{xx} is the Poisson ratio for transverse contraction to transverse extension. σ_{xz} is the Poisson ratio of transverse contraction to longitudinal extension, and a_x is the coefficient of thermal expansion in the x direction, S_x is the isothermal swelling and a_z and S_z these values in the corresponding directions.

By using the calculated values of $\delta H_E/\delta n$, values of p' may be obtained from appropriate values of p, and hence the stress-free isotherm constructed (see Fig. 4.8). It is seen that this calculated isotherm has no point of inflexion, so that according to this theory the characteristic sigmoid shape is due entirely to the heat of swelling. The data obtained from this isotherm may be interpreted in terms of the statistical multimolecular absorption theory using the two general equations,

$$(A-X)(B-X) = \beta X^2$$

and $p'/p_0 = (A-X)/A$.

Values of B and β may be calculated from two points on the isotherm, so that values of X may be obtained for all values of A. It was found that these values are identical with values determined directly from

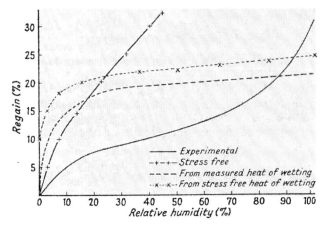

Fig. 4.8.—Regain-relative humidity curves for keratin comparing experimental and stress-free isotherms (No entropy change is assumed in deriving the broken curves).[18, 31]

the isotherm and from the second general equation above: Cassie therefore concluded that this multimolecular adsorption theory accounts for the stress-free isotherm completely.

The swelling theory was investigated and tested by other means, in particular by calculations of heat of wetting. As it is assumed in deriving the two fundamental equations that the only change produced by mixing water with keratin is the absorption of water on localized sites (see p. 99), $-\Delta H_m = (E_1 - E_L)X$. On the other hand, the heat of mixing is given by the integral from 0 to the desired regain corresponding to X of the expression,

$$\left\{ \frac{\delta \Delta H}{\delta n} - \frac{\delta H_E}{\delta n} \right\}$$

as the second term allows for the heat swelling. A constant value for $(E_1 - E_L)$ of 3·5 cal./mol is obtained for all values of X. As the value of B (1·12 mol/100 g.) corresponds closely to the number of peptide keto groups (1·1 mol/100 g.)[61] localized absorption is thought to occur on these groups. The value of 3·5 cal./mol leads to a value of 6·6 cal./mol for the energy of bonding of the water to the fibre

SORPTION AND SWELLING 111

which is close to the value for the hydrogen bonding of CO and NH groups in formic acid.

Finally the occurrence of internal stresses has been used to explain the phenomenon of hysteresis already described (p. 95). Calculation of[26]

$$\left(\frac{\delta H_E}{\delta n}\right)_{ABS} - \left(\frac{\delta H_E}{\delta n}\right)_{DES} \equiv RT \log_e \frac{h_1}{h_2}$$

where h_1 and h_2 are the relative humidities at a given regain on either side of the hysteresis loop, show that the heat differences tend towards large values as regains approach zero representing the decreased tendency of wool to recover after deformation at such regains.

Hydrate formation[3]

In spite of the divergent theories which have been advanced to explain quantitatively the absorption of water vapour and other swelling liquids on wool and other polymers, it is now universally agreed that the initial absorption is confined entirely to specific groups. The solution theory as first advanced by Katz[15] involved this assumption, and although the multimolecular adsorption theory as envisaged by Brunauer, Emmett and Teller[38] is based on physical adsorption only, the more recent interpretations require a highly localized attachment. Thus, as mentioned previously, the increase in the density of wool, and the high heat of absorption at low regains, indicate that the initial water is chemically bound to the keratin. In addition, the classical work of Fisher[62] on the rate of drying of a mass of fibres leads to a similar conclusion. It was found that the rate of drying was constant if the wool contained more than 30 per cent. of water, which was explained by the evaporation of mechanically held water from the surface of the fibres. The rate decreased sharply at c. 30 per cent. regain,[63] corresponding to the saturation value at 100 per cent. relative humidity[20, 34, 64] (see p. 93), and below this value was found to decrease linearly with regain. At low regains, however, the rate falls more rapidly, which is thought to be due to the removal of bound water. From these experiments, a value of 5 per cent., by weight, was attributed to the bound water in keratin, which agrees approximately with the value obtained for the monolayer calculated on B.E.T. theory.

In the case of some polymer systems, e.g. cellulose, definite crystalline hydrates have been detected,[65] and part of the water in gelatin and casein cannot be frozen out.[66] There is considerable doubt, however, concerning the nature of the specific groups responsible for this primary sorption in wool. Jordan Lloyd and Phillips[48] who were among the first workers to consider the problem, came to the conclusion that the water molecules are co-ordinated to oxygen, nitrogen and hydrogen atoms of side chains, and only to a lesser extent to amido and keto groups. They point out that the water-soluble proteins contain high proportions of serine and other hydroxy-amino acids, lysine, arginine and proline.

A more quantitative analysis has subsequently been given by Pauling,[67] who drew attention to the parallelism between the number of side chain polar groups and the amount of water which is adsorbed as a monolayer.

The results given in Table 4.6 show a general parallelism between number of moles of water absorbed and number of polar groups,

TABLE 4.6

Comparison of the number of water molecules held by proteins in the initial absorption, and the number of polar groups[67]

Protein	Water in first layer	(1) Number of polar groups per 10 g.		(2) Total reported amino acids
Silk	226	219	228	107·0
Ovalbumin (crystallized)	329, 342, 344	277	313	75·7
Wool	336	303	341	71·4
Gelatin, collagen	485, 529	328	609	108·8
C Zein, B Zein	290, 228	305	390	106·0
Salmin	592	611	707	110·5
Serum albumin	374	424	424	86·8
δ-Lactoglobulin (crystallized)	370	472	508	115·8

(1) Excluding the carbonyl groups of proline and hydroxy-proline.
(2) Including water added on hydrolysis. Complete analysis corresponds to 110–116.

indicating that one water molecule is associated with each group. These values have been modified recently by Green[68] to allow for the low attraction by amide groups, which gives better agreement for casein and collagen. In the absence of highly reactive side groups, the primary water combines with the polypeptide chains.[69] Thus, the limited absorption of water by nylon presumably occurs on keto and imido groups of adjacent separated chains.[67, 70]

In contrast to the proteins, simple peptides show greatly reduced values for absorption in the monolayer for a given number of polar groups.[71] This indicates that close packing of the chains, and/or high crystallinity severely restrict the initial combination. Thus di- and tri-glycine do not absorb water at all below $0·93\ p_0$ whereas tetra-glycine will absorb even at low x.[71] From the observation that benzoylated tetra-glycine absorbs less than the unsubstituted compound, Mellon concluded that both the amino and carboxyl groups act as absorption centres. Simple polypeptides such as polyglycyl-*dl*-alanine can absorb considerable quantities of water, which is presumably attracted initially to the peptide groups. Experiments with vinyl polymers containing only one species of substituted group show that the sorption capacity varies with the nature of the group, and also with its position within the polymer.[25, 28] In addition, the crystallinity is found to affect the sorption considerably, as also in the case of nylon. Considerable uncertainty exists at present as to the effect of the crystalline to amorphous ratio in proteins on the sorption. The available evidence, which of course is extremely incomplete, indicates that crystallinity does exert a considerable steric influence in the case of the simpler polymers, the units of which can be packed together very closely. On the other hand, Mellon, Korn and Hoover[56] have since shown that dissolution of silk in cupriethylenediamine and wool in thioglycollate solution (see p. 362) followed by reprecipitation gives products which have identical adsorption isotherms to those of the original fibres. As the precipitated powder is probably highly amorphous, the amorphous to crystalline ratio cannot influence the adsorption isotherms. In addition it has been found that freeze-drying of ovalbumin and plakalbumin which affects the crystalline habit completely, has no effect on the isotherm.[21] Further, McMeekin *et al.* have shown that the degree of hydration of lactoglobulin is the same in solution as in the crystal.[72] These observations strongly favour the theories involving preliminary chemical hydration[73] on specific groups followed by multimolecular sorption or solution at higher regains.

In contrast to the results of experiments on modified amino groups in casein,[14,29] Speakman[74] found that deamination of wool has no effect on the absorption curve. Later Blackburn and Phillips[75] showed that acetylation and methylation similarly have no effect. It has also been found, however, that alkylation of nylon does not decrease the sorption so that modification in contrast to complete removal of amino groups need not necessarily lead to a decrease in sorption,[76,77] although the water may combine with them specifically. The results of Speakman, and experiments with polypeptides, suggest that the keto and imido groups of the polypeptide chains are mainly responsible for the primary sorption. In support of this alternative view, Eilers and Labout[78] have shown that the number of localized sites in collagen is approximately equal to the number in wool in spite of the very different side-chain structures.

There is no doubt therefore that the polypeptide groups can hold a considerable portion of the bound water. Thus, in casein Mellon et al.[71] have estimated that 46 per cent. of the water is held on such groups and 25 per cent. on the amino groups, whereas zein accommodates 70 per cent. of the water on peptide bonds. Similarly in the case of silk and wool the weight of evidence suggests that most of the water is localized on the peptide chains.

Heat of absorption

The absorption of swelling liquids, and water in particular, leads to a considerable liberation of heat when dry fibres are used. The thermodynamics of the process were first investigated by Shorter[79] who derived a simple relation between heat of absorption and regain. Experimentally, the heat of wetting ΔH is measured, i.e. the heat evolved (per 100 gram. of dry wool) when a sample at a given regain is competely saturated. Theoretically the differential heat of sorption is of greater significance as this may be related directly to the partial molar free energy of the process. The differential heat of sorption is defined as the amount of heat developed when a very great amount of partially swollen substance absorbs 1 g. of water, i.e.

$$\overline{\Delta H} = \frac{\delta H}{\delta n}$$

$$\text{or} \quad \Delta H = \int_{n_1}^{\infty} \overline{\Delta H} \, . \, dn$$

Table 4.7 records the data on the heat of wetting, determined directly by Hedges.[5]

These results may be expressed by the following equation,

$$\Delta H = 2410 - 184n + 3 \cdot 5 \, . \, n^2 + 0 \cdot 0007 \, . \, n^4$$

which on differentiation gives the following equation for the differential heat of sorption,

$$-\overline{\Delta H} = 184 - 7 \cdot n - 0 \cdot 003 \cdot n^3$$

where n is the regain. Values of the heat of wetting were not determined above 20 per cent. regain, as the heat changes become too small to measure with accuracy.

TABLE 4.7
Heat of wetting $(-\Delta H)$[5]

Initial water content. Percentage of dry weight of wool	Heat of wetting cal./100 g. of dry wool.
0·0	2,410
3·0	1,880
6·4	1,380
9·5	1,010
13·1	630
15·0	470
17·8	330

The differential heat of sorption may be obtained alternatively from the effect of temperature on the sorption isotherms. If p_1 and p_2 are the vapour pressures above a fibre at constant regain at temperatures T_1 and T_2, the heat change is given by the following general relation.

$$\log_e \frac{p_1}{p_2} = \frac{\overline{\Delta H}}{R}\left[\frac{1}{T_1} - \frac{1}{T_2}\right]$$

The required isotherms which have been determined by Hedges[5] up to a relative humidity of 88 per cent. and by Bull[20] up to 95 per cent. lead to values of $\overline{\Delta H}$ which are compared with values determined from Hedges' heat of wetting data in Table 4.8. The agreement is fairly satisfactory except at high regains, as the values of $\overline{\Delta H}$ determined from the heat of wetting data tend towards high negative

TABLE 4.8

Differential heat of wetting $(-\overline{\Delta H})$[5, 20, 80]

Regain percentage	Relative humidity at 17° C.	Relative humidity at 29° C.	$-\overline{\Delta H}$ from isotherms cal./g.	$-\overline{\Delta H}$ from heat of wetting cal./g.
6	10·4	12·2	129·0	141·4
7	13·6	15·8	121·0	134·0
9	22·0	25·1	106·0	118·8
12	37·3	42·0	95·7	94·8
16	59·5	65·4	76·2	60·0
18	69·4	75·0	62·5	40·5

values, whereas the values determined from the isotherms tend towards zero at saturation. This has led to considerable confusion as Cassie[18] considers the heat change to become positive at high regains leading to positive entropy changes (p. 117). In support of this contention he quotes the data of Speakman, Stott and Chang[81] who found that the saturation swelling reaches a maximum at 37° C.

The temperature coefficient of the isotherms determined by Bull[20] up to 95 per cent. relative humidity show, however, that $\overline{\Delta H}$ remains negative, and these values are to be preferred to the values estimated by extrapolation of Hedges' data.

The differential heat changes may be combined with the partial molar free energy of the process given by,[20, 82, 83]

$$\frac{\delta.\Delta F}{\delta n} \equiv \overline{\Delta F} = RT.\ln.\frac{p}{p_0}$$

so that

$$\Delta F = RT.\int_{n1}^{n2} \ln.\frac{p}{p_0}.dn$$

to give values of the entropy change at various regains. Fig. 4·9 shows that at low regains a large decrease in entropy, corresponding to the large evolution of heat, is obtained.[82] This is in harmony with the conclusion that the initial water is bound very strongly to specific groups in the form of water of hydration. The heat of wetting is found to be similar to the corresponding values for other proteins (Table 4.9).

TABLE 4.9

Comparison of heat of sorption of a monolayer E_1 and differential heat of wetting[3]

Polymer	E_1 (B.E.T.)	$-\overline{\Delta H_0}$ cal./mol.	$-\overline{\Delta H_1}$ (average up to n_1) cal./mol.
Nylon	934	4,500	2,260
Silk	1,570	4,000	2,600
Wool	1,450	3,300	2,760
Egg albumin	1,490	3,000	1,500
Collagen	1,840	6,600	5,200
Casein	1,400		
Polyglycine	1,530		

$-\overline{\Delta H_0}$ refers to zero regain; $-\overline{\Delta H_1}$ gives the average over the monolayer.

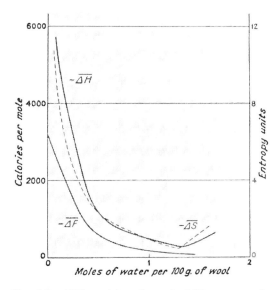

FIG. 4.9.—Differential net heats ($-\Delta H$), net entropies ($-\Delta S$) and free energies ($-\Delta F$) of sorption of water vapour by wool.

The negative entropy change decreases regularly as the regain increases and, as already mentioned, Cassie[32] considers that it may become positive at very high regains, as in the case of cellulose[84] for which more accurate data are available near saturation. Cassie explains this increase in terms of an entropy of mixing of the water which is supposed to exist in discrete clusters. This view has, however, been severely criticized mainly because the most suitable data indicates no such positive entropy change (see Fig. 4.9). According to Davis and McLaren[82] the positive entropy change is quite definite for the soluble proteins but is never shown by the fibrous proteins. Thus, although gelatin shows a well-defined positive entropy change as it approaches solution, collagen does not[77] in common with silk and wool.

Finally, it may be observed that the heat change calculated from the parameter c given by the B.E.T. equation (see p. 100) is much less in all cases than the initial heat of hydration and the initial heats averaged over values of n required for one monolayer (Col. 4, Table 4.9). This shows that the B.E.T. equation cannot be used to derive the initial heat of sorption.

Influence of swelling on electrical properties

The absorption of a swelling liquid has a considerable effect on the electrical properties of wool fibres. Thus the D.C. conductivity κ increases very rapidly with increasing regain m according to the logarithmic relation,

$$\kappa = A \cdot m^b$$

where A and b are empirical constants. According to Murphy and Walker[85] who used wool yarn, $b \sim 16\cdot 4$, although Marsh and Earp[86] give a somewhat lower value ($15\cdot 0$) for individual fibers. Until recently, this was the only well-established observation to have been made, and no satisfactory data were available on which to base a mechanism for the conduction process. The original explanation, which considered the conduction to be due to ionized water held in the pores, has to be modified as the pore hypothesis has already been discounted (p. 91). In addition, it is found that the specific conductivity of fibres with a regain of only 22 per cent. is similar to that of conductivity water.[86] Similar results were obtained by Stamm[87], who investigated the wood-water system, and who concluded that the conductivity of the adsorbed water was two to three times the conductivity of the pure liquid. In view of the recent work of King and

Medley[88] however (see p. 121), this is most likely due to traces of impurities in the fibres, which could have a considerable effect when the wool is highly swollen.

It is a difficult matter to establish with certainty the mechanism by which the conduction proceeds, particularly as the conductivities are very small. Thus, both electrolytic and electronic mechanisms have been proposed, although the recent work of King and Medley supports the former mechanism strongly. Many insulators conduct electricity to a limited extent by means of electron jumps of a very small number of free electrons within the solid. According to the modern theory of metals, all the valency electrons in a true conductor are relatively free to move through the lattice without the application of an external force. As the solid becomes more semi-metallic the motions of the electrons become more restricted and in order to conduct electricity, they must overcome potential energy barriers. Thus the activation energy E for the conductivity of semi-conductors is usually high, and the effect of temperature on conductivity is given by the well-known equation,

$$\kappa = B \cdot e^{-E/RT}$$

where B is a specific constant. The temperature coefficient of electrolytic conduction on the other hand is usually small, although this also obeys the above general equation. The two types of conduction may also be distinguished by studying the Hall effect, which characterizes semi-conductors. If a current is passed through an electronic conductor which is subjected to a magnetic field acting at right angles, an E.M.F. is induced mutually at right angles to these forces. By measuring the magnitude and direction of this force, semi-conductors may be differentiated from metallic conductors. This method is always used when a crystalline or compact solid is under consideration, but is obviously difficult to apply to textile fibres.

Baxter[89] favoured the electronic mechanism mainly because of the very high activation energy (approximately three times that of liquid water, and almost equal to that of ice), which remains approximately constant as the regain increases (Fig. 4.10). He considers the dry wool to be a perfect insulator, the current being carried by electronic jumps between adjacent water molecules. It was found that except at very low regains, collagen, wool and silk give the same activation energy (1·35 e.v., i.e. 32 k. cal./mol).

The results of King and Medley[88] cast considerable doubt on the electronic mechanism, as a direct current has been shown to produce electrolysis in agreement with Faraday's laws. Thus using carefully purified fibres in order to exclude ionic impurities, the quantity of

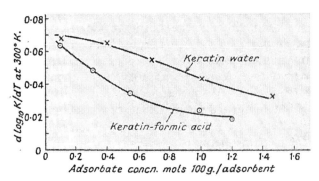

FIG. 4.10.—Activation energy of electrical conduction, K, at different water and formic acid contents.[92] (*Reproduced by permission.*)

hydrogen liberated was found to be proportional to the quantity of electricity passed (Table 4.10).

TABLE 4.10

The electrolysis of swollen wool fibres[92]

Quantity of electricity passed (Coulombs)	Pressure of hydrogen developed	Quantity of electricity calculated	$\dfrac{Q.\ \text{calc.}}{Q.\ \text{measd.}} \times 100$
0·26	0·070	0·230	88
0·37	0·106	0·345	93
0·51	0·140	0·455	89
0·33	0·098	0·318	96

Similarly, O'Sullivan has also mentioned the liberation of gas during the conduction of cellulose swollen with water, although he was unable to obtain a similar quantitative relation.[90]

The presence of ionic impurities is found to affect the conductivity of swollen keratin considerably, in the same way as in analogous systems. For example, O'Sullivan[90] has detected the movement of ion boundaries in cellulose containing as little as 10 per cent. water, impregnated with various salts. Similarly, Fuoss[91] has attributed the electrical conductivity of swollen polymers to ionic impurities. In the case of the ampholyte keratin, it is very difficult to remove the last traces of salts held on charged amino and carboxyl groups (see Chapter 5). Thus dialysis reduces the conductivity of swollen keratin,[29] although prolonged washing has a much smaller effect.

King and Medley[92] have studied in considerable detail, the conductivity of the wool-water system impregnated with various salts and find that potassium chloride produces large increases over the

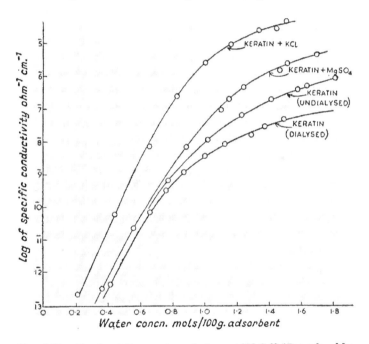

Fig. 4.11.—Conductivity-sorption relations at 25° C.[92] (*Reproduced by permission.*)

whole range of water content. Magnesium sulphate on the other hand, has very little effect in the least-swollen medium, but a considerable effect as the swelling increases (Fig. 4.11). As the conductivity varies by a factor of 10^5 over the 6–18 per cent. water concentration range, whereas the diffusion coefficient of water in keratin varies by only 10^2, distortion of the gel structure plays only a small part in the conduction process. This fact, together with the conductivity data given in Fig. 4.11, indicates that the ions are associated to a large extent at the lower regains.

The change in conductivity with regain is closely parallel to the change in dielectric constant (see p. 123), and King has shown that formic acid has a greater effect on both properties than water.[93] The extent of ion-pair formation can be calculated as a function of dielectric constant by the theory of Kraus and Fuoss,[94] which has been widely applied to strong electrolyte solutions. This treatment has been applied by King and Medley[92] with considerable success to the keratin system, and theoretical curves similar to the experimental curves given in Fig. 4.11 may be derived.

The temperature coefficient is greatest at low regains but decreases regularly to about half the maximum value as the concentration of water increases to 18 per cent. (Fig. 4.10). It will be observed that the extrapolated value for saturation when the salt is almost fully dissociated, approaches a value of the order of 10–12 k. cal./mol for the activation energy, which has been obtained for a variety of diffusion processes within the swollen fibre (Chapter 5, p. 168). The high values at low regains are due partly to the viscous restraint of the medium, but mainly to the increase in dissociation of the ion pairs as the temperature increases. Thus the values obtained by Baxter,[89] which remain high (c. 30 k. cal./mol) as the regain increases, are given approximately by the sum of the energy of migration through the gel structure (c. 10 k. cal./mol), and the energy of dissociation of water (c. 14 k. cal./mol).

As already indicated, the dielectric constant increases considerably with regain both at high and low frequencies (Fig. 4.12).[95] The dielectric constant of the dry fibres has been determined by several workers. Thus Errera and Sack[96] obtained the same value (4·2 at 10^6 c.p.s.), for wool and silk which suggests that this is due largely to the polypeptide chains, as these two proteins have very different numbers and types of side chain groups. Shaw and Windle[97] have obtained the value of 4·4 at 3×10^9 c.p.s. in close agreement with this value and the value of 4·2 obtained by King at audio frequencies.[93]

In contrast to the corresponding values for gelatin[98] however, these values are not equal to the square of the refractive index ($n = 1·547$ for wool,[99] so that $n^2 = 2·39$). The change in dielectric constant with regain varies considerably with frequency (Fig. 4.12) showing that extensive dielectric dispersion occurs. Although the cause of this is not known, King[100] has shown that the dispersion is proportional to the quantity of free water within the fibre.

King has calculated the polarization of the system assuming the Debye relation for isotropic mixtures to hold,[101] and modifying this for the two types of water sorbed. It is found that the total polarization cannot be due to the more labile water molecules alone, and

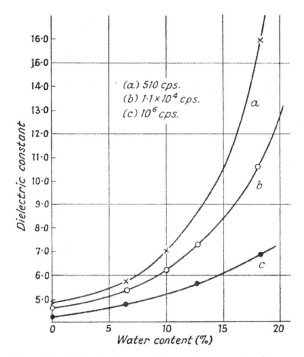

FIG. 4.12.—Relation between dielectric constant and water content of keratin.[100] (*Reproduced by permission.*)

King suggests that the polar groups contribute to the observed dielectric constant. This was also assumed by Fricke[98] in the case of gelatin, where the dielectric constant becomes very large (~ 200) as the water content increases to 50 per cent. There is also a considerable increase in dielectric dispersion. It is considered that in both cases, the adsorbate weakens the hydrogen bonds between CO- and NH-groups of adjacent polypeptide chains, thus allowing these groups freedom of rotation, in agreement with the studies of rotation in polyamides made by Baker and Yager.[102] In support of this deduction, it is shown that formic acid has a much greater effect on the dielectric constant than water or methyl alcohol for a given concentration, although its dipole moment is smaller.[93] Fricke has also pointed out the correlation between dielectric dispersion and mechanical properties, which supports the contention that the increase in dielectric constant and dielectric dispersion are both due to the increased rotation of polar groups in the polypeptide chains.

The effect of water on the high-frequency conductivity of wool has recently been studied by Alexander and Meek.[103] As opposed to the L.F. and D.C. conductivity, the H.F. conductivity does not depend on the transport of charges from one electrode to another, but is the result of the orientation of charged or dipolar molecules in the field. Thus the conductivity is low when the molecules in the system have a small dipole, or when the charged molecules are small and

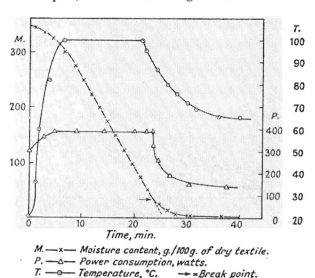

M. —×— Moisture content, g./100 g. of dry textile.
P. —△— Power consumption, watts.
T. —○— Temperature, °C.   → = Break point.

FIG. 4.13.—Change of moisture content, power consumption and temperature with time of drying for wool felt.[103]

mobile, so that they can follow the alternation of field. At 50 mc./s. water molecules, although highly polar, can follow the oscillations without lagging behind, and pure water does not therefore absorb H.F. energy, resulting in a low conductivity. As shown in Fig. 4.13 it was found that the H.F. conductivity of the water held by a pad of fibres remains very high even for water contents of the order of 100 per cent. This suggests that the water is bound in some way and not able to oscillate freely. If the sorbed water dipoles were orientated in a multimolecular film at the surface of the fibres, due to induced polarization which decreases with distance from the surface, the relatively high H.F. conductivity and low dielectric constant could be explained. These observations may have some bearing on the work of Ashpole,[104] who showed that as the relative humidity approaches close to 100 per cent., the quantity of water held by the fibres exceeds the usually accepted saturation value of 33 per cent. by weight, and may reach values of the order of 60 per cent. As swelling data are in agreement with the former value (see Table 4.1) the excess water must be largely accommodated at the fibre surface.

REFERENCES

[1] ROWEN AND GAGLIARDI, *J. Res. N.B.S.*, 1947, **38**, 103.
[2] See KING, *J. Soc. Dy. Col.*, 1950, **66**, 27.
[3] See MCLAREN AND ROWEN, *J. Polymer Sci.*, 1951, **7**, 27.
[4] ROWEN AND BLAINE, *Ind. Eng. Chem.*, 1947, **39**, 1659.
[5] HEDGES, *Trans. Faraday Soc.*, 1926, **22**, 178.
[6] SPEAKMAN, *Proc. Roy. Soc.*, 1931, **132A**, 167.
[7] KING, *J. Text. Inst.*, 1926, **17**, T53.
[8] HERMANS, HERMANS AND VERMAS, *J. Polymer Sci.*, 1946, **1**, 149, 156, 162.
[9] HERMANS, *Physics and Chemistry of Cellulose Fibres*, Elsevier, New York, 1949, p. 180 ff.
[10] ASTBURY, *Nature*, 1930, **126**, 913; ASTBURY AND WOODS, *Proc. Roy. Soc.*, 1933, **232**, 333.
[11] KING, *Trans. Faraday Soc.*, 1947, **43**, 552.
[12] BRUNAUER, *The adsorption of gases and vapours*, Princeton Univ. Press, 1943, p. 5.
[13] HAUSER AND MCLAREN, *Ind. Eng. Chem.*, 1948, **40**, 112.
[14] MELLON, KORN AND HOOVER, *J. Amer. Chem. Soc.*, 1947, **69**, 827.
[15] KATZ, *Trans. Faraday Soc.*, 1933, **29**, 279; *Kolloid Z.*, 1917, **Z**.
[16] DOLE, *J. Chem. Phys.*, 1948, **16**, 25; *Ann. New York Acad. Sci.*, 1949, **51**, 705.
[17] FLORY, *J. Chem. Phys.*, 1942, **10**, 51.
[18] See CASSIE, *Fibrous Proteins Symp.*, *Soc. Dy. Col.*, Bradford, 1946, 86.
[19] GREEN, *Trans. Roy. Soc. N.Z.*, 1948, **77**, 24.
[20] BULL, *J. Amer. Chem. Soc.*, 1944, **66**, 1499.
[21] MCLAREN AND OTTESEN, *Compt. rend. trav. lab. Carlsberg, Ser. chim.*, 1950, **27**, No. 13, p. 325.
[22] GREEN, *Trans. Roy. Soc. N.Z.*, 1948, **77**, 313.
[23] SPEAKMAN, *J. Text. Inst.*, 1936, **27**, T185.
[24] RAO, SUBBA AND RAO, SANSIVA, *Proc. Ind. Acad. Sci.*, 1947, **25**, 221.
[25] COOK, *J. Amer. Chem. Soc.*, 1948, **70**, 2925; HALSEY, *J. Chem. Phys.*, 1948, **16**, 931.
[26] BARKAS, *Trans. Faraday Soc.*, 1942, **38**, 194; *ibid.*, 1946, **42B**, 137.
[27] MELLON, KORN AND HOOVER; *J. Amer. Chem. Soc.*, 1948, **70**, 1144.
[28] MCLAREN AND KATCHMAN, *J. Amer. Chem. Soc.*, 1951, **73**, 2124.
[29] MELLON, KORN AND HOOVER, private communiation, quoted by MCLAREN AND ROWEN, ref. 3.
[30] PEIRCE, *J. Text. Inst.*, 1924, **15**, T501.
[31] SMITH, *J. Amer. Chem. Soc.*, 1947, **69**, 646.
[32] CASSIE, *Trans. Faraday Soc.*, 1945, **41**, 450.

[33] URQUHART, *J. Text. Inst.*, 1929, **20**, T125.
[34] SPEAKMAN, *Trans. Faraday Soc.*, 1929, **25**, 92; *J. Text. Inst.*, 1927, **18**, T446.
[35] CASSIE, *Discussions Faraday Soc.*, 1948, **3**, 11.
[36] SPEAKMAN, *Trans. Faraday Soc.*, 1944, **40**, 6.
[37] GILBERT, *Fibrous Proteins Symp.*, Soc. Dy. Col., Bradford, 1946, 96.
[38] BRUNAUER, EMMETT AND TELLER, *J. Amer. Chem. Soc.*, 1938, **60**, 309.
[39] HILL, *J. Chem. Phys.*, 1946, **14**, 263.
[40] MAYER AND MAYER, *Statistical Mechanics*, Wiley, New York, 1940, p. 63.
[41] GORTER, *Ann. Revs. Biochem.*, 1941, **10**, 619.
[42] NEURATH AND BULL, *Chem. Rev.*, 1938, **23**, 391.
[43] ROWEN AND SIMHA, *J. Phys. and Colloidal Chem.*, 1949, **53**, 921.
[44] HUGGINS, *Ann. New York Acad. Sci.*, 1943, **44**, 431.
[45] BRADLEY, *J. Chem. Soc.*, 1936, 1467.
[46] HOOVER AND MELLON, *J. Amer. Chem. Soc.*, 1950, **72**, 2562.
[47] DE BOER AND ZWICKER, *Z. phys. Chem.*, 1929, **B3**, 407.
[48] JORDAN LLOYD AND PHILLIPS, *Trans. Faraday Soc.*, 1934, **A147**, 382; JORDAN, LLOYD AND MORAN, *Proc. Roy. Soc.*, 1934, **A147**, 382.
[49] GEE AND TRELOAR, *Trans. Faraday Soc.*, 1942, **38**, 147, 418; ORR, *ibid.*, 1944, **40**, 320.
[50] MCBAIN, GOOD, BAKER, DAVIES, WILLAVOYS AND BUCKINGHAM, *Trans. Faraday Soc.*, 1933, **29**, 1086.
[51] BARRER, *Trans. Faraday Soc.*, 1947, **43**, 3.
[52] HAILWOOD AND HORRABIN, *Faraday Soc. Discussions*, 1946, **42B**, 84.
[53] GEE, *ibid.*, 1946, **42B**, 96.
[54] BARRER, *ibid.*, 97.
[55] UBELOHDE, *ibid.*, 99.
[56] MELLON, KORN AND HOOVER, *J. Amer. Chem. Soc.*, 1949, **71**, 2761.
[57] TRELOAR, *Trans. Faraday Soc.*, 1952, **48**, 567.
[58] WHITE AND EYRING, *Text. Res. J.*, 1947, **17**, 523; WHITE AND STAM. *ibid.*, 1949, **19**, 136.
[59] CASSIE, *Trans. Faraday Soc.*, 1945, **41**, 458.
[60] WARBURTON, *Proc. Phys. Soc.*, 1946, **58**, 585.
[61] ASTBURY, *J. Chem. Soc.*, 1942, 339.
[62] FISHER, *Proc. Roy. Soc.*, 1923, **A103**, 139.
[63] PRESTON AND CHEN, *J. Soc. Dy. Col.*, 1946, **62**, 361.
[64] SPEAKMAN AND COOPER, *J. Text. Inst.*, 1936, **27**, T183.
[65] MEYER, MISCH AND BADENHUIZEN, *Helv. Chim. Acta*, 1939, **22**, 59. HERMANS AND WEIDINGER, *J. Colloid Sci.*, 1946, **1**, 185.
[66] BRIGGS, *J. Phys. Chem.*, 1931, **36**, 367; JONES AND GORTNER, *ibid.*, p. 387; see also HOBER, *Physical Chemistry of Cells and Tissues*, Churchill, London, 1935, p. 123 ff.
[67] PAULING, *J. Amer. Chem. Soc.*, 1945, **67**, 555.
[68] GREEN, *Trans. Roy. Soc. N.Z.*, 1949, **77**, 313.
[69] SPONSLER, BATH AND ELLIS, *J. Phys. Chem.*, 1940, **44**, 996; 1947, **69**, 1896.
[70] MAGNE, PORTAS AND WAKEHAM, *J. Amer. Chem. Soc.*, 1950, **72**, 3662.
[71] MELLON, KORN AND HOOVER, *J. Amer. Chem. Soc.*, 1948, **70**, 3040.
[72] MCMEEKIN, GROVES AND HIPP, *J. Amer. Chem. Soc.*, 1950, **72**, 3662.
[73] PERUTZ, *Trans. Faraday Soc.*, 1946, **42B**, 187.
[74] SPEAKMAN, *J. Soc. Chem. Ind.*, 1930, **49**, T209.
[75] BLACKBURN AND PHILLIPS, *Biochem. J.*, 1944, **38**, 171.
[76] BAKER AND FULLER, *Ann. New York Acad. Sci.*, 1943, **44**, 329; *J. Amer. Chem. Soc.*, 1943, **65**, 1120.
[77] DOLE AND FALLER, *J. Amer. Chem. Soc.*, 1950, **72**, 414.
[78] EILERS AND LABOUT, *Fibrous Proteins Symp.*, Soc. Dy. Col., Bradford, 1946, 30.
[79] SHORTER, *J. Text. Inst.*, 1924, **15**, T358.
[80] GOODINGS, *Amer. Dyestuff Reporter*, 1935, **24**, 109.
[81] SPEAKMAN, STOTT AND CHANG, *J. Text. Inst.*, 1933, **24**, T284.
[82] DAVIS AND MCLAREN, *J. Polymer Sci.*, 1948, **3**, 16.
[83] DOLE AND MCLAREN, *J. Amer. Chem. Soc.*, 1947, **69**, 651.
[84] URQUHART AND WILLIAMS, *J. Text. Inst.*, 1924, **15**, T559.

[85] Murphy and Walker, *J. Phys. Chem.*, 1928, **32,** 1761; Murphy, *ibid.*, 1929, **33,** 200, 509.
[86] Marsh and Earp., *Trans. Faraday Soc.*, 1933, 29, 173.
[87] Stamm, *Colloid Symposium Monograph*, 1926, **4,** 246.
[88] King and Medley, *J. Colloid Sci.*, 1949, **4,** 1.
[89] Baxter, *Trans. Faraday Soc.*, 1943, **39,** 207.
[90] O'Sullivan, *J. Text. Inst.*, 1947, **38,** T285.
[91] Fuoss, *J. Amer. Chem. Soc.*, 1939, 61, 2329.
[92] King and Medley, *J. Colloid Sci.*, 1949, **4,** 9.
[93] King, *Nature*, 1946, **158,** 134.
[94] Kraus and Fuoss, *J. Amer. Chem. Soc.*, 1933, 55, 476, 1019.
[95] King, *J. Colloid Sci.*, 1947, 2, 551.
[96] Errera and Sack, *Ind. Eng. Chem.*, 1943, 35, 6, 712.
[97] Shaw and Windle, *J. Appl. Phys.*, 1950, **21,** 956.
[98] Fricke and Parker, *J. Phys. Chem.*, 1940, **44,** 716; Fricke and Curtis, *ibid.*, 1937, **41,** 729.
[99] Fox and Finsh, *Text. Res. J.*, 1940, **11,** 62.
[100] King, *Trans. Faraday Soc.*, 1947, **43,** 601.
[101] Burton and Turnbull, *Proc. Roy. Soc.*, 1937, **158A,** 194.
[102] Baker and Yager, *J. Amer. Chem. Soc.*, 1942, **64,** 9, 2176.
[103] Alexander and Meek, *J. Soc. Dy. Col.*, 1950, **66,** 530.
[104] Ashpole, *Proc. Roy. Soc.*, 1952, **212A,** 112.

CHAPTER 5

Rate Processes within the Fibre

The penetration of neutral molecules into dry wool

ALTHOUGH it has been known for some time that simple molecules can readily penetrate wool fibres, the fundamental rate processes have only recently been considered. It was originally thought that the relatively slow rates of sorption and desorption of water vapour were due to evaporation from the surface and diffusion through the fibre.[1] Cassie and King[2] have shown clearly, however, that under normal circumstances, the rate of evaporation is governed by the rate of diffusion in a stagnant layer of gas at the surface similar to the diffusion layer in solution (see p. 152). If care is taken to remove this layer rapidly as it forms, the observed rate of absorption is due entirely to the temperature changes within the fibre. As the absorption is accompanied by a large evolution of heat, the temperature rises to values of the order of 60°–70° C. in the initial stages. The effect of temperature on the equilibrium regain may be investigated by the Kirchoff relation, shown to hold for the wool-water system by Shorter[3] and Hedges[4] (see Chapter 4, p. 115). It is found that at any given temperature, equilibrium within the fibre is reached almost instantaneously (the maximum possible time being 15 sec), provided that the stagnant layer of gas at the surface is removed. Consequently no information on the structure of the fibres can be obtained from time-regain curves using water vapour.

King[5,6] has also studied the absorption of larger molecules into wool fibres. The temperature effect is still considerable for methyl alcohol, but much less for ethyl alcohol, so that in this case the rate of absorption is governed almost entirely by the rate of diffusion of the molecules through the fibre. On the basis of the old pore hypothesis, Speakman[7] has considered that neutral molecules above a critical size (propyl and butyl alcohols) are incapable of penetrating the fibres because of steric factors. King has shown that these molecules do penetrate to approximately the same extent as the lower alcohols, but at a greatly reduced rate.[6]

In order to understand the processes occurring within the fibre as penetration proceeds, a complete knowledge of the diffusion properties of the solute within the absorbent is required. This is difficult to obtain for the systems under consideration, as not only the rate constant representing the diffusional flow, but also the structure and dimensions of the solid phase change continually during the process. These important conditions have only been examined in detail in recent years, and although work has centred around polymers other than wool, the treatments involved and conclusions reached are general. Consequently it is proposed to give initially a brief account of diffusion in general terms, drawing freely on data from analogous systems. At the same time, particular attention will be paid to results obtained with wool fibres.

In the first part of the chapter, which deals with the diffusion of non-reactive molecules into the initially dry fibre, it is proposed to discuss general methods for investigating the constancy or otherwise of the diffusion coefficient, together with results and conclusions which have been reached for wool fibres and other forms of keratin. To study the concentration dependence of the diffusion of a particular solute molecule, true values of the characteristic diffusion coefficient must be obtained. The available methods may be conveniently divided into three types, the first of which depends on maintaining a steady state within the medium, and is preferred to all others because of the simplicity of interpreting the data. This method is used extensively for measuring diffusion coefficients in solution by carrying out the diffusion across a porous pad,[8] and for studying the diffusional properties of membranes.

In many cases, however, a steady state cannot be achieved, as for example when diffusion into fibres or other naturally produced particles or colloidal systems is studied. In such cases the diffusion coefficient must be obtained from measurements of the rate of absorption of the solute from solution or from the gaseous phase, although such data may be exceedingly difficult to interpret theoretically.

A more direct approach to the measurement of the diffusion coefficient for a varying system involves the measurement of the distribution of solute at various times, so that the concentrations at various distances from the interphase boundary may be ascertained. This procedure provides a complete picture of the diffusion process, and has been used in several important cases. The chief difficulty, apart from the treatment of the varying diffusion coefficient, is that in many systems of considerable technical interest, diffusion is accompanied by swelling, so that the position of the reference boundary also changes with time. The treatment of this system may be extremely involved, but the recent work of Hartley and Crank[42] has done much to clarify the position.

Diffusion in a steady state

The case of steady-state diffusion, which is dealt with first because of its relative simplicity, involves the transport across a thin sheet of thickness (l) which is in contact on one face with a gas at a constant pressure or solute at a constant concentration. The concentration of gas or solute at the opposite face is maintained low and constant during the measurement. The membrane may be pre-conditioned before the actual experiment, so that as $t = 0$, the concentrations within the sheet at the surfaces are in equilibrium with the external phases and remain constant during the process.

The rate of transport P is given by Fick's law in terms of the concentration gradient across the sheet:

$$P = D \cdot dC/dx.$$

As the system is in a steady state, the flux P is independent of x, so that

$$\int_0^l P \cdot dx = P \cdot \int_0^l dx = P \cdot l$$

Thus in general,

$$P = \frac{1}{l} \int_{c1}^{c2} D \cdot dC.$$

This treatment holds for a variable as well as a constant diffusion coefficient and an examination shows that the flux can never decrease with increase in the concentration difference ($c_2 - c_1$) across the membrane for any mode of variation of D with C, so that a low diffusion coefficient at a point within the sheet is compensated by a high concentration gradient.

Although this method has been used extensively for the measurement of diffusion coefficients which were assumed to be concentration-independent, few references are found to its application to a concentration-dependent diffusion process. King.[9] however, has made an exhaustive study of the diffusion of water vapour through a keratin membrane, and Rouse[10] has used a similar method to study the diffusion of water in nylon and polythene. The measurements were further simplified by maintaining the concentration of vapour at the low pressure side at zero by continuous pumping, and collecting the amount of water which had passed through in a given time. By using different initial pressures, average diffusion coefficients over the range $c = 0$ to $c = c_1$ can be obtained, from which values of D corresponding to a given concentration are obtained by graphical differentiation (Fig. 5.1). This general method is described in more

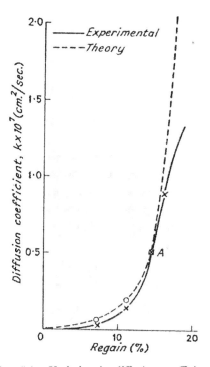

Fig. 5.1.—Variation in diffusion coefficient with regain for the keratin-water system.[9] (*Reproduced by permission.*)

detail on page 141 where methods of obtaining more accurate values are discussed. The diffusion coefficient is found to change rapidly with regain from values of the order of 10^{-9} cm.^{2}/sec. to $> 10^{-7}$ cm.^{2}/sec. at 25° C. as saturation is approached.

Although the actual rate of flow of water vapour was found to be the same at any given pressure as the temperature changed, the diffusion has a positive temperature coefficient owing to the change in regain with temperature. When allowance is made for this difference in surface concentration at the two temperatures, the estimated activation energy increases with decrease in regain from approximately 4·8 k. cal./mol at 16 per cent. regain to 7.5 k. cal./mol at low regains. The low value at high regains indicates that diffusion proceeds through the aqueous phase within the fibre, as the sorbed molecules are probably highly mobile above a certain regain, corresponding approximately to the formation of a monolayer within the fibre (see p. 98).

Diffusion in a non-steady state with constant diffusion coefficient.

The non-steady state is of considerable importance because in many cases the diffusion coefficient has to be determined by measuring the quantity of solute absorbed as a function of time. In almost all cases the concentration at the surface is maintained constant, either by using a large excess of solute in the external solution, or by working in a concentration range where the absorption isotherm enforces an almost constant surface concentration, e.g. in acid dyeing (p. 157). When the volume of the system remains constant, the mathematical treatment is relatively simple in terms of the second law of Fick which in the general case takes the form:

$$\frac{\delta C}{\delta t} = \frac{\delta}{\delta x}\left(D \cdot \frac{\delta C}{\delta x}\right)$$

where D is concentration dependent. In most cases so far studied, D has been assumed independent of C, when the equation takes the simpler form:

$$\frac{\delta C}{\delta t} = D\frac{\delta^2 C}{\delta x^2}$$

This condition may hold approximately when low concentrations of diffusing material are used, and when diffusion proceeds through a porous medium which is initially in the fully swollen state (see p. 157). The simplest case of diffusion into a solid of infinite width is given by the following solution for the boundary condition $C = C_0$, when $x = 0$ for all t:

RATE PROCESSES WITHIN THE FIBRE

$$C_x = C_0\left(1 - \frac{2}{\sqrt{\pi}}\int_0^{x/2\sqrt{(Dt)}} e^{-y^2}.dy\right)$$

where C_x is the concentration at a distance x from the surface. The Gaussian error function may be expanded as follows:

$$C_x = C_0\left(1 - \left\{\frac{2}{\sqrt{\pi}}\frac{x}{2\sqrt{(Dt)}} - \frac{x^3}{3.1![2\sqrt{(Dt)}]^3} + \frac{x^5}{5.2![2\sqrt{(Dt)}]^5} - \ldots\right\}\right)$$

The amount of solute entering area A of solid in time t is given by

$$Q_t = A\int_0^t D\left(\frac{\delta C_x}{\delta x}\right).dt$$

Taking the first term of the above expansion, the following approximation is obtained,

$$Q_t = 2C_0A\sqrt{(Dt/\pi)}$$

This relation only holds in the initial stages of the process, and may be applied under these conditions to diffusion into any solid with a plane boundary. Thus, for a small penetration, this expression may be applied to spheres and cylinders, to obtain an approximate value of D. In the case of a cylinder, which is of importance when dealing with fibres, the total quantity absorbed at equilibrium is given by $Q_\infty = C_0\pi a^2 l$, where a is the radius of a fibre and l is the average length. Thus, the fractional attainment of equilibrium is given by:

$$\frac{Q_t}{Q_\infty} = \frac{4}{a}\sqrt{(Dt/\pi)}$$

The complete solution for radial diffusion[11, 12] into a cylinder from a constant surface concentration is given by:

$$\frac{Q_t}{Q_\infty} = 1 - \frac{2}{a^2}\sum\frac{1}{\alpha_n^2}.\exp-(D\alpha_n^2.t)$$

where α_n is the nth positive root of the Bessel function $J_0(\alpha, a) = 0$. The first four roots are $\alpha_1 = 2{\cdot}405/a$; $\alpha_2 = 5{\cdot}520/a$; $\alpha_3 = 8{\cdot}654/a$; $\alpha_4 = 11{\cdot}7915/a$.

For small values of D and t, this equation approaches the approximate solution given above, and it may be shown that the two solutions are similar for approximatly 40 per cent. of the reaction. For large values of t, only the first term need be considered so that log Q_t is linear in t. The slope of the absorption curve thus approaches $D\alpha_1^2$ from which the diffusion coefficient may be evaluated.

Finally for diffusion into a slab of thickness l, which is useful for non-steady state diffusion into membranes, the corresponding solution is given by,[12]

$$\frac{Q_t}{Q_\infty} = 1 - \frac{8}{\pi^2} \sum_{m=0} \frac{1}{(2m+1)^2} \cdot \exp\{-D \cdot (2m+1)^2 \cdot \pi^2 \cdot t/l^2\}$$

From these solutions, it follows that the time for a particular fractional absorption is independent of the final absorption.[13] Thus, in the most widely used instance where $Q_t/Q_\infty = 0 \cdot 5$,

$$D = 0 \cdot 0494 / \cdot (t_{0.5}/l^2)$$

The constancy of the diffusion coefficient may be investigated by studying the change in $t_{0.5}$ with different constant values of the surface concentration where Q_∞ automatically varies. This criterion has been used by Crank and Park[13] to show the highly variable nature of D for the penetration of chloroform into polystyrene.

Diffusion from a variable surface concentration with a constant diffusion coefficient

In most practical cases, the concentration at the surface cannot remain constant as the process proceeds, as for example in dyeing and most finishing treatments. With a further variable introduced, the mathematical treatment of the general diffusion equation is extremely complex, and formal solutions may be obtained only in the simplest case where the surface concentration within the fibre is equal or proportional to the concentration in solution for all values of t. In other cases where the distribution of solute between fibre and solution is governed by a non-linear absorption isotherm, numerical solutions only can be obtained.[14] Such processes have been considered in recent years, although empirical approaches have been adopted for some time.[12] Wilson and Crank[15] have considered the diffusion of a solute which simultaneously adsorbs on specific sites, such that the mobile solute and adsorbed solute are always in mobile equilibrium, given by a distribution coefficient K. The diffusion equation thus becomes:

$$\frac{\delta C}{\delta t} = \frac{D}{K+1} \cdot \frac{\delta^2 C}{\delta x^2}$$

Solutions have been obtained for diffusion into a slab and into a cylinder, the latter only being considered here owing to its application to fibres. Paterson[16] has given a solution for a sphere, which has recently been extended by Barrer.[17] The latter has shown that under some circumstances, the quantity diffused varies as the square root of time according to the approximate relation,

$$\frac{Q_t}{Q_\infty} = \frac{6}{a} \cdot \frac{K'+1}{K'} \sqrt{\left(\frac{Dt}{\pi}\right)}$$

where a is the radius of the sphere and $1/K' = K \cdot v_\text{I}/v_\text{II}$, v_I and v_II being the volumes of the external gas or liquid phase, and sorption medium respectively, i.e. $1/K'$ = amount of solute absorbed/amount of solute not sorbed. The above expression holds only for small enough values of $\sqrt{(Dt/\pi)}/K'a$. This explains why a parabolic rate law is obtained in many cases where the surface concentration is not constant. This condition is recognized by $t_{0.5}$ for Q_t/Q_∞ being independent of initial concentration (see p. 161), whereas if the parabolic law denoted diffusion from a constant surface concentration, $t_{0.5}$ would be proportional to the square of the concentration (see p. 159).

For diffusion into a cylinder, the general equation for diffusion with adsorption is transformed into polar co-ordinates,

$$\frac{\delta c}{\delta t} = \frac{D}{K+1} \frac{1}{a} \frac{\delta}{\delta a} \cdot \left(a \frac{\delta c}{\delta a}\right)$$

Wilson[15] has shown that for diffusion into a cylinder of radius a and infinite length, surrounded by solution which is imagined to be enclosed in a cylindrical element of cross-section A excluding the volume occupied by the fibre, the following formal solution may be obtained if adsorption is rapid compared with diffusion,

$$\frac{Q_t}{Q_\infty} = 1 - \sum \frac{4(1+\alpha)}{4+4+\alpha^2 q_n^2} \cdot e^{-q_n^2 \cdot \beta t}$$

where q_n is the nth positive root of the Bessel[18] equation $\alpha q_n \cdot J_0(q_n) + 2 J_1(q_n) = 0$, and the constants α and β given by:

$$\alpha = \frac{A}{\pi a^2 \cdot (K+1)} \quad \text{and} \quad \beta = \frac{D}{a^2 \cdot (K+1)}$$

This solution is convenient only for moderate values of t and α, but may be transformed into alternative forms.[19] The most useful solution for small values of t is:

$$\frac{Q_t}{Q_\infty} \frac{1+\alpha}{1-\tfrac{1}{4}\alpha}[1-e^{y^2}.erfc.y] \quad \text{where} \quad y = 2(1+\tfrac{1}{4}\alpha)(\beta t)^{\frac{1}{2}}/\alpha$$

These expressions were originally developed for application to dyeing from a limited dyebath and met with no success, which will be discussed later. The rate of absorption of acids by wool, where instantaneous equilibrium is maintained between absorbed acid and acid in solution, can however be represented satisfactorily in this way.

Variable diffusion coefficient

Measurement from distribution of solute

The most complete procedure for studying a diffusion process is to measure the distribution of solute within the polymer at a series of times. This demonstrates directly the constancy of the diffusion coefficient or the empirical relation between D and concentration. Treatment of the general diffusion equation for variable D is considerably simplified if D is represented as a continuous function of c only, and this assumption has been made in all of the few treatments of a variable D in a non-steady state so far recorded. Boltzmann[20] originally showed that under these conditions the concentration at any point may be represented in terms of a combined parameter $xt^{\frac{1}{2}}$, so that if x is measured from the boundary between the two phases,

$$\left(\frac{\delta x}{\delta t}\right)_c = f.(D, C).t^{\frac{1}{2}}$$

By means of the Boltzmann theorem, the general diffusion equation may be transformed into a simple differential equation. Thus if the new parameter y is expressed as

$$y = x/t^{\frac{1}{2}}$$

it follows that,

$$\frac{y}{2}\cdot\frac{dC}{dy} = \frac{d}{dy}\cdot\left(D\cdot\frac{dC}{dy}\right)$$

This method has been investigated in detail by Crank and Henry[21] for several cases in which D is a single-valued function of C, mainly by studying the shapes of the C–y curves. In most cases the diffusion gradient becomes convex to the distance axis, as for the steady state, except when D increases with concentration at a steadily decreasing rate. The case of an exponential increase in D with C is reproduced in Fig. 5.2, and is of some importance from a practical point of view.

The penetration of a solvent into a polymer capable of swelling often proceeds by way of a sharp boundary which advances regularly into the medium. This must be related to a rapidly changing diffusion gradient, although Crank and Henry[21] consider that it cannot be

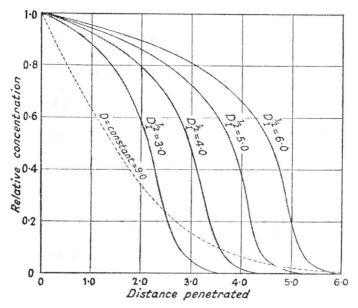

FIG. 5.2.—Theoretical concentration-distance curves for an experimental variation in diffusion coefficient[31] ($D = e^{bc}$), for various times, t given by $D_I \equiv D_0 t$. (*Reproduced by permission.*)

explained completely in this way. Such an advancing front was noticed for the diffusion of organic solvents into cellulose acetate by Hartley and Robinson[22], who attributed this to the effect of a rapidly changing diffusion gradient on the refractive index in the neighbourhood of the front. A similar observation has been made by Hermans[23] for the diffusion of water into cellulose, and very recently Crank and Robinson[24] have examined this type of system in detail, paying particular attention to the refractive index-concentration relationship. In some cases the front is due to an associated chemical reaction which removes solute from the diffusion front. This effect has been treated by Hermans and Vermas[25] and will be discussed at a later stage (p. 163). This can hardly explain the sharp front formed by the penetration of non-reactive solute as in the swelling of polymers with organic solvent, but may have some effect on the penetration of water into cellulose and proteins due to extensive hydration which no doubt occurs.[23]

The diffusion front may be seen in most cases by the microscope, and Hartley[26] has studied its formation and behaviour in some detail with the aid of polarized light. Owing to the uncertainty in the interpretation of the direct optical observations in terms of the concentration relationship to distance, it is advisable to measure the refractive index changes along the diffusion co-ordinate, as already employed by Wiener and Thovert.[27] Use is made of the interference produced by light passing through parallel layers of the diffusion cell, and Robinson[28] has recently evolved an elegant procedure for this purpose.

The appropriate concentration-distribution curves may be calculated from the Boltzmann[20] relation as follows:
Integration of the Boltzmann–Fick equation with respect to y leads to,

$$-\frac{1}{2}\int_0^c y \cdot dc = \left(D \cdot \frac{dc}{dy}\right)_0^{c_1} = \left(D \cdot \frac{dc}{dy}\right)_{c_1}$$

Introducing x and t,

$$D_{c_1} = -\frac{1}{2t} \cdot \frac{dx}{dc} \int_0^{c_1} x \cdot dc$$

This gives the origin from which x is to be measured.

The experimental observations are examined by deriving the concentration-distance curve for a given time from the pre-determined refractive index-concentration relationship.[24] The line where $x = 0$ is found by the above condition that $(D \cdot dc/dy)C_0 = 0$, i.e. the plane where the concentration gradient is zero, so that

$$\int_0^{C_0} y \cdot dC = \int_0^{C_0} x \cdot dC = 0$$

and the diffusion coefficients are evaluated by application of the modified Fick equation, using graphical integration to obtain values of $\int_0^{c_1} x \cdot dC$. The concentration-distance curves for the penetration of chloroform into cellulose acetate,[28] given in Fig. 5.3, show that the diffusion coefficient changes sharply at the boundary as already predicted, although the reason for the peculiar form is not fully understood.[24] In the case of the stretched polymer, direct proof of the Boltzmann relation is obtained thus supporting the validity of the mathematical treatment. According to this requirement, the displacement of a given concentration within the polymer should vary as the

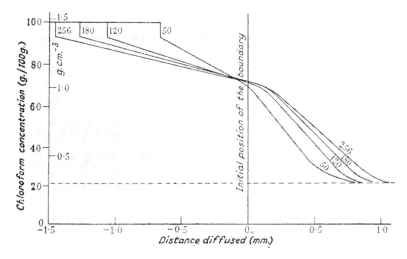

Fig. 5.3.—Concentration-distance distributions for the diffusion of chloroform into cellulose acetate.[28] (Numbers on curves denote time in minutes.) (*Reproduced by permission.*)

square root of time.[20] This is illustrated by Fig. 5.3, and all the curves are seen to intersect at a given point lying on the initial boundary of penetrant and polymer. This follows from the condition given above for determining the boundary conditions, as $x/t^{\frac{1}{2}}$ is constant and zero, so that the concentration is constant for all times at $x = 0$.

In some cases where unstretched, and consequently highly disordered, polymers are used, the Boltzmann relation has been found not to hold,[28] and in such cases the depth of penetration is more nearly proportional to t than to $t^{\frac{1}{2}}$ in the initial stages. This shows that D is not a single-valued function of c and is probably due to diffusion in directions other than across the polymer, which may in turn be due to irregular swelling.

Measurement from rate of absorption

The determination of diffusion rates from rate of absorption data is a subject of considerable contemporary interest, and a brief discussion will be given here. The general case of a variable diffusion coefficient is often treated by making use of a transformation suggested by Eyres, Hartree et al.[29] A modified concentration s defined as follows:

$$s = \int_0^c D \, . \, dc$$

is introduced into the general diffusion equation which then takes the form,

$$\frac{\delta s}{\delta t} = D \, . \, \frac{\delta^2 s}{\delta x^2}$$

In the cases where D is represented as a function of c, this equation has to be solved numerically. Crank and Henry[21] have obtained several solutions using the method of finite difference ratios.[14] Formal solutions have been obtained in the simplest case for diffusion into a semi-infinite solid for a diffusion coefficient which changes in the general way[30]:

$$D = D_0[1 + \beta_0(c - c_0) + \mu^2 \beta_1 (c - c_0)^2]$$

This has been used by King in studies of the rate of absorption of alcohols and water by keratin.[5]

By use of this transformation and also by using the diffusion equation derived from the Boltzmann theorem, the concentration is given as a function of the distance penetrated x. The extent of absorption may then be calculated by the general relation for a sheet of thickness l,

$$\frac{Q_t}{Q_\infty} = \int_0^l \frac{C \, . \, dx}{l \, . \, C_0}$$

and the corresponding equation

$$\frac{Q_t}{Q_\infty} = 1 - \int_0^l \frac{C \, . \, dx}{l \, . \, C_0}$$

for desorption.[21]

From the calculated absorption-time curves for various cases where D changes in a particular way with concentration, several conclusions result.[31] Whenever the diffusion coefficient increases with concentration, absorption is faster than desorption, which is of importance in the conditioning of fibres and other high polymers. Thus the last stages of desorption are extremely slow due to the small value of D, compared with the last stages of absorption where D is a maximum because of the high internal concentration.

The converse is true, so that desorption is faster than absorption when D decreases uniformly with increase in concentration. When the diffusion coefficient passes through a maximum, the two curves may cross, except when the maximum occurs where the concentration is half the initial value. In such cases the system is symmetrical, and as for a constant diffusion coefficient, absorption and desorption curves coincide.

In all cases which have been examined, the actual form of the absorption curve is not affected appreciably by a changing diffusion coefficient, although of course the rate at various times differs considerably. This coincidence enables the diffusion coefficient appropriate to a particular concentration to be determined by a comparatively simple procedure.[21] For a constant diffusion coefficient, theoretical and experimental curves may be compared at any point in order to evaluate D. For a varying diffusion coefficient, the two curves coincide approximately if fitted exactly for 50 per cent. absorption, i.e. where $Q_t/Q_\infty = 0.5$, from which an average diffusion coefficient $\bar{D}$ is obtained. It is found empirically that this is approximately equal to the integrated mean of the diffusion coefficient over the extreme limits of concentration, thus

$$\bar{D} = \frac{1}{C_0} \int_0^{c_0} D \, . \, dc$$

Values of $\bar{D}$ can be obtained in this way for a series of different initial concentrations C_0, which indicate the way in which D varies with concentration. From the graph so obtained, numerical or graphical differentiation leads to approximate values of D at various concentrations. Values of $\bar{D}$ and D obtained as first approximations for the absorption of chloroform on polystyrene are given in Fig. 5.4, from which it is seen that the true diffusion coefficient changes much more rapidly than the average value.[13] Accurate values of D can be obtained by a repetition process, involving successive approximations to the relation between D and $\frac{1}{C_0} \int D \, . \, dc$. Owing to the almost constant form of the rate curves, $\dfrac{1}{C_0} \int D \, . \, dc$ is proportional to $\bar{D}$ under all conditions, so that values of D can be read off directly from a master graph. This procedure has been successfully applied to the diffusion of chloromethanes through polystyrene.[32] It is interesting to find that

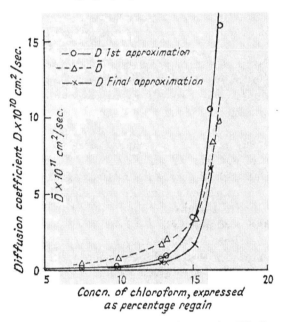

Fig. 5.4.—The effect of concentration on the diffusion coefficient of chloroform into polystyrene determined by rate of absorption measurements.[13] (*Reproduced by permission.*)

the diffusion coefficient varies almost exponentially with concentration, i.e. $D = D_0 \, . \, e^{kc}$, so that the diffusion proceeds according to the theoretical predictions already mentioned (see Fig. 5.2 for the $c-x$ relationship in this case).

The rate of absorption of alcohols into horn and wool fibres, and of water into horn has been studied by King,[5] and the variable diffusion coefficient treated in a similar way. The form of the absorption curves is characteristic of processes of this kind, e.g. the diffusion of water into rubber.[33] The initial limited region corresponds to the diffusion into a semi-infinite solid from constant surface concentration in accordance with Fick's law. Most of the absorption, however, occurs almost linearly with time (Fig. 5.5), with a sharp decrease to zero near equilibrium. This behaviour is a consequence of the greater ease of penetration with increase in swelling, responsible for the variable diffusion coefficient. In accordance with the Boltzmann equation, it is found that the amount of absorption increases as $\sqrt{t}$ for thick slabs.

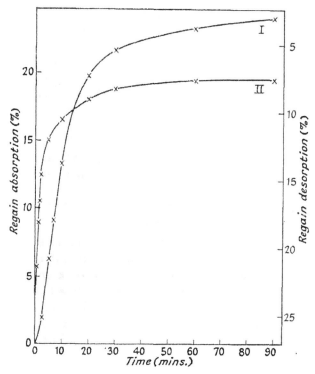

FIG. 5.5.—The rate of sorption of methyl alcohol on wool at 25° C.; I: absorption, II: desorption.[5] (*Reproduced by permission.*)

From the linear slopes of experimental and theoretical curves, approximate values of the minimum diffusion coefficients were evaluated (Table 5.1). It is interesting that the calculated diffusion coefficients for the absorption of methyl alcohol into wool and horn are very similar, the small difference being consistent with the effect of the temperature rise within the fibre.

The absolute values of the diffusion coefficients are probably not highly significant, although the comparison is valid as all were calculated

TABLE 5.1

Minimum values of the diffusion coefficient of water and alcohols in keratin[5]

System	Fibre dia. cm.	Slab thickness cm.	Temp. C.	Diffusion coeff. cm.^{2}/sec.
Water—horn	—	$3 \cdot 0 \times 10^{-2}$	25	$7 \cdot 5 \times 10^{-9}$
Water—horn	—	$6 \cdot 8 \times 10^{-2}$	25	$9 \cdot 0 \times 10^{-9}$
Methanol—wool	2×10^{-3}	—	25	$1 \cdot 7 \times 10^{-10}$
Methanol—horn	—	$7 \cdot 6 \times 10^{-3}$	25	$2 \cdot 2 \times 10^{-10}$
Ethanol—wool	2×10^{-3}	—	20	$3 \cdot 6 \times 10^{-12}$
Ethanol—wool	2×10^{-3}	—	25	$7 \cdot 2 \times 10^{-12}$
Ethanol—wool	2×10^{-3}	—	30	$11 \cdot 6 \times 10^{-12}$

by the same procedure. Thus, the values are somewhat greater for water than the values obtained from measurements of the transport across a keratin membrane[9] (p. 131) which are probably more accurate.

The mechanism of the diffusion process

The variation in diffusion coefficient with concentration given in Fig. 5.1 shows that it increases very sharply within a relatively small concentration range.[9] This behaviour seems to be general for the penetration of solute into a swelling medium. Thus, the diffusion coefficient changes in a similar way for the penetration of chloromethanes into polystyrene[32] and chloroform into cellulose acetate.[24] In the case of wool, the sharp change occurs between 8–10 per cent. regain, which corresponds approximately to the formation of a monolayer on localized sites within the fibre (see p. 111). This indicates that the diffusion proceeds mainly via the more mobile molecules which are in constant equilibrium with those on the localized sites.

The activation energy is usually assumed to be determined by the energy of hole formation in the lattice,[34] so that for diffusion in an elastic medium, low values are observed which are largely independent of the size of the diffusing molecules. The rapid increase in diffusion coefficient may therefore be attributed to the increase in elasticity with regain. King[6], however, rejects this view, as it is possible to vary the elastic constants at constant regain, owing to the hysteresis effect, without affecting the rate of diffusion. For this reason, King interpreted the diffusion in terms of the two types of molecule within the fibre. By applying the Lennard–Jones theory[35] of surface diffusion and the Knudsen cosine law[36] to give the rate of flow of the mobile molecules, the following equation was obtained,

$$D = \kappa \cdot e^{-\epsilon/RT} \cdot \frac{1}{(B-X)^2} \cdot \frac{\delta(A-X)}{dA}$$

where κ is a constant, ϵ the activation energy of the diffusion process. A is the number of molecules absorbed, of which X occupy the low energy sites, the available number being B. Using the values previously obtained by Cassie[37] this equation accounts satisfactorily for the observed variation of D with concentration (see Fig. 5.1). The deviations for large regains are probably due to the assumption that the diffusion of the mobile molecules is unrestricted except for encounters with the localized sites. This cannot be true as D must approach a limiting value, which is given by the self-diffusion coefficient of water,[38] $2\cdot4 \times 10^{-5}$ cm.^{2}/sec. at 25° C., and is probably less than this due to the influence of the keratin lattice.

As the concentration of absorbate decreases to zero, a value of the minimum diffusion coefficient is obtained[6] for the conditions $(B-X) \to B$ as $A \to 0$:

$$D_0 \simeq K \cdot \frac{A}{B^3} a \cdot e^{-(w+\epsilon)/RT}$$

where w is the heat evolved when one mole of liquid is absorbed on to localized sites, ϵ is the energy of activation for the diffusion of mobile molecules and a is a constant given in the Cassie isotherm[37] (p. 110).

It was also found[39] that $\log D_0$ varies proportionally with the polarizability of the sorbed molecule calculated from the appropriate refractive index,[40] i.e.

$$\log D_0 \propto \frac{n^2-1}{n^2+2} \cdot \frac{M}{\rho}$$

A similar relation between $\log D$ and M/ρ has been obtained by Crank and Park for the diffusion of halomethanes into polystyrene.[32] This suggests that the minimum diffusion coefficient for each member of the series of alcohols and water is given by the activation energy only,

i.e.
$$D_0 = D_0' e^{-(w+\epsilon)/RT}$$

where D_0' is a constant throughout the series. As ϵ is independent of the lattice constants, it should depend merely on molecular interaction as for a pure liquid, in agreement with the B.E.T. theory.[41] Thus for water, $\epsilon \sim 4 \cdot 6$ cal./mol.[38] Assuming similar values for the alcohols, corresponding values of w can be calculated from the appropriate values of D_0, assuming Cassie's value of 3·5 cal./mol for water, in order to determine D_0'. The values of w obtained in this way agree approximately with values obtained from heat of wetting data (p. 110), allowance being made for the elastic constraint on the molecular sorption. This agreement was taken to support the proposed diffusion mechanism outlined above.

Diffusion in a swelling medium[42]

If high concentrations of solute penetrate an initially unswollen polymer, considerable changes in dimensions occur (see p. 87), so that the position of the surface from which diffusion proceeds is continually changing. Thus the rate of transfer of solute from the external liquid is dependent on two processes, the intrinsic diffusion of the solute into the solid, and the mass flow of the polymer and solute due to the swelling. The boundary conditions to be applied to the general diffusion equation therefore become dependent on the solution of the equation. This complex system may be examined by adopting an alternative reference for the boundary. Although the volume may change, the total mass will remain constant, and Crank and Harley[42] have suggested fixing a reference plane such that the mass on either side remains constant. The variable dimension ξ is chosen so that a unit increment in its value always includes unit mass of the material of the sheet. Then if the concentration is also defined in terms of this variable, i.e. amount per unit mass, the general diffusion equation can be employed,

$$\frac{\delta C_A}{\delta t} = \frac{\delta}{\delta \xi}\left(D \cdot \frac{\delta C_A}{\delta \xi}\right)$$

where the subscript A refers to the species diffusing into the polymer B.

It is convenient to define ξ in terms of length so that D is then given in the conventional units cm.^{2}/sec. As the concentration of A was originally defined as the amount of A per given weight of B, this transformation is completed simply by multiplying the weight of B by a specific volume which may be taken to be that of the pure compound B. Thus the concentration of A is given as the amount of A per unit basic volume of B, and consequently unit ξ contains unit basic volume of B per unit area.

As in the case of a system of fixed total volume, it can be shown that for the above system where the diffusion is referred to a particular section fixed with respect to the mass of the system, the diffusion coefficients of the two species are equal. As no net transfer occurs across the arbitrarily fixed section,

$$D_A^M \cdot \frac{\delta C_A^M}{\delta \xi} + D_B^M \cdot \frac{\delta C_B^M}{\delta \xi} = 0$$

and

$$C_A^M + C_B^M = 1/V_B^\circ$$

where V_B° is the specific volume of the component B, which may be arbitrarily chosen.

Thus

$$D_A^M \equiv D_B^M$$

The diffusion coefficient determined in this way is therefore a mutual coefficient, dependent on the large diffusing rate of the penetrating species compared with that of the polymer together with the compensating mass flow of the polymer. When no volume change occurs on mixing, this diffusion coefficient is related to the conventional coefficient which can be applied only when the two components do not change in volume, D^V, as follows,

$$D^V = D^M (C_B^V V_B^\circ)^{-2}$$

where C_B^V is the concentration of the polymer given as mass per unit overall volume.

Intrinsic diffusion coefficient

Although the mutual diffusion coefficient D^M must be used in systems where considerable swelling occurs, it has no fundamental significance in terms of the molecular motions of the penetrating molecules. It is important therefore to consider a diffusion coefficient which is characteristic for a given solute and independent of the mass flow within the polymer. Hartley and Crank[42] have shown that D^M

is related to the intrinsic diffusion coefficients of the two components, although in the general case the individual values cannot be evaluated. When a high polymer is the second constituent, however, the second intrinsic coefficient is negligible so that the mutual diffusion coefficient may be related directly to the intrinsic diffusion coefficient of the penetrating solute,

$$D_A^i = D^M(V_B C_B^V)^{-1}(V_B^\circ C_B^V)^{-2}$$

where V_B is the partial specific volume of polymer B at the concentration C_B. If as a first approximation it is assumed that no volume change occurs on mixing, it follows that

$$D_A^i = D^M(1-\bar{V}_A)^{-3}$$

where $\bar{V}_A$ is the volume fraction of the penetrant.

This treatment has been followed by Crank and Park[13] for the diffusion of chloroform into polystyrene and by Park[32] for the diffusion of

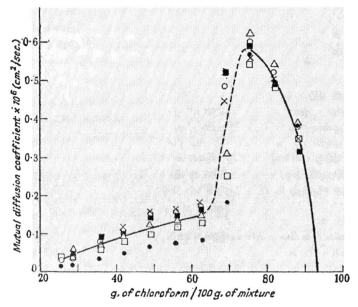

FIG. 5.6.—Mutual diffusion coefficient for chloroform and cellulose acetate[24] at 25° C. (*Reproduced by permission.*)

	time in min.		time in min.
●	200	□	256
○	120	■	154
△	180	×	92
—	mean curve		

other halomethanes. For low regains, the mutual and intrinsic diffusion coefficients, which are very similar in most cases, increase with concentration in much the same way as previously found by King,[5] although the latter made no correction for the change in the position of the phase boundary with swelling. As in the case of wool the total increase in diameter is relatively small (c 16 per cent.), and the diffusion of penetrant is much more important than the opposing mass transfer, the latter may be neglected to a first approximation.

At high regain, however, the mutual diffusion coefficient falls rapidly with increasing concentration, although the intrinsic diffusion coefficient continues to increase (see Figs. 5.6 and 5.7). Both for the

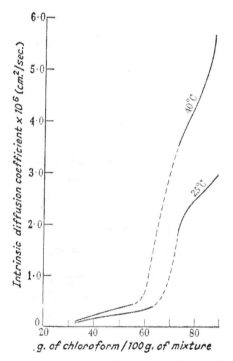

FIG. 5.7.—Intrinsic diffusion coefficients for chloroform and cellulose acetate[24] at 25° and 40° C. (*Reproduced by permission.*)

penetration into cellulose acetate and into polystyrene, the activation energies are found to increase with increasing absorption, although no explanation has been given. This is difficult to understand and is the exact reverse of the observations of King for wool, neglecting the effect of swelling on the position of the phase boundary. In addition it follows that extreme care is required in interpreting experimental activation energies obtained from rate of penetration measurements, where the dimensions of the fibre or polymer are continually changing.

Rate processes involving the fully swollen fibre

The reactions involved in textile finishing and dyeing, almost without exception, involve the treatment of the fibres in aqueous solution, and consequently it is surprising that very few kinetic studies of such processes have been made. As the wool fibres usually remain fully swollen during the process, the system is considerably simpler than those already described where the penetrating molecules cause considerable swelling as diffusion proceeds, and the diffusion coefficient may be assumed to be constant as a first approximation, except when the penetration of very large ions (dyes) is considered. In such cases it is probable that D decreases with increasing concentration in contrast to the effect of swelling on D.

Reaction in a porous solid in contact with liquid involves the following steps.

(a) The transport of reactants to, or products away from, the surface.
(b) The sorption of reactant at the surface.
(c) The diffusion of reactant and/or products through the solid phase.
(d) Chemical reaction at specific sites.

The slowest process governs the measured rate and is referred to as the rate-determining step.

In many of the process recently studied, the rate of the chemical reaction with particular groups, or the rate of sorption in the case of dyeing and acid absorption, is very rapid compared with the other stages. In the oxidation of the fibre where the most prominent reaction occurs with the disulphide bond, chemical modification of the wool does not affect the overall rate of the process.[43] In addition it is known that the rate of reaction with the individual amino acids is extremely rapid,[44] so that the overall rate is controlled by the migration of reactant to reaction sites.

Thus, in many processes an interpretation of the kinetics depends primarily on differentiating between diffusion processes in solution and within the fibre. In some cases such as oxidation with peracetic acid or hydrogen peroxide and, in some reactions where free halogen is slowly generated, there is evidence to show that the actual rate of the chemical reaction is slower than the other steps involved. At present, there are insufficient data available for this to be discussed further.

The rate-determining process of diffusion in solution may be differentiated from a rate-determining process within the fibre in various ways, although in some cases, e.g. acid absorption, the diagnosis is extremely difficult. The following factors may be used to determine which of these two processes is the slower, i.e. rate-determining.[45]

(1) An increase in temperature tends to make the process with the smallest activation energy, and therefore the smallest temperature coefficient, rate-determining. In some cases diffusion in solution and through the solid phase are controlled by an identical activation energy, but in the case of diffusion through textile fibres, high energies are usually observed if the overall rate is governed by diffusion through the fibre.

(2) Agitation affects the overall rate of the process when governed by diffusion in solution, but not when diffusion through the fibre is the controlling factor. Greater agitation tends to make the fibre-diffusion mechanism rate controlling by increasing the supply of reactant to the surface.

(3) Concentration affects the relative values of the two rates in the two phases owing to the change in the distribution coefficient. Usually an increase in concentration renders the fibre-diffusion mechanism rate-controlling.

(4) Particle size affects the rate of the two processes to different extents, and an increase in size strongly promotes the fibre-diffusion mechanism. This test is extremely useful when dealing with synthetic particles such as ion-exchange resins, but is difficult to apply in the present case.

It must be realized that a change in mechanism is not necessarily produced by varying the above conditions, as the diffusion coefficient in the fibre may be very much smaller than the diffusion coefficient in solution.

It is proposed to discuss the rate-controlling diffusion through solution first, in view of the considerable technical importance. As the diffusion through solution is greatly affected by the degree and mode of agitation, the rate is liable to be highly variable. In additio , particularly when fabric is employed, the liquid film is probably variable in width, resulting in uneven treatment and increased risk of damage to the fibres.[46] This is undesirable both in dyeing and finishing, as for example the resistance to felting is greatly decreased by an uneven treatment. It is advantageous therefore to use conditions where the overall rate is governed by a process within the fibre, when the reaction becomes independent of external influences.

Diffusion across a liquid layer at the surface[47]

This rate-determining mechanism is common to all reactions between a liquid and a solid phase, and has been examined mainly in connexion with eletrochemical processes,[48] rate of solution[49] and the reaction of acids with metals.[50] According to the classical approach of Noyes and Whitney,[51] Nernst,[52] and Brunner,[53] diffusion proceeds across a stationary layer of liquid close to the surface, according to the simple law of Fick. Thus the rate of transport P across unit area of the layer is given by

$$P = D\Delta c/\delta$$

where Δc is the concentration difference across the layer of thickness δ. It is found in most cases that δ is of the order of 0.001 cm. even when rapid agitation is employed, whereas Fage[54] observed experimentally that the liquid close to the surface is actually in motion. The limitations of the simple theory led Euken to develop a strictly hydrodynamical theory.[55] A complete treatment is extremely complex and formal solutions have been obtained only in a few simple systems. A more general semi-empirical treatment has been developed by Colburn et al.[56] and recently applied to rate processes at an electrode by Agar and Levich.[48]

These detailed treatments show that the width of a hypothetical diffusion layer depends not only on the rate of stirring, but also on the viscosity and density of the solution and the diffusion coefficient of the migrating species. At a given temperature, for a given solvent however, it is found empirically that in general[57] $\delta = \text{const.}/U^a$, where U is the stirring rate and a varies from 0.5 to 1.0 depending on the mode of stirring and type of surface.

Most of the reactions between wool and a solute in aqueous solution so far studied are highly dependent on the rate of stirring. The rate of oxidation[47] and dyeing[58] becomes more stirring-dependent as the concentration in solution decreases (Fig. 5.8). Using a relatively high concentration of dye in solution (c. 0·001 M) the rate of absorption is found to be independent of stirring even at very low stirring rates, because of the low rate of penetration of dye into the fibre. In some cases however, e.g. the absorption of acids,[59] the rate of migration through the fibre is extremely rapid, so that the process becomes highly stirring-dependent, and it is difficult to obtain experimental conditions such that the overall rate is governed by diffusion through the fibre. In most cases is is found that the rate of stirring is inversely proportional to the thickness of the diffusion layer (Fig. 5.9), so that $a = 1$ in the above empirical equation.[47]

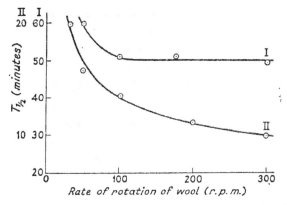

FIG. 5.8.—The effect of stirring on the rate of dyeing of wool with acid orange II[58] at 25° C.
I: 500 mg. dye/l.   II: 125 mg. dye/l.

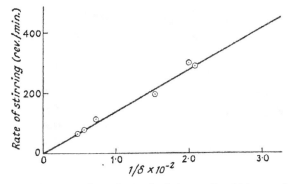

FIG. 5.9.—The effect of rate of stirring on the thickness of the diffusion layer, calculated from rate measurements on the chlorination of wool.[47]

If the overall rate of the process is governed entirely by diffusion across a liquid layer, the rate of transport in the general case is given by,

$$\frac{d(xV)}{dt} = \frac{Dl \cdot A}{\delta}[c-x-C^s]$$

where c is the initial concentration of solute and $c-x$ the concentration after time t, V is the volume of solution associated with a surface of area A and C^s the concentration in the solution adjoining the surface.

In the case where the reactant is removed instantaneously by chemical reaction, C^s may be assumed to be zero, so that

$$D_l = \frac{\delta V}{A \cdot t} \log_e \frac{c}{c-x}$$

This relation is found to hold for low concentrations and/or low

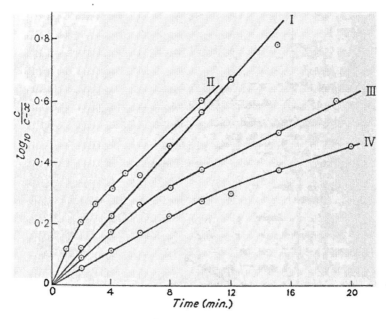

FIG. 5.10.—Graphs of $\log_{10} \frac{c}{c-x}$ against t for low stirring rates at 0° C.[47]
  I: 3 per cent. and 0·5 per cent. Cl_2 solutions at 100 rev./min.
  II: 10 per cent. Cl_2 at 300 rev./min.
  III: 10 per cent. Cl_2 at 200 rev./min.
  IV: 10 per cent. Cl_2 at 115 rev./min.

stirring rates (Fig. 5.10) both for diffusion-controlled chemical reactions and for dyeing even when the reaction bath is almost completely exhausted.

Finally, the values of the experimental activation energies associated with diffusion across a liquid film require some comment, as in most cases it is found that they are substantially higher than the corresponding values for diffusion in aqueous solution[38] (c. 4·2 kcal./mol in the 0–35° C. temperature range). Examination of the effect of temperature on the various hydrodynamical factors which govern the rate of transport, show that this difference is probably due to a change in δ. Thus under the experimental conditions usually employed, E is likely to be of the order of 6 k. cal./mol, so that such activation energies are quite possible for processes controlled by diffusion across a liquid film, and do not necessarily imply that the rate is controlled by a chemical process or by diffusion through the solid.[47]

Diffusion through the fibre

As the degree of agitation or the concentration in solution is increased, the rate of the process becomes independent of stirring rate[43] (Fig. 5.8) and simultaneously the activation energy is found to increase considerably (Fig. 5.11). At the same time, the shape of the

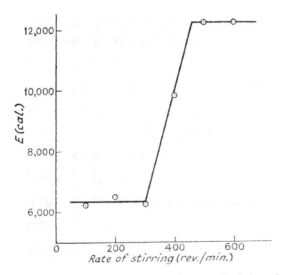

FIG. 5.11.—Relationship between the rate of stirring of the sample and the energy of activation for the reaction of wool and an aqueous solution of chlorine at pH 2.[47]

rate curves changes from an approximately logarithmic form to a parabolic form, which is characteristic of many diffusion processes through solids.[11, 17] In the case of oxidation of the wool with acid chlorine[43] and permanganate solutions,[60] the rate is not controlled by the chemical reaction, for the following reasons:

(1) The reaction in acid and alkaline solution between chlorine or potassium permanganate and the amino acids cystine and tyrosine, which are preferentially oxidized in wool, is extremely rapid. All the other amino acids with the exception of glycine also react very rapidly.[44] It is to be expected that if accessible, the available side chains react at similar rates.

(2) The rate of reaction with wool in which all the disulphide bonds have been modified to lanthionine linkages is the same as for the untreated wool in which the disulphide bond is oxidized preferentially.[61]

(3) The observed activation energy is approximately constant for a wide variety of processes, e.g. dyeing,[60] ion exchange,[63] and oxidation with potassium permanganate[58] and chlorine.[43] The one common feature of all these processes is the rate-controlling diffusion within the fibre.

The diffusion-controlled nature of the reaction is demonstrated by interrupting a kinetic run in order to allow the diffusion gradient within the fibre to decrease.[61] On continuing the reaction, it is found that the rate is considerably greater than the rate immediately prior

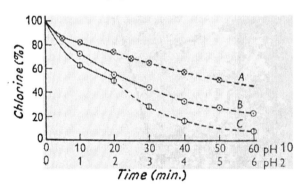

FIG. 5.12.—Rate curves for experiments at 600 rev./min. and 0° for the following treatments:[61] *A*, 4 g. Cl_2/100 g. wool at *p*H 10 interrupted after 20 min.; *B*, 0·5 g. Cl_2/100 g. wool at *p*H 10 interrupted after 30 min.; *C*, 10 g. Cl_2/100 g. wool at *p*H 2 interrupted after 2 min.

to the interruption, due to the re-establishment of a large diffusion gradient (Fig. 5.12). If the rate were controlled by chemical reaction no such effect would be observed.

Diffusion from a constant surface concentration

The general problem of diffusion from a constant surface concentration for a constant diffusion coefficient has already been considered (p. 132) and simple relations between the absorption and time have been derived. There is little direct information to show whether D remains constant when a solute diffuses through a swollen fibre. As the dimensions and internal structure remain constant in these processes, D probably remains fairly constant, particularly when dilute solutions are used (see however p. 167).

Most industrial processes employ an exhausting bath, although continuous processing is widely advocated to-day particularly in treatments to reduce shrinkage, in order to confine reaction to the surface and promote even treatment.[62] In such processes, the bath concentration is maintained constant, so that diffusion proceeds from a constant surface concentration. The kinetics under these conditions have, however, been studied by Hudson and Schmeidler,[63] who investigated ion exchange on wool fibres as a model for dyeing. Fibres initially saturated with sulphurous acid were immersed in a solution containing an excess of salt of the exchanging anion at the appropriate pH to prevent any change in the hydrogen ions combined with the wool. The exchange was almost complete and proceeded from a constant surface concentration, as the change in concentration of the exchanging ion in solution was negligible. From the following observations it was concluded that the overall rate was controlled by diffusion within the fibre.

(*a*) The activation energies were in the range 10–14 k. cal./mol.
(*b*) A linear relation was obtained between Q_t and $\sqrt{t}$ for 50–60 per cent. of the exchange (Fig. 5.13).
(*c*) The rate is almost independent of concentration in solution varying from 1·0 to 0·5 M. This shows that the rate-controlling process is not affected by the Donnan membrane equilibrium, as the concentration of free hydrogen ions and other cations within the fibre changes considerably with salt concentration. The following diffusion coefficients and activation energies, calculated from the Arrhenius equation, were obtained (Table 5.2).

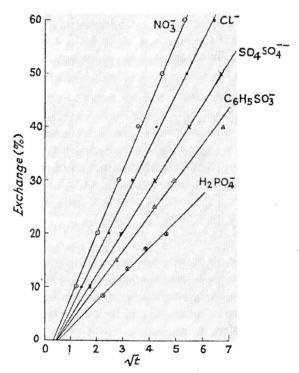

Fig. 5.13.—Rate of exchange between bisulphite ions initially in the wool and 1N solutions of various ions at 18° C.[63]

TABLE 5.2

The kinetics of anion exchange on wool

Exchanging ion N solution	Temp. C.	$10^9 D_f$ cm.^{2}/sec.	Temp. C.	$10^9 D_f$ cm.^{2}/sec.	E k.cal./mol
NO_3	18	2·97	2	1·10	9·92
Cl	16	1·89	3	0·97	10·20
SO_4	18	1·25	—	—	—
$C_6H_5SO_3$	18	0·87	3	0·27	12·40
H_2PO_4	18	0·47	3	0·13	14·00
OH (pH 13)	19	3·24	2	1·10	12·00

The exchange medium, wool, initially treated with 150 ml. of 0·02N H_2SO_3 per gram. Exchanged with N. solutions of salt at pH 2; 1 g. fibre in 50 ml. of solution.

The exchange process involves the simultaneous entry and expulsion of ions of like sign to preserve electrical neutrality, so that D_f is a mutual diffusion coefficient (see p. 147). In the present case the mobility of sulphurous acid is known to be high,[59] so that the overall rate may be governed mainly by the penetrating ion. The anions enter the fibre initially as gegenions to the protein and are highly mobile in the aqueous phase, probably residing in an electrical double layer surrounding the micelles.[65] Thus all the ions are mobile and free to diffuse, so that the values of D_f are probably close to the free-diffusion coefficient of the anion under these conditions. The low values of D_f and the high activation energies show that the diffusion is subject to considerable restriction within the fibre (see p. 168).

In some cases, the surface concentration may remain constant for most of the process, although the concentration in the bath decreases considerably, owing to the nature of the absorption isotherm. This has been observed in dyeing[58] and in the oxidation with potassium permanganate,[60] when high concentrations are used. Under these conditions, it follows from the approximate solution for diffusion into a semi-infinite solid that

$$(x/c)Vc = 2C^s A\sqrt{(D_s t)/\pi}$$
$$\text{so that} \quad t_{\frac{1}{2}} \propto c^2$$

where $t_{\frac{1}{2}}$ is time required for half-exhaustion of the bath. The symbols used have the same significance as on p. 154. This relation is obeyed in the oxidation with permanganate above a certain concentration (Fig. 5.14), and it is also found that x is proportional to $\sqrt{t}$ in each individual experiment.

Similarly in dyeing at relatively high concentration, Speakman and Smith[66] have found that Qs is proportional to $\sqrt{t}$ for most of the reaction. This alone does not necessarily imply a constant surface concentration, as it is found that the linearity holds for much greater absorptions than predicted theoretically. Thus Speakman and Smith found the linearity to persist up to 70 per cent. absorption, although with higher dye concentrations or smaller liquor ratios the linearity holds for only c. 40 per cent. absorption as required by the theoretical equation (see p. 133). Assuming all the dye within the fibre is mobile, the calculated diffusion coefficient varies to only a limited extent with concentration (Table 5.3).

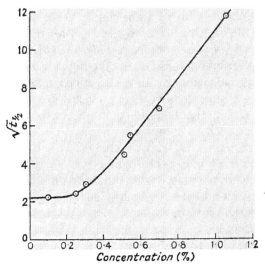

Fig. 5.14.—The relation between $\sqrt{t_{\frac{1}{2}}}$ and c for the reaction with potassium permanganate and wool at 25°C.[60] 1 g. of wool in 3 per cent. H_2SO_4.

TABLE 5.3

Calculated diffusion coefficients for dyeing with Naphthalene Orange II

Conc. mg./l.	D_s values cm.^{2}/sec.		E (25°–35°) k. cal./mol
	25° C.	35° C.	
250	0.700×10^{-11}	1.40×10^{-11}	12.7
500	0.845×10^{-11}	1.95×10^{-11}	15.6

1 g. fibres in 140 ml. solution.

Speakman and Menkart[67] have interpreted the action of mercuric acetate on the elastic properties of wool and hair in terms of diffusion of the reagent into the fibre. The increase in work required to stretch a fibre to a constant extension was found to increase proportionally to the square root of the time, using a constant concentration bath. The assumption is made that the extent of reaction is proportional to the quantity of solute which has diffused into the fibre, and also to the change in elasticity. This was tested qualitatively by studying the penetration, detected by H_2S staining of a cross-section, as a function of time. In support of the diffusion mechanism, it was found that the rate of penetration into Lincoln wool is twice as fast as into human hair in agreement with the ratio of the rates of dyeing of these two fibres with Acid Orange II G.[66]

Diffusion from a varying surface concentration

In most cases the surface does not remain saturated during the process, and the mathematical treatment of the rate curves becomes extremely complex. Variations in the surface concentration may be recognized by studying the effect of changing the concentration of the

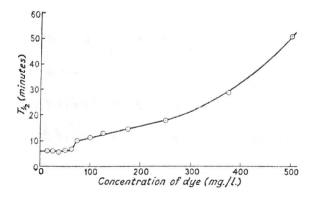

FIG. 5.15.—The effect of dye concentration on the rate of dyeing with acid orange II at 25° C., stirring the wool at 200 r.p.m.

external solution on the rate of the process. In dyeing it is found that $t_{\frac{1}{2}}$ increases linearly with concentration over a considerable range (Fig. 5.15). The process is not governed by diffusion in solution as shown by the high activation energies (Table 5.4).[58]

Two distinct problems can be recognized. Firstly, processes in which the solute reacts with the specific sites thus affecting the concentration of diffusing substance and reactive centres simultaneously, and secondly, processes involving absorption in which a mobile equilibrium is continually preserved within the fibre between the free diffusing concentration and concentration of absorbed solute. Owing to the different treatments which have been described, these two general classes of reaction will be considered separately.

TABLE 5.4

Activation energies (k. cal./mol), calculated from $t_{\frac{1}{2}}$ values at low and high stirring rates, for the dyeing of wool with Acid Orange II G (Colour Index No. 27)[58]

Initial conc. of dye (mg./l.)	Fast stirring						Slow stirring				
	$t_{\frac{1}{2}}$ (0° C.)	$t_{\frac{1}{2}}$ (25° C.)	$t_{\frac{1}{2}}$ (35° C.)	E (0–25° C.)	E (25–35° C.)		$t_{\frac{1}{2}}$ (0° C.)	$t_{\frac{1}{2}}$ (25° C.)	$t_{\frac{1}{2}}$ (35° C.)	E (0–25° C.)	E (25–35° C.)
37.5	17.5	5.5	4.5	7.3	4.0		6.0	25.0	—	—	—
62.5	25.0	6.5	5.0	8.5	4.5		100.0	33.5	—	6.4	—
125.0	78.0	12.0	5.5	12.0	13.0		135.0	42.0	—	6.4	—
250.0	—	18.0	9.0	—	12.0		—	57.5	24.5	—	14.0
500.0	—	50.0	22.0	—	13.0		—	77.5	41.0	—	13.0

Diffusion accompanied by reaction within the fibre[43]

This process has been studied in connexion with many biological systems for many years, but in most cases a constant external concentration was maintained. The classical work of A. V. Hill[68] involving the incorporation of a specific rate constant for the chemical process in the general diffusion equation later extended by Hermans, Dankwerts[69] and Vreedenberg,[70] is limited to diffusion from a constant surface concentration. It is often found experimentally that processes involving diffusion with reaction or absorption obey approximately the simple parabolic rate law applicable to diffusion from a constant surface concentration, and the conditions under which this approximation may hold have already been considered (p. 133).

The process under consideration bears a close resemblance to the well-known tarnishing reaction of metals where the reactant is removed during the course of the diffusion.[71] The chief difference lies in the density of the product formed, as in metals all the atoms in the lattice are reaction centres. In a highly porous medium such as a fibre, the reaction centres are more widely spaced, but if they are assumed to be randomly distributed the system becomes essentially the same. In studies of the oxidation of metals it has been found that the thickness of the product layer increases proportionally to $\sqrt{(2At)}$, so that, if the reaction is complete within the layer, x is proportional to $\sqrt{t}$. Similarly in the reactions of silver with sulphur[72] and in certain oxide reductions,[73] this parabolic law is found to hold. The rate equation has been derived by applying the simple law of Fick. In a detailed analysis of this rate equation, however, Booth has pointed out an inconsistency in the usual derivation, as the diffusion is not in accord with a steady-state treatment.[74] The usual assumption of a linear gradient within the solid leads to the anomaly that

$$\delta^2 c/\delta x^2 = 0$$

This can hold only for a stationary state, i.e. where the width of the diffusion layer does not vary with time. However, by means of a rigorous solution, Booth showed that the diffusion gradient is almost linear when the diffusion concentration is small and the density of the product formed is high. This condition is satisfied for most of the processes occurring within the fibres, and although mathematically unsatisfactory, application of the pseudo-steady-state concept gives results in close agreement with experiment.[43, 58]

The rate of transport of reactant at any particular time across the layer of product of width l is given approximately by:

$$\frac{d(xV)}{dt} = D \cdot A \frac{C_1 - C_2}{l}$$

where C_1 and C_2 are concentrations at the inner and outer boundaries at time t.

It will be assumed that reaction is virtually complete within the product layer as in the case of tarnishing reactions. If C_f is the concentration of product formed, the distance penetrated is given by $l = xV/AC_f$

so that $$\frac{d(xV)}{dt} = \frac{D_s A^2 (c-x)}{xV} C$$

i.e. $$D_s = \left[\frac{V}{A}\right]^2 \cdot \frac{c}{C_f t}\left(\ln\frac{c}{c-x} - \frac{x}{c}\right)$$

It is seen that in this case the half-life value is proportional to initial concentration of reactant in solution.

This simple expression, which reduces to a parabolic law in the initial stage of the process, has been seen to hold in the oxidation of wool with chlorine solutions where the chemical reaction within the fibre is known to be extensive (Fig. 5.16).[43] It is found, however, that the increase in $t_{\frac{1}{2}}$ with concentration is not as large as predicted, which is attributed to incomplete reaction within the penetrated

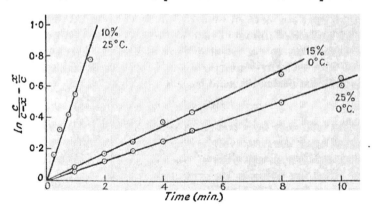

Fig. 5.16.—Graphs of $\ln\frac{c}{c-x} - \frac{x}{c}$ against t for 10 per cent., 15 per cent and 25 per cent. Cl_2 at pH 2.[43]

diffusion element. In the case of dyeing,[58] however, a linear increase is observed (see Fig. 5.15), and this will be considered further when the mechanism of dyeing is discussed.

Diffusion accompanied by absorption

Diffusion processes proceeding from a varying surface concentration, where the solute is not removed by chemical reaction within the fibre, are more amenable to a theoretical treatment. When the distribution coefficient between fibre and solution is high, it may be supposed that most of the solute in the solid phase is physically sorbed on localized sites, and only a small fraction of the molecules are mobile at any given time and responsible for diffusion. This kind of process has been considered by various workers, and applied in particular to the dyeing of cellulose.[75] The concentration of absorbed solute within the fibre is assumed to remain in mobile equilibrium with the free concentration, such that as the diffusion gradient changes, the distribution of absorbed molecules changes automatically. The external equilibrium was assumed by Wilson[15] to be governed by the same law governing the concentration of absorbate within the fibre for a given external concentration at equilibrium. The formal solution of the general diffusion equation in the case of a linear absorption isotherm, has already been discussed on page 135, and will now be applied to the experimental data.

The only process to which this treatment has been applied satisfactorily is the absorption of acids on wool, initially in the isoelectric condition.[59, 76] Here the desired mobile equilibrium exists between combined hydrogen ions on carboxyl groups and free acid, as the combination and dissociation of acids are instantaneous. The rate of acid absorption was found to be highly stirring-dependent, and with efficient agitation becomes very rapid. Further, the process is found to have a low temperature coefficient and consequently it is more difficult in this case to determine whether the rate is governed by diffusion across a liquid layer.

The fibre diffusion mechanism is supported by the fact that the observed rate is slower than that to be expected for diffusion in solution under these circumstances, given by the equation (p. 154).

$$\frac{d(xV)}{dt} = \frac{D_l \cdot A}{\delta}[C - x - C^s{}_1]$$

In acid absorption, the concentration in solution, $C^s{}_1$, is related to the concentration within the fibre by the Langmuir absorption isotherm,[65] i.e.

$$C{^s}_1 = K \cdot \left(\frac{\theta_i}{1-\theta_i}\right) \equiv K\frac{xV}{0 \cdot 88W - xV}$$

where θ_i is the fraction of the total number of absorption sites available occupied by ions of species i, and W is the weight of wool in contact with a volume V of solution. The thickness of the layer, δ, under the given stirring conditions is known from the work on dilute chlorine solutions (see Fig. 9). Using this value, and the value for D_l for the given acid, it is found that the calculated rate is much greater than the observed rates.

This conclusion is supported by the form of the rate curves which in all cases are predicted accurately by the appropriate equations of Crank[19] (see Fig. 5.17). In the present case, the absorption is of the Langmuir type, but may be taken to be approximately linear, under certain conditions over a limited concentration range. This procedure is also justified as the form of the theoretical curves is not highly dependent on the value of K. It has been assumed that the concentration of free acid within the fibres equals the external concentration. It is observed, however, that the calculated values of D_f are extremely small if diffusion proceeds through the aqueous phase within the fibre. Thus, for the diffusion of neutral molecules at high regain, King[9] has found the true diffusion coefficient to be greater than 10^{-7} cm.^{2}/sec. It is found by extrapolation that D at saturation is of the order 5×10^{-7} cm.^{2}/sec.)

TABLE 5.5

Internal diffusion coefficient of HCl in the fibre[59]

Initial conc.	$t_\frac{1}{2}$ sec.	$10^8 D_f$ cm.^{2}/sec.	Conc. of HCl within the fibre*	$10^6 D'_f$ cm.^{2}/sec.
0·092	17	2·760	0·00300	0·850
0·020	110	0·880	0·00200	0·880
0·010	135	0·825	0·00100	0·825
0·005	160	0·610	0·00035	0·870

*Assuming maximum acid-combining capacity of 88 milliequivs./100 g.[77]

In addition, the calculated values of D_f change significantly with concentration. No account has been taken of the Donnan membrane effect on the internal concentration of free acid, although this has been shown to have a considerable influence on the titration curve. Using the calculated values of internal free acid concentration given by Peters and Speakman,[64] the values of D_f^i become independent of external concentration, and approach the value for diffusion in aqueous solution ($c.\ 2\cdot5 \times 10^{-5}$ cm./sec. for HCl at 18° C.).[78]

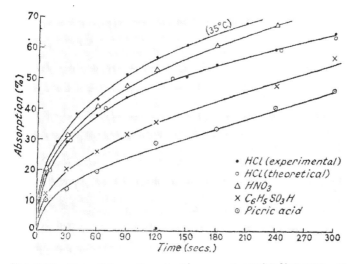

Fig. 5.17.—Rate of absorption of acids on wool at 19° C.[59] (0·01N acid solutions.)

The rate of absorption of several acids is shown in Fig. 5.17 for the same concentration and temperature. Calculations of approximate values of D_f show that the rate of diffusion decreases with complexity of the ion, the order being identical with those of the diffusion coefficients in aqueous solution.

The effect of temperature on the fibre diffusion processes

The experimental activation energies for anion exchange[63] are observed to increase with size of ion except in the case of the phosphate ion, which may be attributed to its hydrogen-bonding characteristics reflected in the very high viscosity of aqueous solutions.[78] In addition, the values are very high compared with corresponding values for other ion exchange processes, which are in some cases of the same order as for diffusion in solution.[45, 79, 80, 81] In general, the value of the activation energy can be correlated with the average pore size of the swollen medium and the size of the hydrated ions. Thus, in the cation exchange on sulphonated resins, a sharp increase in E at a critical ion size is observed.[45] In this case the high activation energy has been attributed to temperature-dependent steric factors, and not to the additional energy required to force the ions through solvent in the pores.

Part of the water in fibres is combined with the macromolecules as shown by the high heat of wetting at low regains.[4] Thus, only some of the water may be considered to be free and even these molecules are probably subject to more restricted motion than in liquid water. Thus the pore size of 40 Å, calculated by Speakman[7] on the assumption that all the swelling is inter-micellar, is probably greater than the average diameter of channels containing free water. A similar value of 30 Å has been obtained for the pore size of cellulose which swells to a similar extent.[25] However, if the pores were of these dimensions, the relatively small diffusing molecules would proceed through free liquid with a low energy of activation. The high values for the activation energy which have been observed led to the suggestion that diffusion proceeds through a uniform gel of low water content. The high activation energy is then due to the deformation necessary to provide holes for the diffusing molecules. In addition, the pore surfaces will be highly irregular due to the morphological complexity and to the considerable number of polar side chains. Thus it has been suggested that diffusion also involves the deformation of the macromolecular chains.[58]

It is found, however, that some processes require much smaller activation energies for diffusion of the solute through the swollen fibres. In the absorption of acids, values of the order of 5–6 k. cal./mol are obtained.[59] Similarly King[9] has found that, at high regains, water absorption (into horn) requires a small activation energy (4·8 k. cal.) compared with the corresponding value at low regains (7·5 k. cal.). In both systems diffusion is accompanied by swelling of the fibre to accommodate the sorbed molecules. At the same time the diffusion coefficient is found to approach the value in aqueous solution (see Fig. 5.1, p. 131). Thus, it appears that the diffusion of the penetrating molecules proceeds as in aqueous solution, in cases where simultaneous elimination of molecules does not occur.

In the cases of anion-exchange, dyeing and diffusion-controlled reactions, the rate proceeds according to a coupled mechanism[80] involving the simultaneous flow of molecules or ions in opposite directions. This flow is regulated by the relative mobilities of the diffusing species, and by the conditions of electroneutrality. The high activation energies under these conditions are then attributed to the interaction of the two diffusion species confined in a volume of molecular dimensions, or to the energy necessary to distort the wool lattice to permit the flow in opposing directions, by a place-exchange mechanism.

These views are supported to some extent by the observation that the diffusion coefficients for acid absorption and anion exchange are governed almost entirely by the activation energies. Assuming the migration to involve activated diffusion, calculation shows that the temperature-independent factor A of the Arrhenius equation $D = Ae^{-E/RT}$ is of the same order for both processes and for the diffusion of the appropriate solute in water (Table 5.6). This substantiates the conclusion that the diffusion coefficients are the true values for the rate-determining transport in these various widely differing processes.

TABLE 5.6

Comparison of the rate parameters for acid absorption and ion exchange on wool, and diffusion in water.[59]

D cm.^{2}/sec.	Temp.° C.	E k.cal./mol	$-\log_{10} A$	Process
10^{-6}	18	5·5	2·00	HCl absorption
2×10^{-9}	18	10·0	1·50*	Cl—HSO$_3$ exchange
3×10^{-5}	20	4·0	1·58	HCl in water

* This slightly high value probably arises from the accelerating influence of the diffusion of H_2SO_3 on the coupled diffusion.

Rate of dyeing

The predominant process in dyeing with simple acid dyes is the anion-exchange mechanism already discussed, although some molecular absorption may occur because of the very high affinity of the ions[77] (see p. 222). The high anion affinity is due to specific attractions between the dye and particular groups within the fibre resulting in the formation of weak bonds, due mainly to dispersive forces. Thus, if these forces are sufficiently great, the dye will have a low mobility within the fibre in contrast to simple anions. For this reason, the calculated diffusion coefficient for the exchange from constant surface concentration, assuming all the dye ions to be mobile, is found to be extremely small (p. 160, Table 5.3). A constant surface concentration may be assumed because of the form of the absorption isotherm even when the concentration in solution decreases considerably during the process.

Over a considerable concentration range, however,[58] the rate is governed by diffusion from a variable surface concentration as shown by the linear increase in $t_{\frac{1}{2}}$ with initial bath concentration (Fig. 5.15). Application of the appropriate equation of Wilson[15] for a linear absorption shows a complete lack of agreement with the experimental curves. As already mentioned, this treatment requires instantaneous equilibrium between free and adsorbed dye as the dyeing[39] proceeds, which necessarily involves rapid desorption of dye from the specific sites. Owing to the bonding between dye and site, desorption may well be slow, as indicated by low rates of levelling, so that thermodynamic equilibrium is not attained during the adsorption. The dyeing may well proceed in two well-defined stages. An initial, almost complete absorption of dye may occur on the external layers of the fibre, followed by a slow re-distribution. According to this picture, the diffusion of free dye ions through the small micropores filled with solvent is accompanied by extensive sorption on the internal surface such that the sorbed ions can no longer participate in the diffusion. The process may then be treated as a chemical reaction in which the reactive sites within the volume element penetrated may have completely or partially reacted (see p. 163). According to this view, the diffusion species advances in the form of a well-defined front at which reaction occurs rapidly. Photomicrographs of dyed fibres show that the dye does indeed spread in the form of a well-defined band,[82] particularly at the lower temperatures, giving rise to the so-called ring dyeing,[83] whereas at higher temperatures the dyeing is more even throughout the fibre.

Apart from such visual observations, little work is recorded on the effect of the structure of the dye on the rate of penetration. With dyes containing a long hydrocarbon chain (e.g. Carbolan Crimson B.S.), an initial rapid absorption is observed followed by a very slow change due to the gradual penetration of the dyes into the fibres[84] (Fig. 5.18). This effect is not observed with other highly aggregated dyes, e.g. Sky Blue FF, Polar Yellow, Benzopurpurine and Coomassie Milling Scarlet. The initial absorption was related to the surface activity of

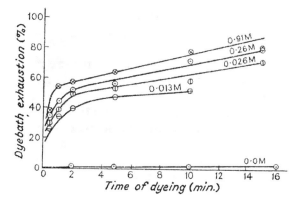

FIG. 5.18.—Rate of dyeing of wool with Carbolan Crimson (75 mg./l.) and different amounts of ammonium acetate at 50° C.[84]

the dyes, the amount absorbed being given by the Langmuir absorption isotherm. In addition, the dyes which exhibited this effect were found to reduce the interfacial tension of water against light petroleum considerably.

As the complexity of the dye is increased in order to render it more resistant to washing, the tendency towards direct dyeing also increases and the dyeing process becomes more complex in every way. Acid dyeing is essentially an ion-exchange process in the presence of excess salt (see p. 231) so that the distribution of electrical charge within and at the surface of the fibre remains essentially constant. For this reason the diffusion coefficient will be assumed to be constant as for simple ion exchange processes, in the absence of direct data. Direct dyeing has been studied mainly in connexion with cellulose, and little agreement between experimental results and theoretical predictions has been obtained.[85] Firstly, the diffusion coefficient is found to increase with dye concentration[86] according to the expression $D_c = D(1+4\cdot8c)$. Secondly, the rate is affected considerably by the salt concentration in solution, so that D at first increases and then decreases at high salt concentrations.[87]

The initial increase in D with salt concentration has been attributed by Hartley[88] and Neale[89] to the effect of salt on the electrical double layer surrounding the fibres, thus reducing electrical repulsion. The subsequent decrease has been explained by Standing et al.[90] in terms of a diffusion with absorption theory (see p. 165) where the absorbed dye has a limited mobility.

Further, the shape of the absorption curve is not predicted by the diffusion with adsorption theory of Crank[14] and Wilson,[15] but resembles closely that for diffusion from a constant surface concentration. Crank[85] has explained this anomaly by allowing for the effect of the changing surface potential due to adsorbed dye ions on the effective concentration at the surface. This treatment shows that the concentration at the surface remains almost constant during the dyeing process, although the concentration in the bulk of the solution decreases considerably. This may be a further reason why parabolic rate laws are frequent under conditions where a simple diffusion theory would not predict such a law.

It is impossible at this stage to give any definite explanation of the difficulties which have been found in studies of direct dyeing. A further factor which must be taken into account is the state of aggregation of the dye.[91] The morphological structure of the fibre may have a pronounced effect on the dyeing process. It has been observed that the penetration of simple molecules and ions proceeds as through a homogeneous medium, except under some circumstances which are discussed below.

The influence of the morphological structure

The penetration of dyes into the fibre can be examined visually by taking photomicrographs of cross-sections at suitable intervals. As a result of the extensive work of Royer, Millson and their collaborators,[83] two distinct effects have been recognized, namely the limited penetration under mild conditions to give ring-dyed fibres, and uneven penetration within the bulk of the fibre on prolonged treatments leading to specks of dye in the cross-sections.

The first behaviour, which is particular pronounced with complex dyes, suggests that considerable resistance to the penetration of dye is experienced at the surface and near surface. Speakman and Smith[66] have shown that the cuticle offers some resistance to the simple dye Acid Orange II, as descaling human hair increased the rate of dyeing to a greater extent than could be due to the decreased radius of the fibres. This is supported by the observations of Lemin and Vickerstaff[92] that in all cases the rate is increased by pre-treating the fibres with various reagents which opened up the surface structure. The effect was particularly pronounced with alkali-treated wool and somewhat less for chlorinated wool. This retardation is, however, relatively small when simple acid dyes are used, but is much greater with more complex dyes. Thus direct cotton dyes penetrate the fibre with great difficulty but readily enter the cortex after extensive modification or removal of the scales. In a detailed examination of lightly treated fibres, Millson and Turl[93] found that the distal edges of the scales are always rapidly dyed, followed by a slow migration to the base of the scales. Mechanical damage of the surface, bending of the fibre, and rapid chemical treatments (acid chlorine solutions) were found to increase the rate of dye uptake considerably. These workers think that the epicuticle provides the main surface resistance.

Lindberg[76] has investigated the effect of the epicuticle on the rate of acid absorption using chopped fibres of virgin wool. It was found that pretreatments known to disrupt or remove the epicuticle increased the rate of absorption considerably. These data are not entirely conclusive, as the rate is observed to increase with the extent of the pretreatments, which indicates that the acceleration may be due at least in part to an opening of the cuticle structure, although Köpke and Nilssen[93a] have shown the existence of a cuticular barrier to penetration of dyes into the fibre. No similar effect was observed, using commercially scoured fabric, in the acid absorption nor in the reaction with hypochlorite ions (Table 5.7). The epicuticle is easily damaged, and in commercial processes it may be sufficiently open to have no effect on the rate of penetration of simple molecules or ions.

It is difficult to distinguish the influence of the epicuticle from that of the cuticle as a whole. Watkins, Royer and Millson[94] have shown clearly that the dyeing behaviour changes sharply at various temperatures depending on the type of dye employed. With a typical milling dye, penetration is slow until a critical temperature is reached when all the fibres become ring dyed. At a second critical temperature in the neighbourhood of 80–100° C. the ringed fibres become completely penetrated. This is in agreement with the contention that the cuticle offers considerable resistance to the penetration of dyes.

Under these conditions the diffusion proceeds through a composite cylinder of varying porosity. For this condition, Crank[95] has evolved a mathematical treatment which, owing to the tedious mathematical computations, has not yet been applied to experimental data.

The influence of the cuticle on the rate of diffusion has been examined by Alexander, Gough and Hudson,[61] who studied the rate of chlorination at various pH's. The rate is found to remain almost constant over a considerable pH range (Fig. 5.19) and is controlled by diffusion through the fibre. Above pH 7, the rate decreases sharply and the shape of the curve changes together with the activation energy. Similar effects are observed in the alkaline reaction with potassium permanganate. Interruption tests show that the rate of reaction in the pH 8–11 region is not governed by uniform diffusion through the fibre. Above pH 8 the solution is composed almost entirely of hypochlorite ions (Fig. 5.19) so that the change in rate with

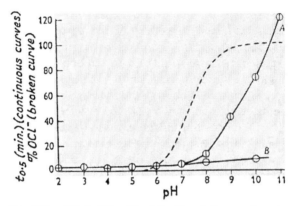

Fig. 5.19.—Effect of the pH of the Cl_2 solution on the rate, given as half-life value ($t_{0.5}$) for untreated (A) and KCN-treated wool fibres (B).[61] Broken curve: effect of pH on the concentration of OCl^- ions, given as percentage of oxidizing agent present. Stirring rate 600 rev./min.; temp. 0° C.; 10 g. Cl_2/100 g. wool.

pH cannot be explained by the decreasing concentration of HOCl in solution, but is probably due in some way to the increased negative charge on the fibre (see Chapter 6). Considerable increases in rate were produced by preliminary treatments, but boiling with acid or potassium cyanide which causes considerable hydrolysis, and treatment with chlorine to open the disulphide bonds, increased the rate in alkaline solution only. Pretreatments with alkaline permanganate or hypochlorite, however, were found to have no effect.

As already mentioned, treatment with KOH in butanol had no effect, so the increases were not due to the removal of the epicuticle.

TABLE 5.7

Effect of pretreatment on the rate of chlorination at 0° C. and pH 10[61]

Treatment	Half-life period (min.)
Untreated	71·0
N HCl 80°C for 8 hr.	5·2
5% Cl_2 pH 2	10·4
10% Cl_2 pH 2	10·0
KOH in butanol (20 sec.)	64·0

These results can be explained by assuming that the cuticle offers resistance to hypochlorite ions because of the negative charge acquired in alkaline solution. If the cuticle is disrupted by preliminary treatments, the area A_m permeable to anions is increased, so that the rate of transport P across the membrane of thickness m is given by

$$P = A_m D (C_1 - C_m)/m$$

where C_1 and C_m represent the concentrations of free hypochlorite ions on either side of the membrane. If it is assumed that diffusion and reaction within the cortex are rapid processes, $C_m \sim 0$ and if C_1 is assumed to be equal to the external concentration in solution,

$$D = \frac{mV}{A_m t} \cdot \ln\frac{C_0}{C_0 - x}$$

where V is the volume of the solution, C_0 the initial concentration and $C_0 - x$ the concentration at time t. This logarithmic relation is obeyed approximately (Fig. 5.20), although in very dilute solution an initial rapid absorption is observed due to saturation of the cuticle, in contrast to the approximately parabolic curves, obtained in acid solution which can be explained by diffusion through a homogeneous medium.

It is usually found that the diffusing species can penetrate the cortex readily and evenly, due to the considerable lateral swelling (p. 87). Speakman and Clegg[96] considered the wool micelles to offer resistance to the penetration of dyes, although in acid solution penetration is greatly facilitated by the increased swelling due to salt-link breakdown. According to Astbury, however, the swelling of the cortex is mainly intermicellar (see p. 90) so that penetration of complex dyes into the individual micelles is doubtful. Swelling is shown to play a negligible part in the diffusion-controlled rate of reaction of wool with chlorine solutions, as the rate of reaction is only slightly decreased as the pH of the solution increases from 2 to 6.

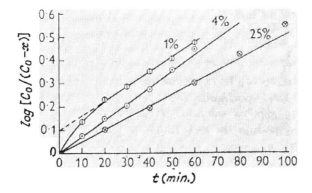

FIG. 5.20.—Graphs of log $[C_0/(C_0-x)]$ against t at 0° for 1, 4 and 25 g. Cl_2/100 g. wool at pH 10[61]; stirring at 600 rev./min. C_0 is the initial concentration of OCl^- ion and (C_0-x) the concentration at time t.

This may be responsible to some extent for the numerous 'specks' of dye which are found in cross-sections of dyed fibres.[97] Speakman and Smith[66] noted an irregular deposit of dye suggesting that solid dye had accumulated at particular points when mohair was dyed with Solway Blue S.E.N. Such deposits are normally absent in fibres dyed with simple acid dyes. This conclusion is supported by the X-ray data of Astbury and Dawson,[98] and the detailed investigation of Royer and Maresh[99] who found this phenomenon particularly marked with chrome dyes and fast to washing dyes of the thiocyanine type. According to these authors, the dye is precipitated or trapped in solution in voids caused by local irregular swelling as the dye penetrates between the cortical cells. Observations with the polarization microscope show that the dye is not in the form of a single large crystal, but in powder form due to precipitation or subsequent drying of the fibre when water readily diffuses out leaving a high local concentration of dye.

REFERENCES

[1] FISHER, *Proc. Roy. Soc.*, 1923, **103A**, 139; TROUTON, *Proc. Roy. Soc.*, 1906, **77A**, 292.
[2] CASSIE AND KING, *Trans. Faraday Soc.*, 1940, **36**, 445; CASSIE, *ibid.*, 453; CASSIE AND BAXTER, *ibid.*, 458.
[3] SHORTER, *J. Text. Inst.*, 1924, **15**, T320.
[4] HEDGES, *Trans. Faraday Soc.*, 1926, **22**, 176.
[5] KING, *ibid.*, 1945, **41**, 325.
[6] KING, *ibid.*, 1947, **43**, 552.
[7] SPEAKMAN, *Proc. Roy. Soc.*, 1931, **132A**, 167.
[8] See WILLIAMS AND CADY, *Chem. Rev.*, 1934, **14**, 171; GORDON, *Ann. N.Y. Acad. Sci.*, 1945, **46**, 211.
[9] KING, *Trans. Faraday Soc.*, 1945, **41**, 479.
[10] ROUSE, *J. Amer. Chem. Soc.*, 1947, **69**, 1068.
[11] BARRER, *Diffusion in and through Solids*, Cam. Univ. Press, 1941, Chap. 1.
[12] BARRER, *Fibrous Proteins Symp.*, Soc. Dy. Col., Bradford, 1946, 108.
[13] CRANK AND PARK, *Trans. Faraday Soc.*, 1949, **45**, 244.
[14] CRANK, *Phil. Mag.*, 1948, **39**, 140.
[15] WILSON, *ibid.*, 1948, **39**, 48.
[16] PATERSON, *Proc. Phys. Soc.*, 1947, **59**, 50.
[17] BARRER, *Trans. Faraday Soc.*, 1949, **45**, 358.
[18] WATSON, *Theory of Bessel Function*, Cambridge, 1922, p. 78.
[19] CRANK, *Phil. Mag.*, 1948, **39**, 362.
[20] BOLTZMANN, *Ann. Phys. Lpz.*, 1894, **53**, 959.
[21] CRANK AND HENRY, *Trans. Faraday Soc.*, 1949, **45**, 636.
[22] HARTLEY, *Trans. Faraday Soc.*, 1946, **42B**, 6; ROBINSON, *ibid.*, 12.
[23] HERMANS, *Contributions to the Physics of Cellulose Fibres*, Elsevier, 1946, p. 31; *Physics and Chemistry of Cellulose Fibres*, Elsevier, 1949.
[24] CRANK AND ROBINSON, *Proc. Roy. Soc.*, 1951, **204A**, 549.
[25] HERMANS AND VERMAS, *J. Polymer Sci.*, 1946, **1**, 149.
[26] HARTLEY, *Trans. Faraday Soc.*, 1949, **45**, 820.
[27] WIENER, *Ann. Physik.*, 1890, **41**, 675; THOVERT, *Compt. Rend.*, 1901, **133**, 1197; 1902, **134**, 594, 826; 1902, **135**, 579; 1903, **137**, 1249; 1904, **138**, 481; 1910, **150**, 270; *Ann. de Phys.*, 1914, **2**, 369.
[28] ROBINSON, *Proc. Roy. Soc.*, 1950, **204A**, 339.
[29] EYRES, HARTREE, INGHAM, JACKSON, SARJANT AND WAGSTAFF, *Phil. Trans.*, 1946, **240**, 1.
[30] HOPKINS, *Physic. Soc. Proc.*, 1938, **50**, 703.
[31] CRANK AND HENRY, *Trans. Faraday Soc.*, 1949, **45**, 1119.
[32] PARK, *ibid.*, 1950, **46**, 684.
[33] DAYNES, *ibid.*, 1937, **33**, 531.
[34] BARRER, *ibid.*, 1939, **35**, 646.
[35] LENNARD-JONES, *ibid.*, 1932, **28**, 349.
[36] KNUDSEN, *The Kinetic Theory of Gases—Some Modern Aspects*, Methuen.
[37] CASSIE, *Trans. Faraday Soc.*, 1945, **41**, 450.
[38] ORR AND BUTLER, *J. Chem. Soc.*, 1935, 1273; PARTINGTON, HUDSON AND BAGNALL, *Nature*, 1952, **169**, 583.
[39] KING, *Trans. Faraday Soc.*, 1947, **43**, 601.
[40] FOX AND FINSH, *Textile Res. J.*, 1940, **11**, 62.
[41] BRUNAUER, EMMETT AND TELLER, *J. Amer. Chem. Soc.*, 1938, **60**, 309.
[42] HARTLEY AND CRANK, *Trans. Faraday Soc.*, 1949, **45**, 801.
[43] ALEXANDER, GOUGH AND HUDSON, *ibid.*, 1949, **45**, 1109.

[44] ALEXANDER, CARTER AND EARLAND, *J. Soc. Dy. Col.*, 1951, **67**, 17.
[45] KRESSMAN AND KITCHENER, *Discussions of the Faraday Soc.*, 1949, **7**, 90.
[46] ALEXANDER, *Amer. Dyestuff Reporter*, 1950, **39**, 420.
[47] ALEXANDER, GOUGH AND HUDSON, *Trans. Faraday Soc.*, 1949, **45**, 1058.
[48] AGAR, *Faraday Soc. Discussions*, 1947, **1**, 26; LEVICH, *ibid.*, p. 37; *Acta Physicochim.*, 1942, **17**, 257; *ibid.*, 1944, **19**, 117, 133.
[49a] KING et al., *J. Amer. Chem. Soc.*, 1935, **57**, 828; *ibid.*, 1937, **59**, 1375; *Ind. Eng. Chem.*, 1937, **29**, 75; [b] GUTMAN et al., *Rocz. Chem.*, 1928, **8**, 445; 1932, **12**, 9; JABLCZYNSKI, GUTMAN AND WOLCZNK, *Z. anorg. Chem.*, 1931, **202**, 403; WILDERMAN, *Phil. Mag.*, 1929, **18**, 558.
[50] JABLCZYNSKI, *Z. anorg. Chem.*, 1929, **180**, 184; KING AND BRAVERMANN, *J. Amer. Chem. Soc.*, 1932, **54**, 1744; YAMASAKI, *Int. Cong. Pure Appl. Chem.*, 1909, **10**, 172; see also MOELWYN HUGHES, *Kinetics of Reaction in Solution*.
[51] NOYES AND WHITNEY, *Z. physik. Chem.*, 1897, **23**, 689.
[52] NERNST, *ibid.*, 1904, **47**, 52.
[53] BRUNNER, *ibid.*, 1904, **47**, 56.
[54] FAGE, *Proc. Roy. Soc.*, 1932, **135A**, 828.
[55] EUKEN, *Z. Elektrochem.*, 1932, **38**, 341.
[56] COLBURN, *Ind. Eng. Chem.*, 1930, **22**, 967; *Trans. Amer. Inst. Chem. Eng.*, 1933, **29**, 164; CHILTERN AND COLBURN; *Ind. Eng. Chem.*, 1934, **26**, 1183.
[57] HIXSON AND CROWELL, *ibid.*, 1931, **23**, 1160; VAN NAME, *Z. physik. Chem.*, 1910, **73**, 97; WILDERMAN, *ibid.*, 1909, **66**, 445; KING AND SHACK, *J. Amer. Chem. Soc.*, 1935, **57**, 1212.
[58] ALEXANDER AND HUDSON, *Text. Res. J.*, 1950, **20**, 481.
[59] HUDSON, *Discussions of the Faraday Soc.*, 1954, **12**, 15.
[60] ALEXANDER AND HUDSON, *J. Phys. Chem.*, 1949, **53**, 733.
[61] ALEXANDER, GOUGH AND HUDSON, *Biochem. J.*, 1951, **48**, 20.
[62] FRISHMAN AND HARRIS, B.P. 621,989.
[63] HUDSON AND SCHMEIDLER, *J. Phys. Chem.*, 1951, **55**, 1120.
[64] PETERS AND SPEAKMAN, *J. Soc. Dy. Col.*, 1949, **65**, 63.
[65] ALEXANDER AND KITCHENER, *Text. Res. J.*, 1950, **20**, 203.
[66] SPEAKMAN AND SMITH, *J. Soc. Dy. Col.*, 1936, **52**, 121.
[67] MENKART AND SPEAKMAN, *ibid.*, 1947, **63**, 322.
[68] HILL, *Proc. Roy. Soc.*, 1929, **104B**, 39.
[69] HERMANS, *J. Colloid Sci.*, 1947, **2**, 387; DANKWERTS, *T.F.S.*, 1950, **46**, 300, 701; *T.F.S.*, 1951, **47**, 1014.
[70] VREEDENBERG, *Rec. Trav. Chim.*, 1948, **67**, 839.
[71] MOTT AND GURNEY, *Electronic Processes in Ionic Crystals* (O.U.P., 1940) Ch. VIII.
[72] WAGNER, *Z. physik. Chem. B*, 1933, **21**, 25.
[73] WAGNER, *ibid.*, 1938, **40**, 197.
[74] BOOTH, *Trans. Faraday Soc.*, 1948, **44**, 796.
[75] MORTON, *Trans. Faraday Soc.*, 1935, **31**, 262; NEALE, *ibid.*, 1935, **31**, 282; VALKO, *ibid.*, 1935, **31**, 278; STANDING, *ibid.*, 1945, **41**, 410; WILLIS, WARWICKER, STANDING AND URQUHART, *ibid.*, 1945, **41**, 506; MORTON, *J. Soc. Dy. Col.*, 1946, **62**, 272.
[76] LINDBERG, *Text. Res. J.*, 1950, **20**, 381.
[77] STEINHARDT, FUGITT AND HARRIS, *J. Res. N.B.S.*, 1940, **24**, 335. STEINHARDT AND HARRIS, *ibid.*, 1940, **24**, 335.
[78] International Critical Tables.
[79] BOYD, MYERS AND ADAMSON, *J. Amer. Chem. Soc.*, 1947, **69**, 2836.
[80] HALE AND REICHENBERG, *Discussions of the Faraday Soc.*, 1949, **7**, 79.
[81] KUNIN AND MYERS, *J. Phys. Chem.*, 1947, **51**, 1111.
[82] MORTON, *Textil-Rundschau*, 1949, **4**, 39.
[83] ROYER, FIDDEL AND MILLSON, *Amer. Dyestuff Reptr.*, 1948, **37**, 116.
[84] ALEXANDER AND CHARMAN, *Text. Res. J.*, 1950, **20**, 761.
[85] CRANK, *J. Soc. Dy. Col.*, 1948, **64**, 386; *ibid.*, 1950, **66**, 366.
[86] GARVIE AND NEALE, *Trans. Faraday Soc.*, 1938, **34**, 335.
[87] NEALE AND STRINGFELLOW, *ibid.*, 1933, **29**, 1167; NEALE AND PATEL, *ibid.*, 1934 **30**, 905.

[88] HARTLEY, *ibid.*, 1935, **31**, 281.
[89] NEALE, *J. Soc. Dy. and Col.*, 1936, **52**, 252.
[90] STANDING, WARWICKER AND WILLIS, *J. Text. Inst.*, 1947, **38**, T335.
[91] VICKERSTAFF, *The Physical Chemistry of Dyeing*, Oliver and Boyd, London, 1950, p. 36 and p. 57 ff, ALEXANDER AND STACEY, *Proc. Roy. Soc.*, 1952, **212A**, 274.
[92] LEMIN AND VICKERSTAFF, *Fibrous Proteins Symposium, Soc. Dy. Col.*, Bradford, 1946, 129.
[93] MILLSON AND TURL, *Amer. Dyestuff Reporter*, 1950.
[93a] KÖPKE AND NILSSEN, *Second Quinquennial Wool Text. Res. Conf. (J. Text. Inst.)*, Harrogate, 1960, T.1398.
[94] WATKINS, ROYER AND MILLSON, *ibid.*, 1944.
[95] CRANK AND GODSON, *Phil. Mag.*, 1947, **38**, 794.
[96] SPEAKMAN AND CLEGG, *J. Soc. Dy. Col.*, 1934, **50**, 348.
[97] MILLSON AND ROYER, *Amer. Dyestuff Reporter*, 1940, **29**, 697.
[98] ASTBURY AND DAWSON, *J. Soc. Dy. and Col.*, 1938, **54**, 6.
[99] ROYER AND MARESH, *Amer. Dyestuff Reporter*, 1943.

CHAPTER 6

Acid-Base Characteristics

As wool contains a high proportion of diamino and dicarboxylic amino acids incorporated in the polypeptide chain, the protein is amphoteric in nature. It is thought that the numbers of free acid and basic groups are approximately equal,[1] so that at neutrality when both types of group are fully ionized, the system is electrically neutral:

$$\text{Wool—COO}^- \quad {}^+\text{NH}_3\text{—Wool}$$

As a result of studies of the change in mechanical properties with pH (see p. 213), it has been concluded that these groups exert an electrostatic attraction on each other which contributes to the strength of the fibre. The cohesion is also demonstrated by the minimum swelling at neutrality.[2]

The electrostatic force is commonly referred to as the salt link, and according to Speakman[1] the two oppositely charged groups are close together forming a true chemical compound. This arrangement does not appear to be very probable on steric considerations, and a more random distribution of the groups seems more likely, although a certain degree of order must prevail so as to produce a resultant electrostatic attraction.

Wool combines with acids and alkalis according to the generally accepted mechanism for zwitterions, involving back titration of acid and base.[3]

Thus, an ampholyte which is originally in the *isoelectric condition*, i.e. with no net electrical charge, contains ionized acid and base groups. Titration with acid involves the combination of protons with the weak acidic groups, *although the total amount of acid combined is determined by the number of basic groups*. However, the total number of acid groups can be obtained from the titration curve by measuring the acid uptake in the pH range 5·5 to 2·5 in the presence of N/10 KCl. This is made clear when a protein is considered which possesses an unequal number of acid and base groups. For example, a protein with 50 amino groups per 100 carboxylic groups at the isoelectric point would carry 50 charged amino groups, 50 charged carboxyl groups and 50 neutralized carboxyl groups. Thus, on titration with acid 50 more carboxyl groups are neutralized, i.e. the same number as the total number of basic groups. The position of the isoelectric point is of course determined by the strength of the acid and basic groups as well as the relative proportions.

Similarly, combination with base involves the back titration of charged amino groups with hydroxyl ions, and the two processes may be summarized as follows:

$$-COO^- \ldots\ldots ^+NH_3- \xrightarrow{HCl} -COOH \quad ^+NH_3- \atop Cl^-$$

$$-COO^- \ldots\ldots ^+NH_3 \xrightarrow{NaOH} -COO^- \atop Na^+ \quad NH_2-$$

The heat changes produced by the addition of acid and alkali are in agreement with this mechanism.[4] Thus, from the titration curves at various temperatures, it is found that the heat evolved in acid is very small, decreasing with increasing temperature to reach zero at about 50–60° C. The observed value and temperature-dependent behaviour are characteristic of the titration of carboxylic acids.[5] On the other hand the heat of titration of wool with base is high, in agreement with the high heats of dissociation of hydrogen ions from imidazole and substituted ammonium bases,[6] ranging from 6 k. cal./mol. for the former to 9–14 k. cal./mol. for the latter.

These reactions have been studied in detail notably by Speakman, and by Harris *et al.*, who showed the titration curve with acid to be surprisingly reproducible and almost independent of the type and source of the wool employed. Thus, for many simple acids a maximum combining capacity of *c.* 82 milliequivalents/100 g. of fibres has been obtained after allowing for the HCl imbibed by the fibres. This figure corresponds approximately to a 1N. solution within the fibre, which demonstrates the considerable affinity of the acid. The determination of the exact figure is exceedingly difficult because of the experimental procedure employed. Usually samples of wool initially in the isoelectric condition are allowed to come to equilibrium for many hours with a known volume of acid, which is analysed before and after treatment. The change in concentration is very small when high concentrations are employed. Further, some of the absorbed acid probably remains in solution in the free water within the fibre, the volume of which is difficult to estimate (see Chapter 4).

In addition, it is known that some of the acid and basic groups are combined in the form of an amide (see p. 307) which is hydrolysed in highly acidic solution, thus modifying the titration curve at low pH.[7]

It is usually claimed that the number of free carboxyl groups agrees closely with the values calculated from the quantities of glutamic and aspartic acids,[8] although the most likely analyses suggest that some free carboxyl groups arise from other residues. This comparison is of some importance in deriving the state of combination of cystine in keratin, and is discussed further in Chapter 8.

Titration data

In contrast to soluble proteins, wool and silk absorb no acid or base within a wide range of pH (c. 4–8), where they remain in the isoelectric condition although histidine ionizes in this region. Originally much controversy reigned over the significance of the isoelectric region. Some workers[9] denied the existence of such a region of zero potential, and considered the fibre to possess an isoelectric point which is well defined by electrophoretic measurements.[10] This apparent anomaly is a consequence of the two-phase system and as electrophoresis is essentially a property of the surface, the isoelectric condition of the fibre as determined by such measurements need not necessarily be the same as that determined by titration (see p. 190). The titration curve of wool is compared in Fig.6.1 with that of a similar soluble protein, egg albumin, chosen because of the similarity in composition, particularly with respect to the diamino acids. The characteristic sigmoid form of the titration curves for wool are entirely different in shape from the corresponding curves for a soluble protein, which shows that considerable opposition is offered to the penetration of acid near the isoelectric region. This opposition has been explained qualitatively in terms of a general Donnan membrane effect, and also on the basis of the effect of the electrical double layer surrounding the particle. Early explanations attributed the opposition to the stability of the salt links in the isoelectric region which require a minimum concentration of acid or base to produce their breakdown.[1, 2, 12] Lloyd and Bidder[13] even attributed the lack of absorption to covalent bonds between polypeptide chains and the polar side groups, rendering the latter inaccessible until broken by strong mineral acid or base.

ACID-BASE CHARACTERISTICS

The most significant observations on acid absorption involve the effect of salts on the titration curves. Neutral salts affect the position considerably, moving both acid and alkaline curves in such a way

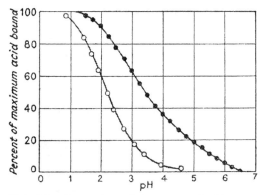

FIG. 6.1.—Comparison of the curves of acid combination of wool (O)[10] and of egg albumin (●)[10] at 25° C. in the absence of added salt. (*Reproduced by permission.*)

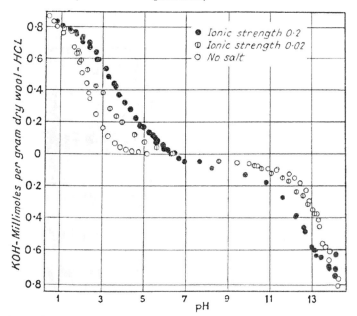

FIG. 6.2.—Combination of wool with hydrochloric acid and with potassium hydroxide as a function of pH.[14] (*Reproduced by permission.*)

as to reduce the isoelectric region (Fig. 6.2), so that at high salt concentrations the wool behaves as a soluble protein, the acid and alkaline curves meeting at the isoionic point.[14]

The influence of salt on the acid-titration curve is due to the anion alone, as the cation has no effect on the amount of acid combined. That the anion affects the titration considerably is shown by the many observations on the combination of different acids. It is found that the position of the curve depends on the nature of the acid,

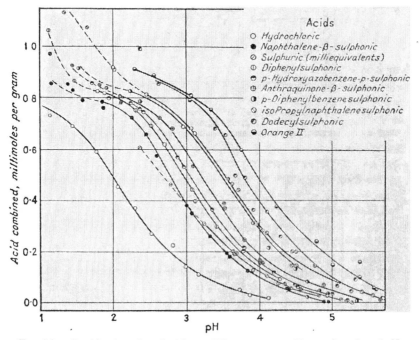

Fig. 6.3.—Combination of wool with ten different strong acids as a function of pH at 50° C.[7] (*Reproduced by permission.*)

although the shape is largely unaffected except with the more complex acids.[7, 15] In most cases the titration curves lie closer to the isoelectric region than the hydrochloric acid curve (Fig. 6.3), which indicates that the anion itself is exerting a specific affinity for the fibre in addition to the high affinity of the proton for the carboxyl groups. When this affinity is very high, as for highly complex anions, suprastoichiometric absorption is observed. Thus the maximum acid-combining capacity obtained by titration with certain dye acids is much greater than the usual (total) value of 88 millimol./100 g.

Finally, the shape of the titration curve and the maximum acid-combining capacity are affected by various chemical treatments. Thus alkali-treated and chlorinated wool absorb greater quantities of acid in the pH 2–5 region although the maximum acid-combining capacity is only slightly affected. This was attributed by Lemin and Vickerstaff[16] to disulphide bond breakdown accompanied by greater swelling. On the other hand, wool which has been carbonized by immersion in conc. sulphuric acid absorbs less HCl both in the pH 2–5 region and at very high acidities. Thus the maximum combining capacity was 77 milliequivalents/100 g. after a 15-min. treatment,

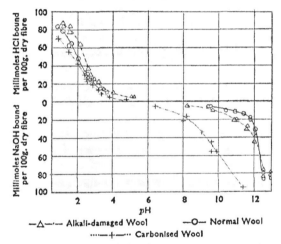

FIG. 6.4.—The effect of chemical pretreatments on the acid and alkaline titration curves.[16] (*Reproduced by permission.*)

and the isoionic point had fallen to 5·5, which is attributed to an increase in the number of strongly acidic groups, due to the formation of sulphamic acid. On the alkaline side, the oxidized and alkali-treated wool absorbs a little more base than untreated, whereas a considerable increase is detected after carbonization (see Fig. 6.4).

The effect of temperature

In acid solution a very small decrease in the amount of hydrochloric acid combined at a given pH may be observed from the two curves[4] at 0° and 25° C. although the results at 25°, 40° and 50° C. are so close as to be indistinguishable (Fig. 6.5). This shows firstly that the heat of the process is small and secondly that it reaches zero at some maximum temperature, above which it probably decreases. This is entirely characteristic of the dissociation of carboxylic acids, the heat of which is invariably small. Recorded values[6] are close to zero, both positive and negative depending on the temperature range employed. For the 0–25° C. range, the heat absorbed per mole of the significant acids is as follows: aspartic 1·6, 2·1 k. cal.; glutamic 1·9, 1·04 k. cal. These values are close to the average value of 2·5 k. cal./mol. for wool, although the quantitative agreement is probably not significant owing to the heterogeneous nature of the latter.

On the other hand, the heats of dissociation of hydrogen ions from imidazole and substituted ammonium bases are large, ranging from 6 k. cal./mol. for the former to 9–14 k. cal/mol. for the latter.[5] The combination of alkali with wool is more difficult to determine accurately than acid combination, and is also more complex theoretically. A more detailed discussion of the nature of the alkali process will be given later (p. 210). Even at low temperatures, considerable reaction with the disulphide group,[17] in which one sulphur atom is eliminated from each cystine linkage, may occur (see p. 254).

Values for the heat changes on absorption of acid and alkali have been obtained from the equilibrium constant K associated with each process. This is related directly to the pH shift with temperature of a given amount of acid or base combined, as will be shown in later quantitative treatments, by the simple relation,[18]

$$\Delta p\text{H} = -\Delta \log K$$

A heat change ΔH may then be obtained simply from the reaction isochore,

$$\Delta H = 2 \cdot 303\, R \frac{T_1 T_2}{T_1 - T_2} . \Delta \log K$$

Values obtained in this way for various acids and under various conditions are compared in Table 6.1.

The average value of ΔH for the absorption of HCl ($\sim$2 k. cal./mol.) is close to the calorimetric value of Speakman and Stott[19] 3·5 k. cal./mol. at 0° C. for soluble proteins and amino acids. The two cases are not strictly comparable, as the heat of mixing may be significant in the heterogeneous process with wool. Approximate calculations of the heat of transfer from one phase to another suggest

ACID-BASE CHARACTERISTICS

TABLE 6.1

Heat absorbed on titration with acids and sodium hydroxide[4]

Acid or base	Salt	Temperature range	ΔH (average) k.cal./mol
Hydrochloric acid ..	Nil	0–25°	2·38
,, ,,	,,	25–50°	±0·50
,, ,,	0·2M	0–25°	2·60
,, ,,	0·2M	25–50°	0·53
,, ,,	0·5M	0–50°	1·28
Sulphuric acid ..	Nil	0–50°	0·53
Naphthalene β-sulphonic acid ..	Nil	0–25°	3·64
,, ,,	,,	25–50°	2·22
Dodecyl sulphonic acid ..	,,	25–50°	3·18
Picric acid ..	,,	0–25°	6·55
Orange II acid ..	,,	25–50°	>11·30
Sodium hydroxide	,,	0–25°	13·10
,, ,,	,,	25–50°	13·72
,, ,,	0·2N	0–25°	12·20
,, ,,	0·2N	25–50°	14·43
,, ,,	0·5N	0–50°	12·90

that this effect may explain the whole of the observed heat change, so that the actual heat of dissociation of the acid groups in wool may be very small indeed.

It is observed from Fig. 6.5 that the curves are almost parallel over a wide range, so that the above method of Wyman[18] may be used. This leads to an average value of the heat changes as several different basic groups ionise within this region. The values of ΔH given in Table 6.2 are found to be relatively constant, and lie within the range 10,000 to 13,000 cal./mol. which is characteristic of the dissociation H^+ from amino or guanidino groups. The values are about twice as great as the value for tyrosine,[20] and this supports the belief that the hydroxyl groups do not titrate except at high alkalinities[14] (see p. 212). The experimental values agree with the figure for soluble proteins, e.g. horse haemoglobin (11,500 cal./mol.) and casein (14,000 cal./mol.).

These values are close to the heat of dissociation of water[21] (13,480 cal./mol. at 25° C.), which supports the contention that the

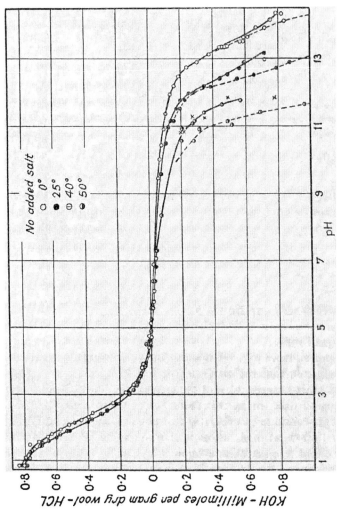

Fig. 6.5.—Combination of wool with hydrochloric acid and with potassium hydroxide as a function of pH and temperature in the absence of added salt.[4] (The dotted lines show the correction by Harris and Rutherford for the reaction of alkali with cystine.) *(Reproduced by permission.)*

alkali titration is due to the removal of hydrogen ions from the ionised amino and guanidino groups:

$$-NH_3^+ \longrightarrow NH_2 + H^+$$

$$-C{\overset{\displaystyle NH_2}{\underset{\displaystyle NH_2^+}{\diagup}}} \longrightarrow -C{\overset{\displaystyle NH_2}{\underset{\displaystyle NH}{\diagup}}} + H^+$$

TABLE 6.2

Effect of the quantity of base sorbed on the heat of sorption[4]

Base bound per gram-millimol	$-\Delta pH$ 0–20° C.	ΔH cal./mol.	Base bound per gram-millimol	$-\Delta pH$ 25–50° C.	ΔH cal./mol
0·1	0·75	11,190	0·1	1·47	25,820
0·2	0·81	12,070	0·2	1·04	18,310
0·3	0·85	12,640	0·3	0·84	14,800
0·4	0·88	13,100	0·4	0·78	13,720
0·5	0·89	13,240	0·5	0·84	14,800
0·6	0·83	12,380			
0·7	0·81	12,030			

It must be realized that the calculated heat change also depends on the affinity of the corresponding gegen-ion. Thus the results of Table 6.1 show that ΔH increases considerably with the complexity of the anion. The heat changes accompanying the absorption of the simpler acids may be explained entirely by the reaction of the proton with the carboxyl group, and by the concentration difference of the acid in the two phases. It follows therefore that the chloride ion and sulphate ion have no specific affinity for the fibre and that no heat change, other than that of mixing, accompanies their entry into the fibre. This conclusion is confirmed by quantitative treatments of the titration process (see p. 195 *ff*).

It has already been mentioned that many ions exhibit a specific attraction for the fibre, as indicated by the greater extent of absorption of their acids. Correspondingly, the heat of absorption increases, so that with simple dye acids the absorption of the anion is accompanied by a considerable heat change. This shows that a considerable decrease in entropy occurs, consistent with the supposition that complex anions are held on localized sites, unlike simple gegen-ions, which are completely mobile within the fibre (see p. 197).

The difference in behaviour between soluble and insoluble proteins

Although the chemical processes of acid and alkaline titration of wool are known to be essentially the same as for soluble proteins and individual amino acids, striking differences in the titration curves have already been pointed out. Thus the extensive isoelectric region and the different effect of salts imply that the titration processes in the two cases proceed by entirely different mechanisms.

A satisfactory quantitative interpretation of the titration of soluble proteins has been known for some time. Although the shape of the titration curve for a soluble protein is similar to that of the individual amino acids, and no isoelectric region is evident, the titration curve

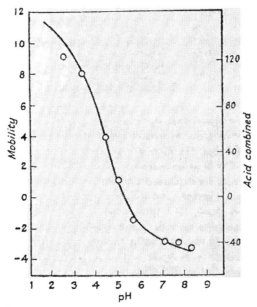

Fig. 6.6.—The effect of pH on the electrokinetic mobility (circles) and acid combination (continuous curve) of myosin.[23]

is spread over a wider pH range. This may be taken to show that the acid or base tends to increase in strength as the number of combined protons or hydroxyl ions is increased. Cannan[22] has explained this behaviour quantitatively in terms of the repulsion between the charged fibre and the proton which increases as the titration proceeds. Owing to the relatively small size and the flexibility of a soluble protein molecule, it may be assumed that all the ionized groups are orientated towards the bulk of the solution and are therefore freely accessible to protons or hydroxyl ions. This is demonstrated by the observation that the electrophoretic mobility,[23] which is controlled by the charge on the outside of a particle, and the titration curves coincide (Fig. 6.6). It is also known that proteins can form monomolecular films on oil/water and air/water interfaces showing that soluble proteins are capable of self-orientation in such a way that polar and non-polar side chains lie in opposite directions.

As fibrous proteins cannot behave in this manner, differences in electrical mobility and titration curves are to be expected (Fig. 6.7). Sookne and Harris[10] have re-determined the isoelectric point by a streaming method and give a value of pH 4·5 against their original value[24] of 3·4. The pH for zero acid combination (isoionic point) is, however, approximately 6·3.[16] This is determined most conveniently by constructing titration curves in the presence of a high salt concentration. It is interesting to mention at this point that the cold drawing of nylon, which increases the crystallinity, changes the electrophoretic mobility, without materially affecting the titration curve, presumably due to the re-arrangement of polar groups into less accessible positions.[25]

The differences between the titration behaviour of soluble and insoluble proteins thus lies in the very great difference in size. Only a limited number of free hydrogen ions can combine with an insoluble protein as the potential thus acquired excludes the further penetration of protons. The potential produced by a given charge decreases as the radius increases, but the number of groups capable of accepting a charge increases with the cube of the radius. The potential is therefore proportional to the square of the radius for a given amount of H^+ combined per unit volume of the fibre. It is readily shown that as the size of the protein changes from molecular to microscopic dimensions only a very small fraction of the total possible number of protons can be absorbed in this way. Further absorption cannot proceed without the simultaneous entry of anions into the fibre during most of the titration, the curve of which is therefore different in form from that of a soluble protein.

The apparent increase in strength of the acid groups is therefore due to the potential at the surface of the fibre and not to a change in the dissociation constant of the carboxyl groups. It is probable, however, that some change in pK is produced by (a) combination of the appropriate amino acid in a macromolecule, (b) mutual interaction of

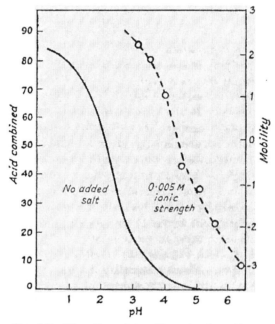

FIG. 6.7.—The effect of the pH on the electrokinetic mobility[10] (circles) and acid combination (continuous curve) of wool.[14]

the fixed carboxyl ions, and (c) the change of dielectric constant within the fibre. The last two factors would however tend to decrease, not increase, the acid strength, and in the light of the quantitative treatments to be described below, it seems likely that the intrinsic strength is not seriously affected by combination.

The influence of the electrical double layer and the effect of salts

The preceding discussion has shown that owing to the preliminary absorption of hydrogen ions, the fibre in acid solution is at a higher positive potential than a soluble protein, differing from it only in size. The apparent increase in acid strength is then not attributed to the presence of positive groups in close proximity to each other, but to the positive potential of the fibre as a whole. In most elementary treatments it is assumed that this potential is uniform throughout the fibre itself with a sharp discontinuity at the surface (Fig. 6.8). This assumption requires that all the acid (or basic) groups are equal in

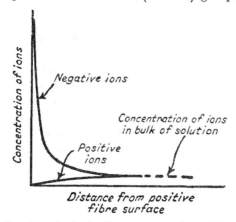

FIG. 6.8.—Ionic distribution near a charged fibre surface.[27]

strength, are randomly distributed throughout the fibre and do not interact the one with the others. In support of these assumptions Cannan[26] has shown that a single dissociation constant may be assigned to a polybasic system with acid groups differing in strength.

The sharp discontinuity of the potential at the fibre surface is due to the electrical double layer which surrounds all colloid systems in solution; a result of the preferential absorption of either anions or cations[27] (hydrogen ions in this case). The double layer surrounding the wool fibre will have a considerable effect on the titration curve, particularly when small amounts of hydrogen ion are combined.

In the case of acid titration, the hydrogen ion, i.e. the potential-determining ion, is bound by strong covalent bonds to the surface. The diffuse layer of gegen-ions held loosely in the neighbourhood of the surface, extends into the solution to a distance controlled by the ionic strength of the medium (Fig. 6.8). For wool fibres of diameter $> 10\ \mu$, the flat-plate theory, which may be assumed to apply approximately, shows that the thickness of the layer varies inversely as the charge and square root of the concentration of ions present.[28] On the addition of neutral salts of low affinity, the double layer is condensed, although the surface charge is unaffected.

The effect of salts on the titration curves of soluble proteins has been studied in detail by Cannan *et al.*[22] who have shown the pH changes with the number of protons combined, h, as follows

$$\Delta p\text{H} = \alpha h + \beta$$

where α is a function of ionic strength only and β is a constant.

The simple double-layer treatment, which satisfactorily explains the titration of soluble proteins, has been combined with the general theory of dual entry of ions into an insoluble protein by Alexander and Kitchener.[28] These authors distinguish between 'inside' and 'outside' absorption sites; the former determine the form of the titration curve, but the latter govern its position on the pH axis, and the electrokinetic properties. This differentiation is not altogether satisfactory as the effective surface is difficult to define owing to the porous nature of the fibres and may vary with the property measured. The size of the 'outside' portion of the fibre is the most important unit to define, and may be taken as the effective range of the inter-ionic forces. As these vary with the liquid medium, the outside layer is best identified with the diffuse double layer.

When the potential-determining charges on a protein are separated by a greater distance than the thickness of the double layer the titration of a fibre becomes similar to that of a soluble protein as the arbitrary distinction between outside and inside has disappeared. Such a condition is satisfied by high pH values, i.e. near the isoelectric region, or by high salt concentrations which collapse the double layer. Under most conditions, in the absence of salt, the double layer is greater than the mean distance between potentially charged sites (*c.* 10 Å for wool). Near the isoelectric point where only *c.* 1 per cent. of the total number of possible ions is charged, the separation is of the order of 30 Å whereas the thickness of the double layer is only 10 Å in the presence of 0·1 N. salt and the fibre becomes completely permeable to ions.[28]

Quantitative theories of acid combination

In all the treatments so far proposed the fibre is considered as a swollen gel containing a definite quantity of free water which acts as a solvent for protons and anions (see Chapter 3). The composition of this internal solution is of the greatest importance in treatments based on the operation of a membrane equilibrium, but of little consequence in more general thermodynamic treatments. In the latter case the chemical potential of the acid within the fibre is almost independent of the free acid dissolved in the internal aqueous phase. Thus for example, at the point of half-neutralization c. 40 milliequivalents of H+ will be combined. If it is assumsd that the average strength of the carboxyl groups is similar to that of a corresponding soluble protein (i.e. pK 4·0), the pH of the internal solution is 4·0, so that a negligible proportion of the hydrogen ions reside in the internal solution. This can therefore be neglected except in a Donnan membrane treatment, where the internal free hydrogen-ion concentration governs the amount of hydrogen ion absorbed.

The equilibrium of acid between solution and fibre may thus be regarded as a simple distribution, in the first instance, governed by the simultaneous penetration of anion and proton into the fibre:

$$H_s^+ + A_s^- \overset{K}{\rightleftharpoons} H_f^+ + A_f^-$$

$$K = \frac{(a_H \cdot a_A)_f}{(a_H a_A)_s}$$

The distribution coefficient K must be related to the strength of the acid groups. The activity of a sorbate is given by $\theta_i/1-\theta_i$ where θ_i is the number of ions of species i within the fibre divided by the total capacity.

Thus in general,

$$\left(\frac{\theta_H}{1-\theta_H}\right)\left(\frac{\theta_A}{1-\theta_A}\right) \equiv \left(\frac{\theta_H}{1-\theta_H}\right)^2 = K[H^+]_s[A^-]_s$$

or

$$\log_{10}\frac{\theta}{1-\theta} = \tfrac{1}{2}\log_{10} K - \tfrac{1}{2}pH + \tfrac{1}{2}\log_{10}[A^-]$$

Fig. 6.9 shows that this relation can predict accurately the influence of added electrolytes on the titration curve except for the region close to the isoelectric point.[29] This suggests that the fibre is titrating as a soluble protein near the isoelectric region, i.e. the 'outside part' only is being neutralized. This is supported by the fact that the electrophoretic mobility tends towards a constant at pH values where only a small fraction of total acid is bound.[10] This behaviour is to be expected also from electrostatic considerations, as the most stable configuration is obtained when the charges are separated as widely as possible. This preliminary charging of the outside of the fibre requires a logarithmic increase in concentration to produce equal increments of sorption, so that acid binding by this mechanism is linear with pH as already mentioned (p. 194). This is found to be the case for most proteins including wool for a limited region in the neighbourhood of the isoelectric point.

However, when the acid concentration has increased to about one hundred times that at the isoelectric point, the much more extensive sorption process controlled by mass action occurs. This process is

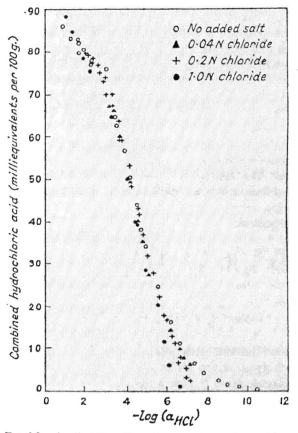

FIG. 6.9.—Applicability of law of mass action to the combination of HCl with wool at different ionic strengths.[29]

independent of the charge, i.e. the potential on the outside electrical double layer, as the electrical work done in introducing two oppositely charged ions into an equipotential region is zero.

The location of the anions within the fibre is more difficult to predict than the extent of combination under given conditions, and here the three alternative theories diverge. According to a thermodynamic treatment, the anions are repelled from negative centres and therefore tend to congregate in the vicinity of positively charged groups. The anions are thus regarded as absorbed on a limited number of sites identified with the amino groups (see p. 237). The opposite view is taken by the Donnan approach, which in the case of the simpler acids at least, considers the anions to be dissolved in the water within the fibre as in a normal aqueous solution. Although this is supported by earlier work on the effect of acids on the swelling of gels (p. 205), it encounters difficulties when the specific effect of complex ions is considered. The thermodynamic treatment[29] merely requires that a particular anion has a given affinity for the absorption site, thus increasing the bonding energy (Table 6.1) and decreasing the mobility and entropy.

The theory of Steinhardt and Harris[4, 14]

Steinhardt and Harris, who evolved the first detailed theory, took an intermediate view of the anion absorption. They proposed the formation of a partially dissociated bond between RNH_3^+ groups and Cl^- ions, governed by mass action, in addition to the dissociation of the acid groups, governed by a single dissociation constant K_H'. Thus, under a given set of conditions the following equilibria control the absorption:

$$\left.\begin{array}{ll} K_H' & WHA \rightleftharpoons WA^- + H^+ \\ K_A' & WA^- \rightleftharpoons W^\pm + A^- \\ K_H & WH^+ \rightleftharpoons W^\pm + H^+ \\ K_A & WHA \rightleftharpoons WH^+ + A^- \end{array}\right\} \text{ where } K_H'K_A' = K_HK_A$$

Two alternative relations may be derived from these equations directly in terms of the hydrogen ion activity, a_H for the conditions (a) $[WH^+] > [WA^-]$ and (b) $[WA^-] > [WH^+]$, corresponding to very low or very high affinity of A for the fibre,

(a) $$\frac{[WHA]+[WH^+]}{[WHA]+[WH^+]+[WA^-]+[W^\pm]} = \frac{1}{1 + \dfrac{K_H'}{a_H}\left[\dfrac{a_{A^-}+K_A'}{a_{A^-}+K_A}\right]}$$

H

(b) $$\frac{[WHA]+[WA^-]}{[WHA]+[WH^+]+[WA^-]+[W^\pm]} = \frac{1}{1+\dfrac{K_A}{a_A}\left[\dfrac{a_H+K_H}{a_H+K_H'}\right]}$$

From these expressions it follows that the titration curve resembles that of the corresponding monobasic acid providing the dissociating groups have the same intrinsic dissociation constants and do not exert any electrostatic interaction on each other. These conditions have been established for soluble proteins by many workers.[30] In addition, however, the apparent strength is observed to change with salt concentration. When a_{A^-} is large, this term approaches unity, so that the position is then determined solely by K_H and the titration curve is identical with that of a similar dissolved acid or protein (see p. 184).

The shift in the titration curve with varying salt concentration may be represented by the position of half-neutralization

$$(pH)_{\frac{1}{2}} = pK_H' + \log_{10} \cdot \left(\frac{a_{A^-}+K_A}{a_{A^-}+K_A'}\right)$$

Electrostatic considerations dictate that $K_A' > K_A$ (and $K_H > K_H'$). Thus in the range $K_A' \gg a_{A^-} \gg K_A$ the logarithmic term approaches $\log a_{A^-}/K_A'$, i.e. $(pH)_{\frac{1}{2}} \propto \log_{10}[A^-]$ which is confirmed experimentally and which has already been deduced in a very simple manner by the application of the law of mass action. In the present instance this relation may be used to evaluate K_A'.

In solution containing no added salt, a_{A^-} is not constant, but equals a_H so that the general relation becomes

$$\Theta = \frac{[WHA]+[WH^+]}{[WHA]+[WH^+]+[WA^-]+[W^\pm]} = \frac{1}{1+\dfrac{K_H'}{a_H}\cdot\left(\dfrac{a_H+K_A'}{a_H+K_A}\right)}$$

As K_A' has been assumed to be very large, and K_A very small

$$\Theta = \frac{1}{1+\dfrac{K_H' K_A'}{(a_H)^2}}$$

which explains why the amount of acid combined in the absence of salt depends on the square of the hydrogen ion concentration.

Using the calculated value of K_A', theoretical and experimental curves are found to be of a similar form although the slopes are considerably different. This causes the theoretical curves to be much broader along the pH axis, which was attributed by Steinhardt and Harris either to the variable nature of K_H, or to a constant dissociation constant for all the groups with electrical interaction varying uniformly with the degree of ionization. The latter explanation was regarded as the more probable, mainly because of the high symmetry in the theoretical and experimental differences (see Fig. 6.10), and also because with high salt concentration the mid-point of the titration curve occurs at pH 4·0–4·1, characteristic for the dissociation of carboxyl groups in proteins.

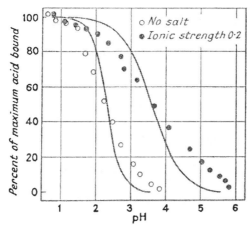

Fig. 6.10.—Experimental data for acid combination as a function of pH compared with theoretical curves for a monobasic acid.[14]

This treatment, therefore, although illuminating in some respects meets with only partial success. The picture involving the various equilibria, all governed by mass action, is difficult to conceive, particularly the dissociation of the salt $RNH_3^+Cl^-$. The most serious disadvantage, however, is the inability of the theory to predict the form of the titration curves, and the very different slopes of theoretical and experimental curves (e.g. for HCl $-1·737$ as against $-3·9$ in the presence of KCl, and $-0·869$ as against $-1·9$ in the absence of salt). To overcome this considerable and uniform difference, Steinhardt and Harris introduce the concentrations quite arbitrarily as square-root terms to make a rough allowance for the electrostatic interaction between neighbouring groups. In the simplified mass action treatment (p. 195), and in the thermodynamic treatment discussed below no arbitrary modification is required, the square-root terms being a direct consequence of the participation of proton and anion in the heterogeneous equilibrium.

The theory of Gilbert and Rideal[31]

The thermodynamic approach has so far proved to be the most satisfactory, and leads to a simple yet comprehensive interpretation of the absorption process. The basis on which this theory rests has already been developed (p. 193) when the distribution of potential within and near to the surface of the fibres was considered in detail. The fibre itself may be considered as a region of uniform potential with a sharp discontinuity at the boundary corresponding to the fixed part of the electrical double layer. Local fluctuations, which are all important in the theory of anion affinity (p. 229) are mutually neutralized, so that the mean potential within the fibre is that of the highest contour of the surrounding double layer. It must be mentioned here however, that the recent measurements of keratin membrane potentials by Baxter[32] are not in entire agreement with the Gilbert-Rideal theory.

As a first approximation, the positive groups are assumed to interact with anions in a similar manner, the anion being free to occupy any positive site. Similarly all the carboxyl groups are assumed to have the same affinity for the proton. Thus, anions and protons may be absorbed independently, except in so far as they affect the net charge of the fibre.

The problem then consists of the calculation of the chemical potentials of the two species of ions within the fibre and in the external solution. The chemical potential μ of an uncharged absorbed substance, distributed at random among a limited number of sites, has been deduced statistically by Fowler[33] in terms of θ, the fraction of sites occupied.

$$\mu = \mu_0 + RT \log_e \frac{\theta}{1-\theta}$$

If the substance has unit charge and the sites are in a region of potential ψ, the chemical potential is greater by an amount ψF. μ^0 is thus defined as the chemical potential when $\theta = 0\cdot5$ and $\psi = 0$. Substituting appropriate values for proton and anion in the fibre,

$$\mu_H = (\mu_H^0)_f + RT \log_e \frac{\theta_H}{1-\theta_H} + \psi F$$

$$\mu_A = (\mu_A^0)_f + RT \log_e \frac{\theta_A}{1-\theta_A} - \psi F$$

Combining these relations with the corresponding relations for the chemical potential of the two ions in solution, i.e. $\mu_i = (\mu_i^0)_s + RT \log_e a_i$, the general relation is finally obtained:

$$2RT \log_e \frac{\theta_{HCl}}{1-\theta_{HCl}} = -(\Delta\mu_H^0 + \Delta\mu_A^0) + RT \log_e (H^+)(A^-) f_A f_H$$

This follows from the general requirement that the free energy of the acid is the same in the two phases at equilibrium. The term $\Delta\mu_i$ represents the difference in the standard chemical potential of the ion of species i in the fibre and in the solution, i.e. $(\mu_i^0)_f - (\mu_i^0)_s$, and is a direct measure of the affinity of the ion for the sorption site.

For the absorption of pure acid the general equation reduces to,

$$\log_{10} \frac{\theta_H}{1-\theta_H} = -pH - \frac{(\Delta\mu_H^0 + \Delta\mu_A^0)}{4 \cdot 6 RT}$$

so that the plot of $\log(HCl)_f/(HCl)_f^s - (HCl)_f$ against pH should be linear with a slope of $-1 \cdot 0$, where $(HCl)_f^s$ is the saturation value. This is found to agree closely with the experimental data (see Fig. 6.11).

On the other hand, in the presence of a constant concentration of a salt, for the sake of simplicity assumed at this stage to have an anion common with that of the acid, the plot of $\log(HCl)_f/(HCl)_f^s - (HCl)_f$ against $\log(Cl^-)(H^+)$ should be linear with a slope of $-0 \cdot 5$ (cf. Fig. 6.11).

This shows clearly that the greater the anion concentration the less the concentration of hydrogen ions required for a given degree of saturation. The value of half-saturation which is frequently used as a standard for investigating the effect of salt concentration on the position of the titration curve, is given theoretically by

$$(pH)_{0.5} = \log_{10}(Cl) - \frac{(\Delta\mu_H^0 - \Delta\mu_{Cl}^0)}{2 \cdot 303 RT}$$

This theory, therefore, agrees closely with the experimental data over a wide pH range, and over a wide concentration range of added inert salt, considering the necessary oversimplifications. Thus besides assigning a single value of K for the carboxyl groups of the different constituent amino acids, interaction of charged amino groups with neighbouring salt links has been neglected. Gilbert and Rideal have made the important contribution of showing that the shape of the titration curve can be predicted by assuming that only a limited number of salt links are available for combination with protons.

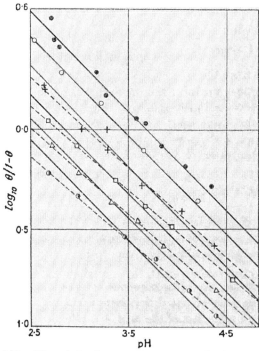

FIG. 6.11.—The relation between the quantity of acid combined and pH according to the Gilbert and Rideal theory[31] (solid lines): points experimental for salt concentrations, 0·20N, 0·10N, 0·05N, 0·02N, 0·01N and 0·005N.

The combined ions may be regarded as occupying separate sites since the protons combine covalently with carboxyl groups, while the anions are probably held loosely (in the case of an ion of low affinity) in the neighbourhood of the charged amino groups by Coulombic forces.

Considerable deviations between theoretical and experimental curves are, however, observed in the region near the point of zero acid combination (Fig. 6.12). This may be due partly to a possible overlap with the titration curve of the lysine amino groups which has not been taken into account, but more probably due to the influence of the electrical double layer, already considered.

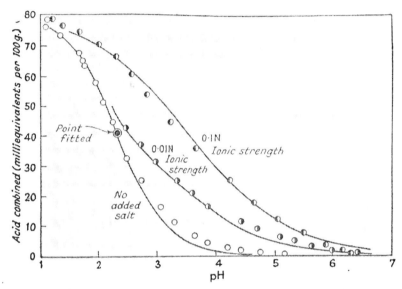

FIG. 6.12.—Theoretical and experimental acid-binding curves at different ionic strengths for wool. Points experimental, curves calculated from theory of Gilbert and Rideal[31], fitted at one point only and corrected for activity coefficient of chloride ion.

The Donnan theory

It has been well established in the above discussions that the peculiar form of the titration curve follows from the two-phase system, and that the position is determined by the conditions at the boundary. Hydrogen ions entering the fibre combine extensively with carboxyl groups, thus rendering the fibre as a whole positive. The accompanying anion is known to be highly mobile within the fibre. Under these conditions the protein may be regarded as a highly ionized salt, and according to the principles laid down by Donnan,[34] the concentrations of free acid in the two phases are unequal. This may be demonstrated very simply as follows. Consider the simple system represented diagrammatically as follows:

	Solid	Liquid
H+	(y)	H+ (x)
Cl−	$(y+z)$	Cl− (x)
WH+	(b)	
W	(a)	

In order to apply a simple membrane treatment it is usually assumed that the free hydrogen ions (y) and the gegen-ions $(y+z)$ are located in the aqueous phase within the fibre, which may be treated as an ordinary solution.

Thus at equilibrium,

$$(\mu_{HCl})_I = (\mu_{HCl})_{II}$$

i.e.

$$\mu_{H^+}^0 + RT\log_e(a_{H^+})_I + \mu_{Cl^-}^0 + RT\log_e(a_{Cl^-})_I = \mu_{H^+}^0 + RT\log_e(a_{H^+})_{II} + \mu_{Cl^-}^0 + RT\log_e(a_{Cl^-})_{II}$$

Thus,

$$(a_{H^+})_I \times (a_{Cl^-})_I = (a_{H^+})_{II} \times (a_{Cl^-})_{II}$$

i.e.

$$x^2 - y^2 = yz$$

This is the generalized expression from which the concentration of free hydrogen ions within the fibre may be calculated for any given external concentration. It is also seen that the concentration of diffusible ions in the gel is greater than in the solution by an amount e where

$$2x + e = 2y + z$$

These general principles were first applied to insoluble proteins by Procter and Wilson[35] in a successful interpretation of the swelling of gelatin in acid and alkali. According to this general theory, the mobile anions tend to diffuse outwards thus exerting a uniform pull on the jelly, which causes an increase in volume. In a very dilute solution $x = y \simeq 0$, so that the concentration of acid on either side of the membrane becomes equal, and as $z \rightarrow 0$, the protein is then in its isoelectric condition. This is the position of minimum swelling. In more concentrated solutions, only part of the acid combines, i.e. $y + z > z > e$, and e is a direct measure of the extent of swelling.

In most dilute solutions e increases almost directly with initial acid concentration, but approaches a maximum as the formation of the salt nears completion, and must then decrease as x becomes larger according to $e = -2x + \sqrt{(4x^2 + z^2)}$, where z has a limiting maximum value. The repression of swelling on the addition of a salt is caused by an increase in the value of x resulting in a decrease in e, assisted to some extent by the repression of the ionization of the protein salt.

The outward force is opposed by the osmotic pressure which prevents the protein from swelling indefinitely. The measurement of the osmotic pressures developed across a protein membrane under different external conditions was performed in considerable detail by Loeb[36] in a verification of the Donnan theory.

The application of the Donnan theory to give an accurate and quantitative interpretation of the titration curves of wool fibres and other insoluble proteins encounters several difficulties.[14] The problem of the phase boundary has already been considered (p. 194) and has been identified with the inner part of the electrical double layer. This arises from the ionization of the carboxyl groups at the surface, which causes the fibre to acquire a negative charge. Such a charge sets up a membrane potential which influences the distribution of acid in the two phases, and may be identified with the membrane of the Donnan theory. The electrical double layer, the significance of which was not appreciated fully in the days of Donnan, Loeb and Procter, thus renders a membrane approach self-evident.

Elöd and Silva[37] and Speakman[1] originally advocated the application of the Donnon treatment to the acid titration of wool. This was considered later by Steinhardt and Harris,[14] although a complete treatment was not given until recently. The chief difficulty lies in assessing the volumes and concentrations within the fibre, and in the distinction between acid bound and acid merely dissolved in the fibre. In particular the volume of free water is difficult to estimate (see Chapter 4, p. 97). It is usual in the absence of a more satisfactory value to consider all the water, except that absorbed as a monomolecular layer, as solvent water. In some cases, however, the maximum regain value has been used. The membrane treatment was applied satisfactorily in its elementary form to the acid combination of silk by Morton[38] and to wool by Peters,[39] who showed that the effect of salt could be explained in this way.

The most complete treatment has been given recently by Peters and Speakman,[40] who calculate the concentration of free hydrogen ions within the fibre for given external acidities, and in this way construct an 'internal' pH curve. This may be derived from the complete expression of Donnan and Guggenheim[34] for the chemical potential of an ion of species i in solution:

$$\mu_i = \mu_i^0 + RT \log_e(i)\gamma_i + p\overline{V}_i$$

where γ_i is the activity coefficient and $\overline{V}_i$ the partial molar volume of the ion. By the equation of the chemical potentials of the acid in the two phases, the following equation for the titration of pure acid is obtained, where v is the volume of free water.

$$p\mathrm{H}_{int.} \simeq 2p\mathrm{H}_{ext} + \log_{10}\left(\frac{b}{v} + x\right)\gamma$$

$$\simeq 2p\mathrm{H}_{ext.} + \log_{10}\frac{b}{v}\cdot\gamma$$

where b is the amount of acid combined with the fibre and x is the external concentration of hydrogen ions. The coefficient γ is impossible to derive with certainty as it includes the activity coefficients of the anion and the activity of water in the two phases, and the molecular volumes of water and anion. Various assumptions were made including the arbitrary equation of all activity coefficients to unity. In all cases the dubious assumption was made that the partial molar volumes of the anion and a water molecule are similar.

By this procedure the 'internal' titration curve was constructed, and shown to be similar to that of a soluble protein with a pK value of 4·3 for both hydrochloric and sulphuric acids.

Peters and Speakman[40] show further that the Donnan membrane treatment could explain qualitatively the effect of adding neutral salt. For a constant external concentration of anion (X) it may readily be shown that:

$$p\mathrm{H}_{ext.} \simeq p\mathrm{H}_{int.} + \log_{10}(X) - \log_{10}\frac{b}{v}$$

Using this relation, all the experimental results are found to lie on the same internal titration curve (Fig. 6.13). In addition it follows that at half-neutralization

$$(p\mathrm{H})_{0.5} \simeq \log_{10}(X) + \mathrm{const.}$$

This relation was first derived experimentally by Steinhardt and Harris and follows directly from the Gilbert–Rideal theory. On using a more exact equation, however, the Donnan treatment predicts that $p\text{H}_{int}$ approaches $p\text{H}_{ext}$ as the salt concentration is increased. The $p\text{H}$-log (X) graph should therefore asymptotically approach a limiting value ($p\text{H}_{int}$.) of about 4·5, as originally observed by Steinhardt and Harris.

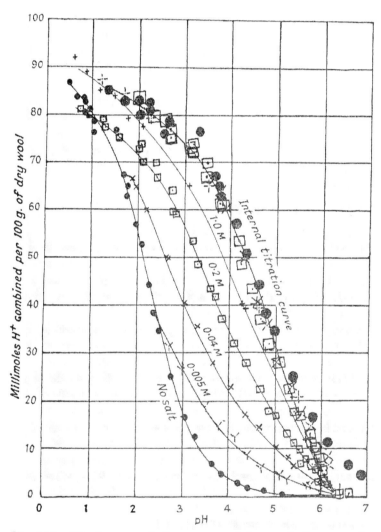

Fig. 6.13.—Titration curves in the presence and absence of salts compared with the internal titration curve calculated on the Donnan membrane theory.[39] (*Reproduced by permission.*)

Alexander and Kitchener[28] consider that although this treatment is fundamentally sound it does not contribute materially to an understanding of the process as claimed by Peters and Speakman. The Gilbert–Rideal treatment can explain most of the observed data without introducing arbitrary and indefinite quantities and the meaningless term 'internal pH' which cannot be measured.

A theoretical examination of the concentration of free hydrogen ions within the fibre is, however, of use when the nature and rate of certain reactions in the highly swollen fibres are considered. Thus an internal pH concept does explain the catalytic effect of salt on the acid hydrolysis of amide groups, and is of value in the interpretation of the rate of absorption of varying concentrations of acid (p. 167).

The chief disadvantage of a membrane treatment is, as already mentioned, the difficulty of interpreting the titration of an acid with a high anion affinity. The Gilbert–Rideal treatment on the other hand allows for quantitative values of affinities, and for this reason is the most commonly employed theory.

The absorption of weak acids

It has so far been assumed that the acid employed is completely dissociated, as most of the available experimental data deals with strong acids. Speakman and Stott[2] found that weak acids gave completely different titration curves in that much greater quantities of the weak acid are absorbed, particularly at low pH, in agreement with the earlier work of Meyer.[41]

This extensive absorption is accompanied by considerable swelling of the fibres (see p. 97), which was considered theoretically by Procter and Wilson.[35] The weak acid forms a highly ionized protein salt which represses the ionization of the free acid within the fibre. Thus y (p. 204) is small and $(x-y)$ and consequently e are large. For this reason the swelling increases with the total concentration of acid, and is not repressed by the addition of an excess, as is the case with strong acids. In the case of acetic acid, the swelling continues up to M solution, but formic acid shows a slight repression. Very weak acids, e.g. boric acid, show very little swelling.

It is important to note that although the swelling in H_2SO_4 is very small compared with formic acid, the latter is only twice as effective in facilitating fibre extension (see p. 213 ff).

The titration curves were examined more closely and extended by Steinhardt, Fugitt and Harris[42] and by Elöd and Fröhlich[43] who showed that the simple aliphatic carboxylic acids exhibit low anion affinities (Fig. 6.14). This can only be detected at high pH where the acids are largely in the ionized form. At low pH however, considerable quantities of acid are absorbed, and no tendency to reach a maximum can be found. Steinhardt *et al.* demonstrated that the

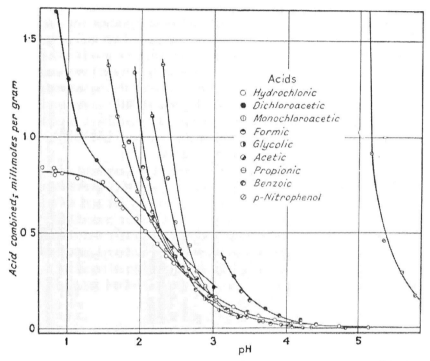

FIG. 6.14.—Combination of purified wool with a number of weak acids at 0° C.[42] The curve for hydrochloric acid is given for comparison. (*Reproduced by permission.*)

excess acid combined is proportional to the concentration of undissociated acid in solution, by (*a*) direct comparison with the hydrochloric acid curve, and (*b*) by varying the amount of acid combined at a given pH by the addition of small quantities of a common ion salt. The distribution coefficient thus derived is found to decrease with increasing absorption, due to the fact that the absorption must be limited as the fibres do not eventually dissolve.

Thus the position and form of the titration curve are determined by (*a*) this partition coefficient and (*b*) the strength of the acid, so that in general no correlation with the specific affinity of the anion exists except at high pH.

Most hydroxylic compounds, in particular phenols are absorbed extensively by wool and other fibres including nylon and cellulose acetate, and consequently the absorption is generally attributed to general solvation of the macromolecules due to hydrogen bonding. This view is supported by calculation of the heat of absorption from titration curves at two temperatures. For monochloracetic acid an extremely low value (200–300 cal./mol.) in comparison with values for strong acids (see p. 187) has been obtained. This indicates that the combination of the undissociated acid molecule involves the replacement of another hydroxylic compound (i.e. water) by selective solvation. Although the heat of hydration is large the replacement of a water molecule by a second molecule, held slightly more strongly, would involve only a small heat change. The partition coefficient thus gives a measure of the affinity of the solvating molecule for the wool compared with that of a water molecule.

The recent work of Elöd and Fröhlich[43] is in agreement with this explanation, but in addition these workers show that the absorption isotherm for weak acids is very similar for wool, silk and nylon, although the hydrochloric acid titration curves differ considerably for the three fibres. The excess absorption is therefore not connected with the salt link, but probably occurs on the carbonyl group of the polypeptide chain. This would explain the very great swelling of the fibres in such solutions, and the considerable effect on the X-ray pattern (see p. 91).

The titration of wool with alkali

In contrast to the extensive studies which have been made on the acid absorption process, relatively little attention has been paid to the alkaline titration. This is largely due to the difficulty of obtaining accurate data particularly at high pH, because of the simultaneous reaction of the disulphide bond. As discussed in some detail in Chapter 8, prolonged treatment with alkali leads to the formation of lanthionine with the elimination of one sulphur atom per cystine linkage. Harris and Rutherford[44] thought that alkali produces thiol and aldehyde groups in the wool with the elimination of sodium hydrogen sulphide. Thus on this basis, the measured uptake of alkali was corrected for the simultaneous degradation assuming the reaction of two sodium hydroxide molecules with each disulphide bond. The discovery of lanthionine and recent work which shows that the fate of the eliminated sulphur atom is unknown (see p. 259), indicate that Harris and Rutherford overcorrected for the removal of alkali by this reaction. These workers had previously suggested, as a result of their corrections, that the absorption of alkali reaches a maximum at *c*. 0·7 millimol per 100 g. of wool, but in view of later work and of the objections raised above, it must be concluded that no tendency to reach a saturation value can be detected (Figs. 6.2 and 6.5).

TABLE 6.3

*p*K values of basic amino groups[45]

Amino acid	Content milliequiv./100 g.	*p*K
Histidine	6·79	5·6– 7·0
Terminal NH₂	2·0–3·0	7·6– 8·4
Lysine	18·9	9·4–10·6
Arginine	59·7	11·6–12·6
Tyrosine	25·7	9·8–10·4

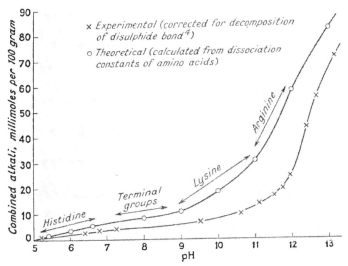

FIG. 6.15.—Comparison of experimental titration curve with computed curve assuming the dissociation constants of the individual basic amino acids.

As in the case of the acid titration curve, the experimental curve on the alkaline side is displaced considerably along the pH axis away from the point of neutrality with respect to the theoretical curve. This is illustrated in Fig. 6.15 which includes a theoretical curve computed from the following basic amino acid contents and pK values[45]: The forms of the two curves are similar as on the acid side, but the slope of the experimental curve is greater, due to the two-phase system. In spite of the extreme complexity, the theoretical curve is similar to the titration curve of a single base.

The ionization of the weakly acid groups of tryosine was omitted in the theoretical curve, as all the uptake of base can be accounted for by the basic amino acids. Anomalous behaviour has been observed with soluble proteins, as spectroscopic observations[46] show that hydroxyl ions do not react with the tyrosine in native protein until the pH exceeds 12·0, although the pK of the phenolic group in the

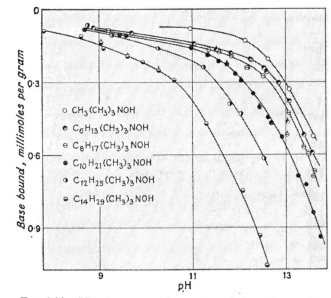

FIG. 6.16.—Titration curves of a number of strong bases with wool at 0° C.[48] (*Reproduced by permission.*)

free amino acid is 10. If the protein is denatured, however, the alkali is absorbed in the correct pH region. It is tentatively suggested that the increased sorption of alkali by wool[47] supercontracted in phenol without the disruption of disulphide bonds may be due to this cause.

Steinhardt, Fugitt and Harris[4] also investigated the effect of added salts on the titration curve and observed similar displacements to those observed in the acid region and it is highly probable that salts influence the position of the titration curve by a mass action effect, similar to that which is now well established in the acid process. In agreement with this conclusion, it has recently been shown by Steinhardt and Zaiser[48] that cations exhibit specific affinities for the fibres. Thus, the titration curves of twelve strong monovalent bases show wide differences in their position with respect to the pH axis (Fig. 6.16). In contrast to the effect of ions on the acid titration curve, it is observed that cations cause a more marked effect on a particular part of the curve than on the remainder. As shown in Fig. 6.16 tetradecyl trimethyl ammonium hydroxide produces the largest shift of 2 pH units in this region, i.e. approximately 1 pH unit more than the displacement in the other regions. Studies of the effect of formaldehyde (p. 327) on the alkaline titration curve have established a considerable displacement in the same region which has been attributed to the reaction with lysine.[49]

Steinhardt and Zaiser[48] observed that symmetrical tetrabutyl ammonium hydroxide is unique in that the usual maximum displacement of c. 2·5 pH units occurs when approximately 0·1 milliequiv. of base is bound, but no displacement is observed with larger combined amounts.

Influence of the "salt link" on elasticity

The general properties of the charged amino and carboxyl groups of wool were first investigated by Speakman and Stott,[2] who studied the changes in mechanical properties with changes in pH. In this way their influence on the structure of the protein was established before the detailed studies of acid absorption were made. Because of the almost linear relation between the amount of acid absorbed and the change in tensile strength,[1] Speakman concluded that the carboxyl and amino groups are close to each other, and form definite electrostatic bridges known as salt links. Similarly Stearn and Eyring[50] regard the salt link as a hydrogen bond between the two ionizable groups. In agreement with the postulate of bond formation, it was found that no change in elasticity occurs over the isoelectric region, pH 4–8, which was attributed to the stability of these electrostatic bonds (Fig. 6.17).

Further studies of the eleastic properties of wool which had been treated by the Van Slyke reagent to give almost complete deamination, supported this picture. The resistance of the deaminated fibres to extension was found to be independent of pH from 1–5 as the salt links were destroyed by the removal of amino groups.[51] At pH 5 however, the liberated carboxyl groups begin to titrate (Fig. 6.17) in a

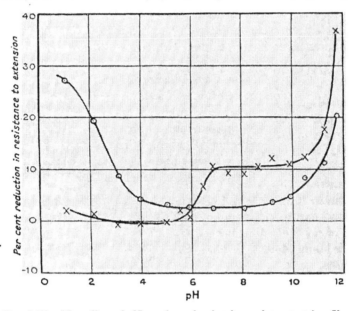

Fig. 6.17.—The effect of pH on the reduction in work to stretch a fibre by 30 per cent. for normal[2] (○) and deaminated hair[51] (×). (*Reproduced by permission.*)

similar way to the free carboxyl groups of gelatin.[52] This is demonstrated by the sharp decrease in strength, which in this case is not attributed to salt link breakdown but to the swelling which follows the combination of carboxyl groups with alkali. The anion has recently been shown by Speakman and Elliott[53] to affect the tensile strength of the fibre. Thus, the more complex the acid the smaller the reduction in work at all pH values (Fig. 6.18). This was attributed to increased cohesion of the chains due to the high affinity of the ion (see Chapter 6), although Speakman and Elliott consider the rapid increases in the 30 per cent. index as the sorption approaches saturation indicate micellar subdivision, which was postulated from independent data. This possibility is discussed in greater detail on p. 238.

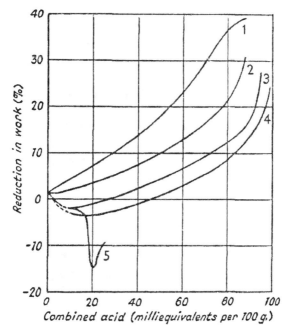

FIG. 6.18.—The effect of absorption of various acids on the reduction in the work required to stretch a fibre 30 per cent. in water.[53]
1. Hydrochloric acid
2. Toluene-p-sulphonic acid
3. Naphthalene-β-sulphonic acid
4. Anthraquinone-β-sulphonic acid
5. Free Acid of Orange II
(*Reproduced by permission.*)

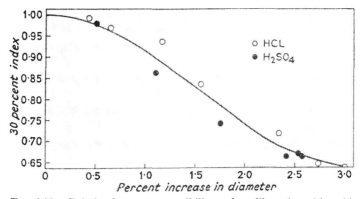

FIG. 6.19.—Relation between extensibility and swelling of wool in acid solutions of different concentrations.[2, 24] (*Reproduced by permission.*)

Sookne and Harris[54] do not agree with the contention of Speakman that definite electostatic bonds are formed in the fibre, and prefer to consider the changes in mechanical properties to be due to swelling. The theory of Procter and Wilson[35] shows that the absorption of acids causes increased swelling (Table 6.4) simply due to osmotic effects (p. 204), and Sookne and Harris, using the data of Speakman and Stott[2] given in Table 6.4, showed that the reduction in work required to stretch a fibre is related directly to the extent of swelling (Fig. 6.19). As, therefore, the quantity of acid or base absorbed, the consequent swelling, and the changes in elasticity are so intimately connected, it is difficult to establish conclusively which of these is the most fundamental factor.

TABLE 6.4

The effect of absorption of hydrochloric acid on lateral swelling and tensile strength[1]

Swelling medium	Percentage swelling	Reduction in work for 30 per cent. extension
Water	100	1·2
pH 5	100	1·2
4	100·5	1·4
3	101	9·4
2	102	24·3
1	103	33·2
0	101	33·8

Harris based his criticism largely on the effect of neutral salts on the swelling of the fibres. Speakman and Hirst[1] showed that the addition of potassium chloride affected the resistance to extension slightly in the complete pH range, and also depressed the lateral swelling. In agreement with the effect of salts on the swelling of hydrophilic colloids in general, which is given by the Hofmeister[55] series, Sookne and Harris[54] found that salts could either increase or decrease the elasticity, depending on the nature of the anion (Table 6.5).

The close relation between elastic changes and swelling was taken by these authors to prove the fundamental nature of the latter in the absorption. The mechanism of the salt action has not been fully explained by the Donnan theory. It is likely that the action is controlled by several individual effects including the extent of hydration

TABLE 6.5
Effect of salts on the tensile strength[54]

Salt N. solution	Change in tensile strength expressed as 30 per cent. index
K_2SO_4	1·03
$KOOC.CH_3$	1·02
KCl	1·01
KBr	0·97
KI	0·90
KCNS	0·83

of the ions, and the specific affinity of the anions for polar groups within the fibre relative to water. The recent results of Speakman and Whewell[56] on the effect of concentrated sodium chloride solutions, demonstrate that the wool fibres are actually weakened by the presence of the salt compared with normal wool. They show that the difference in vapour pressure between water and saturated sodium chloride solution should lead to a 42 per cent. increase in fibre strength, when a fibre is transferred from the first solution to the second, whereas the observed rise is only 11 per cent. Thus the effect of salts on the elasticity cannot be explained entirely by a desiccant effect and is probably due to electrostatic effects.

The complexity of the action was demonstrated by Sookne and Harris,[54] who found that the relation between swelling, caused by neutral salts, and elasticity is entirely different from the relation between the osmotic swelling and elasticity shown in Fig. 6.19. In addition, the presence of concentrated salt solutions has recently been shown to produce considerable hysteresis in the extension-relaxation cycle.[57] These difficulties do not of course support or oppose either theory, but merely show that the swelling caused by acids and neutral salts is controlled by two different mechanisms.

In further support of the swelling theory, Sookne and Harris[54] showed that the reduction in work required to extend fibres by 30 per cent. in N/10 hydrochloric acid solution saturated with potassium chloride is less than in water. Speakman[58] considers, however, that this comparison should have been made with saturated potassium chloride solution and not water, when it is found that the fibres are much more resistant to stretching than in the acid medium (Fig. 6.20).

Further studies on the change in the work of stretching in acid solution, fibres which have been pretreated in various ways support the swelling mechanism. Thus, fibres which have been treated by reagents which destroy the disulphide bonds, thereby increasing the

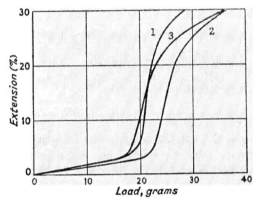

Fig. 6.20.—The effect of acid and salts on the elasticity of wool fibres in water.[58]
1. Water
2. Water saturated with NaCl
3. N/10 HCl saturated with NaCl

TABLE 6.6

The influence of fibre swelling on the contribution of the salt links to the fibre strength[59]

Treatment	Change in tensile strength expressed as 30 per cent. index
Nil	38·0
20% cystine oxidized*	20·0
50% cystine oxidized*	3·0
Fibre in 3% phenol solution	18·0
Fibre in 25% LiBr	11·0

* With peracetic acid.

swelling, are weakened by acid to a much smaller extent than untreated wool.[59] Preliminary swelling with phenol or concentrated solutions of lithium bromide has a similar effect (Table 6.6), and it is difficult to understand why neutral salts can disrupt a particular bond of the type envisaged by Speakman. It was also found by Alexander[59] that strong solutions of potassium chloride increase the work required to stretch the fibre, but as the concentration of salt increases the difference between the elasticity at pH 1 and pH 7 gradually disappears (Table 6.7).

TABLE 6.7

Influence of salt concentration on mechanical properties[59]

Concentration of KCl N.	Change in tensile strength expressed as 30 per cent. index	
	KCl at pH 7	KCl in 0·1N . HCl
Nil	98	60
1·0	97	63
2·0	101	69
4·0	105	87
6·0	109	96

These observations are more readily explained by assuming a more or less statistical distribution of NH_3^+ and COO^- groups which is probably the case with soluble proteins,[60] so that high salt concentrations reduce the interaction of charged groups through the breakdown of the double layer between them. In the absence of salt or in the presence of only dilute solutions, the two charged groups are closer together than the thickness of the double layer (*c.* 20 Å) and therefore mutually attract each other. In high salt concentrations, the double layer is reduced to the thickness of the order of magnitude of the average distance between COO^- and NH_3^+ groups. Then as the interaction between them is reduced, the effects resulting from the combination of COO^- groups and protons is also reduced (see Tables 6.6 and 6.7).

REFERENCES

[1] SPEAKMAN AND HIRST, *Nature*, 1931, **127**, 665, *Trans. Faraday Soc.*, 1933, **29**, 148; see also SPEAKMAN, *J. Text. Inst.*, 1941, **32**.
[2] SPEAKMAN AND STOTT, *Trans. Faraday Soc.*, 1934, **30**, 539.
[3] ADAMS, *J. Amer. Chem. Soc.*, 1916, **38**, 1503; BJERRUM, *Z. physik. Chem.*, 1923, **104A**, 147.
[4] STEINHARDT, FUGITT AND HARRIS, *J. Res. N.B.S.*, 1940, **25**, 519.
[5] HARNED AND OWEN, *Chem. Rev.*, 1939, **25**, 31.
[6] COHN, *Ergeb. Physiol.*, 1931, **33**, 781.
[7] STEINHARDT, FUGITT AND HARRIS, *J. Res. N.B.S.*, 1942, **28**, 201.
[8] ASTBURY, *J. Amer. Chem. Soc.*, 1944, **66**, 339.
[9] See the discussion between SPEAKMAN, ELÖD, J. LLOYD, PETERS AND NEALE, *Trans. Faraday Soc.*, 1933, **29**, 165.
[10] SOOKNE AND HARRIS, *J. Res. N.B.S.*, 1939, **23**, 471.
[11] KEKWICK AND CANNAN, *Biochem. J.*, 1936, **30**, 227.
[12] SPEAKMAN AND STOTT, *Trans. Faraday Soc.*, 1935, **31**, 1424.
[13] LLOYD AND BIDDER, *ibid.*, 1934, **31**, 864.
[14] STEINHARDT AND HARRIS, *J. Res. N.B.S.*, 1940, **24**, 335; *Text. Res. J.*, 1940, **10**, 269.
[15] STEINHARDT, FUGITT AND HARRIS, *J. Res. N.B.S.*, 1941, **26**, 293.
[16] LEMIN AND VICKERSTAFF, *Fibrous Proteins Symp.*, *Soc. Dy. & Col.*, 1946, 129.
[17] HARRIS AND RUTHERFORD, *J. Res. N.B.S.*, 1939, **22**, 535.
[18] WYMAN, *Chem. Rev.*, 1936, **19**, 213; *J. Biol. Chem.*, 1939, **127**, 1.
[19] SPEAKMAN AND STOTT, *Trans. Faraday Soc.*, 1938, **34**, 1203.
[20] DALTON, KIRK AND SCHMIDT, *J. Biol. Chem.*, 1930, **88**, 589. WINNEK AND SCHMIDT, *J. Gen. Physiol.*, 1934–35, **18**, 889.
[21] HARNED AND HAMER, *J. Amer. Chem. Soc.*, 1933, **55**, 2194.
[22] CANNAN, KILBRICK AND PALMER, *Ann. N.Y. Acad. Sci.*, 1941, **41**, 243. CANNAN, *Chem. Rev.*, 1942, **30**, 395.
[23] ERDAS AND SNELLMAN, *Biochim. et Biophys. Acta*, 1948, **2**, 642.
[24] HARRIS, *J. Res. N.B.S.*, 1932, **8**, 779.
[25] HARRIS AND SOOKNE, *J. Res. N.B.S.*, 1941, **26**, 289.
[26] CANNAN, *Cold Spring Harbour Symposia in Quantitative Biology*, 1938, **6**, 1.
[27] VERWEY, *Chem. Rev.*, 1935, **16**, 363; VERWEY AND OVERBEECK, *Theory of the Stability of Hydrophobic Colloids*, Elsevier, Amsterdam, 1948.
[28] ALEXANDER AND KITCHENER, *Text. Res. J.*, 1950, **20**, 203.
[29] ALEXANDER AND KITCHENER, *J. Soc. Dy. Col.*, 1949, **65**, 284.
[30] SIMMS, *J. Amer. Chem. Soc.*, 1926, **48**, 1239; WEBER, *Biochem. Z.*, 1927, **189**, 381; VON MURALT, *J. Amer. Chem. Soc.*, 1930, **52**, 3518.
[31] GILBERT AND RIDEAL, *Proc. Roy. Soc.*, 1943, **182A**, 335.
[32] BAXTER, *J. Colloid Science*, 1949, **2**, 495.
[33] FOWLER AND GUGGENHEIM, *Statistical Thermodynamics*, Cambridge, 1939.
[34a] DONNAN, *Z. phys. Chem.*, 1934, **168A**, 369; [b] DONNAN AND GUGGENHEIM, *ibid.*, 1932, **162A**, 346.
[35] PROCTER AND WILSON, *J. Chem. Soc.*, 1916, **109**, 307; PROCTER, *ibid.*, 1914, **105**, 313.
[36] LOEB, *Proteins and the Theory of Colloidal Behaviour*, N.Y., McGraw Hill, 1922.
[37] ELÖD AND SILVA, *Z. phys. Chem.*, 1928, **137A**, 142.
[38] MORTON, *Fibrous Proteins Symposium. Soc. Dy. Col.*, Bradford, 1946, p. 155.
[39] PETERS, *ibid.*, p. 138.
[40] PETERS AND SPEAKMAN, *J. Soc. Dy. Col.*, 1949, **65**, 63.
[41] MEYER AND FIKENTSHER, *Melliand Textilber.*, 1927, **8**, 781.
[42] STEINHARDT, FUGITT AND HARRIS, *J. Res., N.B.S.*, 1943, **30**, 123.
[43] ELÖD AND FRÖLICH, *Melliand Textilber.*, 1949, **30**, 239, 405.
[44] HARRIS AND RUTHERFORD, *J. Res. N.B.S.*, 1939, **22**, 535.
[45] COHN AND EDSALL, *Proteins, Amino Acids and Peptides*, Reinhold, New York, 1943.

[46] CRAMER AND NEUBERGER, *Biochem. J.*, 1943, **37,** 302; BEVAN, HOLIDAY AND JOPE, *Faraday Soc. Discussions*, 1950, **9,** 406.
[47] ALEXANDER, unpublished.
[48] STEINHARDT AND ZAISER, *J. Biol. Chem.*, 1950, **183,** 789.
[49] STEINHARDT, FUGITT AND HARRIS, *ibid.*, 1946, **165,** 285.
[50] EYRING AND STEARN, *Chem. Rev.*, 1939, **24,** 253.
[51] SPEAKMAN AND STOTT, *Nature*, 1938, **141,** 414.
[52] ATKIN AND DOUGLAS, *J. Soc. Leather Trades Chemists*, 1924, **29,** 148.
[53] SPEAKMAN AND ELLIOTT, *Fibrous Proteins Symposium, Soc. Dy. Col.*, Bradford, 1946, p. 116.
[54] SOOKNE AND HARRIS, *J. Res. N.B.S.*, 1937, **19,** 535.
[55] HOFMEISTER, *Arch. expt. Path.*, 1888, **25,** 13; 1891, **28,** 219.
[56] SPEAKMAN AND WHEWELL, *J. Textile Inst.*, 1950, **41,** 329.
[57] BOGATY, SOOKNE AND HARRIS, *Text. Res. J.*, 1951, **21,** 479.
[58] SPEAKMAN, *Amer. Dy. Rptr.*, 1938, **27,** 168.
[59] ALEXANDER, *Kolloid Z.*, 1951, **122,** 8.
[60] JACOBSEN AND LINDERSTRØM-LANG, *Nature*, 1949, **164,** 411

CHAPTER 7

Ion Exchange and Dyeing Equilibria

Evidence for anion affinity

SPEAKMAN and Hirst, who were among the first workers to determine complete titration curves, found that several simple acids followed approximately the same titration curve. Thus above pH2 the extent of absorption of hydrochloric, phosphoric, sulphuric, hydrobromic and ethylsulphuric acids appears to be almost independent of the nature of the anion. Such behaviour is to be expected from the Donnan theory, which considers the anion in the internal phase to exist in the same state as in aqueous solution. Thus no provision is made for any observed specific affinity of the anion for the fibre. It should be stressed that the term 'affinity' refers strictly throughout this and the preceding chapter to the thermodynamic property controlling the equilibrium, and is not connected in any way with the rate process, although the two factors are frequently confused in the technical sphere.

Meyer[2] observed that different acids may be absorbed to different extents, and the extensive results obtained by Steinhardt, Harris, *et al.*[3] on the same subject show that the quantity of hydrogen ion combined at a given pH is highly dependent on the nature of the anion. From the assembly of curves already given in Fig. 6.3. and Fig. 7.1, it may be seen that the more complex the anion the greater the affinity of the acid for the fibre (see also Table 7.1, p. 227). Thus any theory of acid combination must make provision for the effect of the affinity, specific to the anion, for particular sites within the fibre. The original theory of Steinhardt and Harris was designed partly for the calculation of anion affinities. This was possible because the anion was supposed to combine with amino groups to form a salt which in the case of ions of low affinity was highly dissociated. Thus the reciprocal of the equilibrium coefficient K_A^- for this dissociation gives a measure of the anion affinity.

The thermodynamic treatment developed by Gilbert and Rideal assumes the affinity of both ions contributes to the sorption process. This provides the simplest method of representing anion affinity in

ION EXCHANGE AND DYEING EQUILIBRIA 223

terms of chemical potentials, and is now widely accepted. It is convenient to adopt a reference standard against which the affinity of any ion may be compared. The chloride ion is usually chosen for this purpose as several considerations indicate that this ion has negligible specific affinity for the fibre. Thus the heat of absorption of hydrochloric acid is extremely small and comparable with the heat of dissociation of carboxyl groups (see p. 186). In addition, the Gilbert–Rideal theory leads to an equilibrium constant for the absorption of

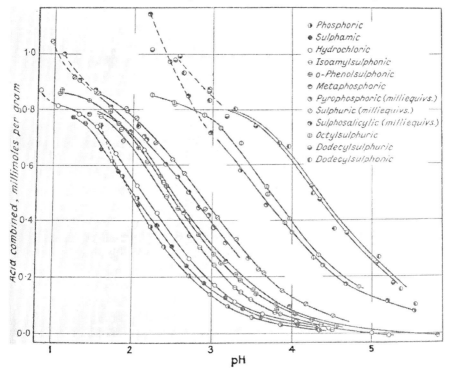

FIG. 7.1.—Combination of wool with twelve different strong acids.[3]
(*Reproduced by permission.*)

hydrochloric acid in close agreement with the value of the dissociation constant of a carboxylic acid ($pK \sim 4\cdot0$). Thus the general expression for the titration curve (p. 201) may be written as

$$RT \log_e a_{\mathrm{H}} a_{\mathrm{A}} \frac{(1-\theta)^2}{\theta^2} = \Delta\mu_{\mathrm{H}}^0 + \Delta\mu_{\mathrm{A}}^0 = RT \log_e K$$

At the point of half-neutralization, i.e. when $1-\theta/\theta = 1$, $a_{\mathrm{H}} a_{\mathrm{A}} = K$

or
$$(p\text{H})_{0.5} = \log_{10} a_A - \log_{10} K$$

For the titration of pure acid, $pK = 2(p\text{H})_{0.5}$.

Assuming that the chloride ion has no affinity for the fibre, the pK of the carboxyl groups will equal $2(p\text{H})_{0.5}$ which in the case of hydrochloric acid, leads to a value of 4·4, in close agreement with the value for single carboxylic acids.

It seems, therefore, justifiable to assume that the chloride ion has no specific affinity for the fibre.[4] A few acids appear to have even less affinity than hydrochloric acid (e.g. phosphoric, formic and sulphamic acids). This may be ascribed to the high affinity of the acid (in the case of the first two at least) for liquid water, with which it is hydrogen bonded. This extensive solvation is expected to be somewhat less within the fibre than in aqueous solution, and leads to an abnormally low anion affinity.

Anion affinity is also demonstrated by the supra-stoichiometric combining capacity of many complex acids (see p. 185). Early results showing a maximum combining capacity in excess of the normal value of 82 milliequivalents/100 g. fibre were usually attributed to experimental error, although similar results were obtained with hide collagen.[5] The existence of a second step in the titration curve at very low pH values (see Figs. 5.2 and 5.3) has been attributed to the titration of more weakly basic groups (amide groups) or to absorption by a different mechanism.[3] The position is, however, further complicated by hydrolytic decomposition of amide groups and possibly the peptide chains, particularly by acids of high affinity.[4] Steinhardt, Fugitt and Harris consider the absorption of Acid Orange II to be too great to be explained in this way, and assume the direct absorption of dye on specific sites. In support of this explanation, Skinner and Vickerstaff[6] observed the sorption of as much as 1·3 milliequivalents/g. of Disulphine Blue V and 2·1 m.eq./g. of Carbolan Blue at pH 2·2. Polar Yellow[7] also exhibits considerable suprastoichiometry. That specific anion affinity is responsible for the increased absorption in such cases is supported by the observations that considerable quantities of dye acid (0·18 and 0·29 m.eq./g. of Disulphine Blue V and Carbolan Blue, respectively) are absorbed at pH 6·1, i.e. close to the isoionic point (the point of zero acid-combining capacity) (see p. 184). Even Acid Orange II is absorbed to some extent by neutral wool, in the presence of salt.[8] In addition, many neutral-dyeing wool dyes are adsorbed at neutrality by cotton which has no basic groups. This is due to the affinity of the large anion for the macromolecule, and by analogy it would follow that such affinity also contributes to the uptake of large acid dyes by wool. These direct dyes exhibit such high affinities (see Table 7.3) that they behave as potential-determining ions independently of the stoichiometric proton combination. The uptake of dye as a leading ion follows entirely different laws and is aided, not prevented, by the addition of a salt (see p. 239).

Electrokinetic evidence also gives some indication of the extent of anion affinity. Although some simple salts, e.g. KCl, influence the electrokinetic properties only by reducing the width of the double layer, salts with larger anions exhibit pronounced specific effects. Sookne and Harris[9] obtained large shifts of the isoelectric point for wool and silk in phthalate buffers and smaller shifts in concentrated acetate buffers.

A lowering of the electrokinetic potential by such salts might possibly be ascribed to the sorption of anions on top of protons in the manner postulated by the Stern theory, with a small proportion of the gegen-ions remaining in the diffuse part of the layer. However, the shift of the isoelectric point of wool which occurs on the addition of large ions cannot be explained in this way. Thus, if phthalate is added to isoelectric wool, direct sorption of the phthalate ion renders the surface negative, and therefore more protons are required (i.e. a lower pH) to restore it to the isoelectric condition. The existence of an isoelectric point for cotton,[10] which only contains acidic groups yet becomes positively charged below pH 2·5, proves that absorption of cations must occur by non-Coulombic forces.

The affinity of small anions is very small compared with the proton affinity, e.g. the displacement of the isoelectric point of wool produced by 0·1N sodium acetate is counterbalanced by the addition of 5×10^{-6}N hydrogen ions. The apparent anion affinity is thus roughly 2×10^4 times smaller than the proton affinity.[42] It is known that the acetate ion has a very low affinity (p. 208), nevertheless anion affinity increases so rapidly with size that for dye ions it is comparable with or even greater than proton affinity. (Table 7.3.)

Determination of anion affinity

Quantitative values of anion affinities were first calculated by Steinhardt and Harris by application of their general equation for the titration curves (see p. 197). If the mid-point of the titration curve only is considered, i.e. where $\theta = 0.5$, the following two relations applying to ions of low and high affinity respectively are obtained.[3, 11] The symbols here have the same significance as in Chapter 6, p. 197.

$$K_A' = a_H \frac{(a_H + K_H)}{K_H'} - a_H = \frac{a_H^2 - a_H K_H'}{K_H'} - \frac{K_H}{K_H - a_H} \quad \text{(low affinity)}$$

and $\quad K_A' = \dfrac{K_H}{K_H'} - \dfrac{a_H(a_H + K_H')}{a_H + K_H} \quad$ (high affinity)

In practice, the factor $K_H/K_H - a_H$ approaches unity for all the acids studied, and as $K_A' > 1$, the first equation may be reduced to

$$K_A' \simeq \frac{a_H^2 - K_H' a_H}{K_H}$$

The simplification of the second equation when $a_A \ll K_H$ is

$$K_A' \simeq \frac{a_H^2 + K_H' \cdot a_H}{K_H'}$$

The evaluation of K_A' representing the reciprocal of the anion affinity as already mentioned, requires a knowledge of K_H', the intrinsic acidity of the carboxyl groups within the fibre. Both equations hold for titration curves whose mid-points lie in a low pH region but above pH 4·2 (i.e. pK′ at 0⁰ C.) the first equation can no longer hold. This equation could therefore be used to calculate the affinities in Table 7.1. These affinities are represented in the logarithmic form to bring them into line with the values obtained from the Gilbert–Rideal theory also included in this table.

A further simplification may be introduced for ions of great affinity when the difference in the positions of the curves produced by a given increase in affinity is so large that ΔpH approaches $-\Delta pK_A$. Under these conditions the affinity constant becomes independent of K_H and is equal to the value of a_H at the mid-point. This is of importance as the affinities of commercial dyes lie in this range, and consequently a comparison of their affinities becomes extremely simple.

Affinities are usually calculated in a much simpler manner from the thermodynamic treatment of Gilbert and Rideal. The foregoing explanation of the Steinhardt and Harris method, together with a collection of their affinity values, has been given, because the most extensive accurate experimental work on acid absorption was carried out by these workers who interpreted all their results in this way.

TABLE 7.1

Affinity of acids for wool at 0° C.[3, 11]

Acid	pH at half saturation	Affinity k. cal./mol.	
		$-RT \log K_A$*	$-\Delta\mu$†
Phosphoric	2·16	0·12	—
Sulphamic	2·22	0·20	—
Hydrochloric	2·32	0·43	—
Ethylsulphuric	2·33	0·44	0·20
Hydrobromic	2·47	0·69	—
Nitric	2·58	0·89	—
Isoamylsulphonic	2·58	0·89	—
Benzenesulphonic	2·63	1·00	0·90
p-Toluenesulphonic	2·66	1·04	1·00
o-Phenolsulphonic	2·66	1·04	—
o-Xylene-p-sulphonic	2·71	1·12	1·20
Metaphosphoric	2·72	1·14	—
Trichloracetic..	2·73	1·16	1·20
o-Nitrobenzenesulphonic	2·86	1·39	1·45
4-Nitrochlorobenzene-2-sulphonic ..	3·07	1·76	2·25
Sulphuric	3·08	1·78	0·76
2, 5-Dichlorobenzene sulphonic ..	3·13	1·87	2·10
2, 4-Dinitrobenzene sulphonic ..	3·17	1·94	2·25
Naphthalene-2-sulphonic	3·24	2·06	—
2, 4, 6-Trinitroresorcinol	3·64	2·73	3·35
Picric	3·86	3·08	3·95
Flavianic	4·24	3·67	—

*By the Steinhart and Harris method. †By the Gilbert and Rideal method.

There are several ways of determining affinity thermodynamically, of which the most obvious is the direct substitution for θ, (H⁺) and (A⁻) in the appropriate equation relating to the titration in the presence or absence of salt. This serves at the same time to examine the validity of the Gilbert–Rideal theory over the complete titration range in the particular case considered. It is found that, except at very low or very high pH values, the calculated affinity for hydrochloric acid and the simpler acids remains almost constant and a similar value is obtained for various initial salt concentrations (Table 7.2). With more complex acids and dyes, however, significant deviations are noted (Table 7.4).

TABLE 7.2

Affinity of hydrochloric acid for wool at 0° C.[17]

	Acid alone		Constant Ionic Concentration, 0·1N		
pH	θ	$-\Delta\mu$ (kg. cal.)	pH	θ	$-\Delta\mu$ (kg. cal.)
1·09	0·927	5·43	1·09	0·929	5·37
1·47	0·940	(6·62)	1·38	0·935	5·83
1·66	0·830	5·83	1·68	0·908	5·78
1·71	0·793	5·68	1·99	0·858	5·64
1·74	0·770	5·63	2·27	0·810	5·62
1·93	0·695	5·68	2·54	0·735	5·48
2·06	0·622	5·65	2·81	0·655	5·42
2·28	0·549	5·88	3·21	0·544	5·42
2·34	0·475	5·70	3·65	0·435	5·50
2·45	0·427	5·75	4·21	0·305	5·57
2·73	0·305	5·86	4·62	0·214	5·56
3·04	0·207	6·10	5·03	0·149	5·60
3·27	0·134	6·10	5·47	0·092	5·55
3·62	0·082	6·38	—	—	—
3·84	0·055	6·45	—	—	—

For this reason and also for convenience, the approximate affinity is usually determined from the mid-point of the titration curve. This also serves as a rough method of compensating for the impossibility of introducing activity coefficients in the majority of cases. Thus from the general equation,

$$(pH)_{\theta=0.5} = \log_{10}(Cl^-) - \frac{(\Delta\mu_H^0 + \Delta\mu_{Cl}^0)}{2\cdot303\,RT}$$

If the pure acid only is employed this reduces further to,

$$-\Delta\mu^0 = 4\cdot6\,RT(pH)_{\theta=0.5}$$

If a solution containing equal concentrations of two acids of low and high anion affinity (e.g. a dye) are allowed to come to equilibrium with a sample of wool, the proportion of each anion absorbed is related directly to the anion affinity of the two ions. Alternatively, if a sample of wool saturated with a given acid is placed in a solution of equivalent acidity of a second acid, the extent of the displacement of the one anion by the other is governed by the relative anion affinity. Estimations of the extent of the resulting displacement can then be used to determine the affinity of one of the ions. The quantitative treatment will become clear after the general process of anion exchange has been considered in detail.

Nature of anion affinity

Little is known of the forces responsible for the affinity of the anion for the fibre, due mainly to the lack of data on the affinities of series of similar ions, and also to the molecular complexities of the ions which exhibit high affinities. It is seen clearly from the data of Table 7.1 that a steady increase with molecular weight and size of ion is prevalent, although various groups exhibit considerable additional effects. It is fairly certain that a uniform relation exists between structure and affinity as, in the few cases so far studied, the influence of a particular group on the affinity is roughly additive. Thus, using the data of Speakman and Clegg,[12] Gilbert showed that the replacement of a benzene ring by a naphthalene residue causes a constant increase in $\Delta\mu$ and, in three out of four cases, the introduction of a second sulphonic acid group again causes a uniform change.[13] Steinhardt and Harris point out several cases where the affinity appears to be dependent on molecular weight only, although the nature of the substituted group is changed considerably. Thus o-phenolsulphonic and p-toluenesulphonic acids have equal molecular weights and affinities, and similarly the affinities of xylene, nitrobenzene, nitrochlorobenzene, and dinitrobenzene sulphonic acids are consistent with molecular weight differences only. Lemin[14] has also found that increasing the length of the carbon chain ($n = 1$ to $n = 7$) of a simple azo dye increases the affinity linearly. In addition, dibasic acids have greater affinities than the corresponding monobasic acids. Similar results were obtained for affinities of cations on synthetic resins,[15] and would be expected on general grounds, due to increased electrostatic attraction.

Steinhardt, Fugitt and Harris[3] have suggested that as dispersive forces are probably the most predominant, the area of the ion which can be presented to the absorption site is also of importance. Thus, in general, planar ions would be expected to be absorbed more readily than the more spherical ions. Examination of the available data reveals a marked tendency for ions which can adopt a planar configuration to be absorbed more strongly. Hydrochloric and hydrobromic acids have small affinities compared with nitric acid. Ethylsulphuric acid has an affinity of the same order as hydrochloric acid whereas benzenesulphonic acid has a high affinity due to the benzene ring. The relative affinities of various stereoisomers support this suggestion.

With dyes of the type ⟨◯⟩—C(H)=C(H)—⟨◯⟩ the *trans*-configuration, which is planar, has a much greater affinity than the *cis*-compound.[13, 16] Many substantive dyes are largely planar in structure, although Vickerstaff[17] points out that a completely planar structure is not essential for high affinity, e.g. Coomassie Milling Scarlet 2G (benzidine 2:2-disulphonic acid).

Although the available data are limited, the conclusion seems justified that both the molecular weight and the geometrical structure influence the affinity. This does not, however, provide a complete explanation, as some groups (in particular hydroxyl and amino groups) contribute specifically to the affinity.

It is found that the specific anion affinities depend on the nature of the solid phase, and different orders are exhibited with various fibres and other absorbents.[17] Other forces probably operating include dipole–dipole interaction and hydrogen bonding. The high affinity of all phenols supports the latter suggestion. In addition, Pfeiffer[18] has isolated stable molecular compounds of diketopiperazine and characteristic dye nuclei such as *p*-hydroxyazobenzene, *p*-aminoazobenzene and simple dyes such as Orange II. Pfeiffer also showed that the free hydrogen atoms of the amino and hydroxyl groups were essential for the bonding. Assuming diketopiperazine to be representative of the polypeptide chain, the association most probably occurs by hydrogen bonding with the CO group:

$$\begin{array}{l} |\\ \text{NH}\\ |\\ \text{C}=\text{O}\cdots\text{H}-\text{O}-\langle\bigcirc\rangle-\text{N}=\text{N}-\\ |\\ \text{CH}_2\\ | \end{array}$$

Finally, dyes with high surface activity have been shown to have considerable affinity for the fibre.[19] Thus dyes of the Carbolan type rely largely on the long paraffin chain for the very rapid absorption on the surface of the fibre at room temperature, no such behaviour being detected with the dye Azogeranine of similar structure but without the side chains. It is unlikely that the very great affinity of Carbolan Crimson over Azogeranine can be explained entirely by Van der Waals' forces. Surface absorption on the large internal surface of the fibre, as a result of interfacial tension lowering, may make a large contribution to the affinity.

Anion exchange and dyeing equilibria

The absorption of a mixture of acids in the presence or absence of salt, although more complex than the absorption of a single anion, may be interpreted on the equilibrium theories developed earlier. It has already been stressed that the extent to which a particular anion is absorbed is controlled entirely by pH, concentration and the relative anion affinities. Thus if an equimolecular solution of hydrochloric and hydrobromic acids is used, almost equal amounts of the two anions will be absorbed owing to their low affinities. If, however, the hydrobromic acid is replaced by an acid of high affinity, e.g. benzene- or naphthalene-sulphonic acid, or particularly by a dye acid, the quantity of this second ion absorbed will be much greater than the quantity of chloride; moreover the total acid combined, at any pH value except at saturation, is increased.

The extent to which the two anions compete independently for available sites on the fibre has been well illustrated experimentally by Steinhardt, Fugitt and Harris.[20] As the pH decreases and the number of hydrogen ions absorbed increases, the amount of the ion of high affinity also increases regularly (Fig. 7.2). The amount of the second ion combined, however, increases at first and then rapidly decreases to zero at high acidities, so that the dye (in this particular case) is absorbed almost exclusively. The pH at which this change occurs and the maximum quantity of the ion of low affinity absorbed, depends on its concentration and affinity. Thus from Fig. 7.2 it is seen that naphthalenesulphonic acid exhibits a much greater effect than hydrochloric acid.

The problem is merely a specific case of the process of ion exchange on a weak acid exchanger,[21] such that the capacity is affected by the pH and the nature of the absorbed ions except at very low pH. The mechanism of ion exchange has been studied in great detail recently, particularly cation exchange on strong acid resins. Boyd, Schubert and Adamson[22] have interpreted the process in terms of the independent absorption of competing ions on specific sites of equal potential according to the Langmuir isotherm. Kitchener and Kressman[15] and others have shown that ion exchange follows the classical laws of mass action, and that an approximately constant value of K may be obtained. This is strictly true only for a limited concentration range

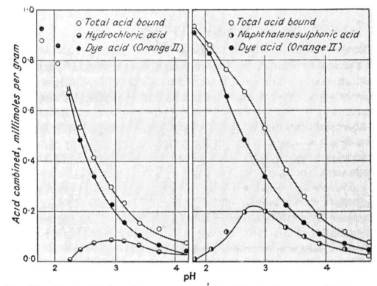

Fig. 7.2.—Total acid, dye acid, and second acid combined with wool at 50° C. when dye acid and second acid are initially present in equimolar proportions.[20]
(*Reproduced by permission.*)

as shown by Duncan and Lister.[23] Recent work by Gregor[24] and by Pepper, Reichenberg and Macauley[25] shows that the law of mass action only holds for slightly cross-linked polymers, considerable deviations resulting on extensive cross-linking. A complete explanation is not known, although it is probably due to the effect of ionic strength on swelling. Glueckauf and Duncan[26] have developed a thermodynamic treatment, whereby the activity coefficients in the fibre are related to the swelling pressure.

Dyeing

The process of acid dyeing may thus be regarded in the first instance as a simple ion exchange process involving two species of anions with very high and very low affinities, respectively. When simple (levelling) dyes are employed, a strongly acid dyebath is always employed (3% H_2SO_4) together with a large concentration of salt (e.g. 10% Glauber's salt). As already mentioned, the cations have no affinity for the fibre, but the excess of inorganic anions introduced by the salt addition enhances the competition of the inorganic ion with the dye anions. The dyeing conditions are thus represented approximately by the conditions given by Steinhardt, Fugitt and Harris[20] at pH $c.$ 1·3. It will be observed that at equilibrium an appreciable concentration of dye ions will remain in solution whereas in the absence of salt the concentration would be negligible. (Fig. 7.4).

The process of dyeing is also highly dependent on the rate of attainment of equilibrium. This has been considered in detail elsewhere (Chapter 5), but for the purpose of illustrating the dyeing mechanism will be considered briefly here. Although (and partly because) the dye has a much greater affinity, it will migrate much more slowly than the chloride ion through the fibre (see p. 160). On immersing wool in the isoelectric condition into a solution of acid, hydrogen ions are rapidly adsorbed followed almost exclusively by the mobile chloride ions until equilibrium is virtually reached. Simultaneously the dye gradually penetrates, exchanging with the chloride ions which are returned to the solution, until a true equilibrium is reached governed by the relative concentration and affinities of the

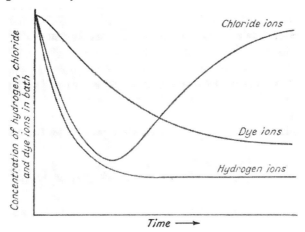

FIG. 7.3.—Representation of the dyeing process based on the data by Elöd.[27]

two ionic species in solution. This process is described completely by an elegant experiment of Elöd[27] who measured the change in concentration of chloride, hydrogen and dye ions from a dyebath containing the sodium salt of the dye (Crystal Ponceau, a levelling dye) and hydrochloric acid (Fig. 7.3). The significant feature of this demonstration is that the preliminary absorption of HCl satisfies the electrical requirements of the titration, but that owing to the difference in free energy of solution and fibre, the slow exchange occurs. This illustrates again the difficulty of applying a Donnan membrane treatment to such a system.

As dyeing is usually performed in the presence of high salt concentrations, the fibre remains saturated with hydrogen ions during the exchange, so that it behaves as an exchange medium of constant capacity. The overall process may be represented simply as follows, where s refers to ions in solution and f to ions in the fibre;

$$H_s^+ + Cl_s^- \rightleftharpoons H_f^+ + Cl_f^-$$
$$Cl_f^- + D_s^- \rightleftharpoons D_f^- + Cl_s^-$$

the mass action equation can therefore be written

$$K = \frac{(a_{Cl^-})_s \cdot (a_{D^-})_f}{(a_{D^-})_s \cdot (a_{Cl^-})_f}$$

The addition of neutral salt decreases the absorption of dye by competing for the available sites (Fig. 7.4). This has the effect of maintaining a greater concentration of dye in solution, which greatly assists the levelling. Neale[28] has proposed that salt decreases the electrical attraction between dye anion and the positively charged fibre, but electroneutrality is maintained throughout the process owing to the presence of anions in the double layer. Although the surface charge can have no effect on the equilibrium, it may exert a considerable influence on the rate of dyeing (see p. 172).

The replacement of one ion by another provides a further method for determining anion affinity if the affinity of one of the anions is known, and usually the chloride ion is chosen for this purpose. The total number of protons combined and present in solution can be determined by titration and the concentration of dye measured colorimetrically; according to the Gilbert–Rideal treatment, the affinity of the dye anion is then given by,

$$\Delta\mu_D = RT \log_e K \quad (\Delta\mu_{Cl} = 0)$$

This method, under carefully chosen conditions, becomes extremely simple and relatively rapid even at low temperatures. The chief disadvantage lies in the great difference in affinity between the two ions, resulting in very low concentrations of dye in solution. For this reason the displacement of one dye with another reference ion of high affinity is often employed, particularly in the case of the milling dyes.[29]

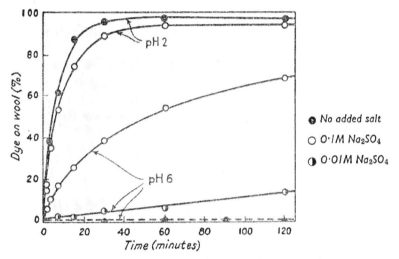

FIG. 7.4.—The effect of salts in increasing the uptake of Orange II at pH 6 and decreasing dye uptake at pH 2. (Temperature 25° C.; well agitated bath; ½ per cent. dye on weight of wool.)[8]

A problem arises here in the calculation of standard affinities for polyvalent dye anions.[13] Considering the general case of two polyvalent ions of valency z and y, respectively, the chemical potential of dye on the fibre is:

$$\mu_D = \mu_D{}^0 + RT\ln\frac{\theta_D}{1-\theta} + zF\psi$$

where $\theta = \theta_D + \theta_x$, and θ_D is the fraction of the total number of sites occupied by dye, and θ_x is the fraction occupied by ion x.

Equating this to the corresponding value in solution leads to,

$$-\Delta\mu_D{}^0 = RT\ln\frac{\theta_D}{1-\theta} + z\psi F - RT\ln(D)_s$$

Similarly for the anion x of charge y,

$$-\Delta\mu_x{}^0 = RTln\frac{\theta_x}{1-\theta} + y\psi F - RTln(x)_s$$

Rearranging these equations to eliminate ψF leads to the final expression for the difference in affinity:

$$(-y\Delta\mu_D{}^0 - z\Delta\mu_x{}^0) = RTln\frac{\theta_D^y \cdot (x)_s^z}{\theta_x^z \cdot (D)_s^y} + (z-y).RTln(1-\theta)$$

This equation can only be applied directly in cases already considered when $y = z$, so that for dibasic dyes, an affinity value may be obtained by displacing the dye ion by the sulphate ion. Lemin and Vickerstaff[29] have maintained, however, that in any unsymmetrical exchange the term $(z-y)RT \ln(1-\theta_x-\theta_D)$ may be ignored, which is probably not true when $\theta \to 1$. These authors quote the following values (Table 7.3), obtained by both chloride and sulphate exchanges, in support of this empirical procedure. The thermodynamic

TABLE 7.3

Affinity of dyes from exchange measurements.[17]

Dye	Basicity	Affinity k. cal./mol	
		By chloride	By sulphate
Naphthalene Orange C	1	4·6	5·2
Solway Ultra Blue B	1	7·2	7·1
Metanil Yellow YK	1	5·0	5·3
Solway Blue BN	2	4·8	5·10

treatment is of course only approximate, as in all cases activities have been replaced by concentrations. This procedure is hardly justifiable when the complex nature of a dye is considered. Thus, over large concentration ranges, the value of $\Delta\mu$ and the distribution constant K are found to vary considerably. The deviations are very marked as the dye employed becomes more complex, due mainly to the comparable affinity of the dye ion and the proton (see Table 7.4). Systems in which the dye ion enters the fibre partly as a gegen-ion and partly as a leading ion, cannot be treated theoretically at present. Even though the dye enters mainly as a gegen-ion, as is the case with most simple (levelling) acid dyes, the simple treatment cannot be expected to hold strictly for a variety of causes. Firstly, the complete neglect of activity coefficients in both phases is unjustifiable, although some cancellation of errors would result from the ratio employed in the general expression. Secondly, considerable electrical

TABLE 7.4

Adsorption of mixtures of Solway Ultra Blue B and Naphthalene Orange G on wool at 60° C.[22]

Concentration of dyes in solution		Concentration of dyes in the fibre		Final pH	Affinities k. cal./mol.	
Naph. Orange G	Solway Blue B	Naph. Orange G	Solway Blue B		Naph. Orange G	Solway Blue B
1.33×10^{-4}	4.11×10^{-5}	3.34×10^{-2}	5.01×10^{-2}	5.46	10.6	11.6
2.67	9.04	6.55	9.95	5.18	10.7	11.7
4.22	1.46×10^{-4}	9.40	1.40×10^{-1}	4.88	10.6	11.6
5.28	1.84	1.32×10^{-1}	1.91	4.70	10.8	11.8
6.40	2.06	1.74	2.49	4.54	11.1	12.1
6.73	2.63	2.25	3.03	4.24	11.2	12.0
9.00	3.41	2.71	3.65	4.07	11.5	12.4
9.30	3.29	3.05	4.05	3.92	11.9	12.8
9.90	3.54	3.46	4.58	3.72	12.8	13.6

Equimolecular mixtures were used in each case.

interaction between like anions is to be expected when they are absorbed on the fibre as the sites are necessarily not far apart. This may be negligible for simple anions, but becomes more significant when dyes are used owing to their greatly increased size.

The distribution of dye within the fibre

It has been assumed in the above interpretation of the dyeing process that the complex dye acid combines stoichiometrically with the wool. The available evidence shows that even with dyes, the maximum combining capacity is rarely exceeded (see p. 181). This is remarkable because the affinity of even simple dyes is exceedingly high compared with inorganic gegen-ions, and indicates that complex anions are held within the fibre in a similar way to ions of very low affinity. Thus in acid solution a large dye anion is attracted primarily to a charged amino group forming a relatively strong bond, and at the same time may be held to an adjacent amide group by secondary forces. The association with amide groups is recognized by the considerable decrease in tensile strength,[30] as in the case of nylon. In this way, the stoichiometric combination may be reconciled with a very high affinity.

It is interesting to consider further the combination of polyvalent ions with the fibre. Speakman and Stott[31] have shown that dyes with as many as five sulphonic acid groups are absorbed stoichiometrically. It is inconceivable that the spatial distribution is such that each of the groups in a compact structure such as phenol trisulphonic acid is exactly adjacent to a positively charged group in the fibre. The monovalent anions of a simple inorganic acid are thought to be highly mobile within the fibre, but distributed in such a way as to remain, on the average, close to each positive group. A polybasic dye of high affinity cannot behave in this way. Because of the considerable decrease in mobility with increased strength in bonding, the anion is most likely to be associated at any given time with one particular positive group, while serving to neutralize the charge on the fibre as a whole. This view is supported by approximate calculations of the energy variations in a regularly alternating potential field. It can be shown simply that a polyvalent ion would be in a position of maximum stability close to a single centre rather than at a mean position between two or more oppositely charged centres. This requires the operation of electrostatic forces over considerable distances which would not appear to lead to a stable system. This is compensated by a readjustment in the composition of the electrical double layer within the fibre.

A similar conclusion is reached when the influence of the complex morphological structure on the stoichiometry of acid combination is considered. It has been established by both X-ray measurements[32] and permanent set data[33] that dye anions cannot penetrate the crystalline phase of the fibre. Speakman and Elliott[34] account for this anomaly by postulating micellar subdivision which allows anions to approach sufficiently close to all the positive amino-groups to satisfy the electrical action, while allowing the crystalline phase to retain its characteristic α keratin structure. Support for this view was mainly drawn from data on the effect of dye penetration on the tensile strength of the fibres. This has been criticized considerably because the extent of subdivision necessary for the chemical combination of anions would probably be far too great to allow a well-defined α-photograph.[35]

It is usually agreed that protons can enter micelles freely and Astbury and Dawson[32] consider that simple anions also penetrate. The possibility that charged micelles can absorb the correct number of dye ions on the surface is difficult to accept in the light of simple calculations of the electrostatic energy developed, assuming the accepted dimensions of the micelle. It is possible however that the diameter of a micelle is much less than the accepted value of 200 Å, and that because of the high degree of order within the micelle all the amino groups may be directed towards the surface (see Chapter 12).

A further observation which has a bearing on the stoichiometric combination is the presence of dye crystallites within the fibre. Dyes with very high affinities such as Solway Blue are highly aggregated in solution, so it may be that the presence of crystallites explains the supra-stoichiometric absorption of this type of dye.[36] The dyeing process could then be imagined as occurring in two stages, firstly the stoichiometric absorption governed by the ion exchange mechanism, and secondly a re-aggregation of dye within the fibre.

Neutral and basic dyeing

If a dye has a sufficiently high affinity, it may become the potential-determining ion (see Table 7.3), and under these conditions dyeing will proceed in the absence of mineral acid. In this instance, cations enter the fibre as gegen-ions, so that the addition of salt increases the uptake of dye by the common ion effect of the cations in the same way as the anion in the acid process.[8] This is clearly demonstrated in Fig. 7.4.

It is found experimentally in this case that the absorption does not follow the Langmuir but the Freundlich isotherm.[37] This is to be expected as the fibre can no longer be treated as a uniform medium containing a limited number of sites of widely varying energy. The data on direct dyeing of wool are limited and difficult to interpret theoretically, and in addition the direct dyeing process has in several cases been confused with the anion exchange process. More comprehensive data have been obtained using cotton, which are found to support the following general mechanism:

$$Na^+_s + D^-_s \rightleftharpoons Na^+_f + D^-_f$$

$$\frac{(Na^+)_f . (D^-)_f}{(Na^+)_s . (D^-)_s} = K$$

In the presence of excess salt $(D^-)_f = A \cdot (D^-)_s^m$
where A and m are empirical constants for the particular case, so that as $(Na^+)_f \simeq (D^-)_f$

$$\frac{A^2 \cdot (D^-)_s^{2m}}{(Na^+)_s \cdot (D^-)_s} = K$$

Thus for a dyebath at constant concentration, $\log(D^-)_f$ is proportional to log (salt), which has been confirmed experimentally.

A more rigorous treatment is difficult to envisage until a clear understanding of the forces governing the specific anion affinity is obtained. Thus Willis, Warwicker, Standing and Urquhart[38] were forced to simplify their approach to such an extent that it became virtually equivalent to a simple thermodynamic treatment. Peters and Vickerstaff[39] have more recently considered the absorption to be a partition between external solution and internal solution. The volume of the internal solution was taken to be the volume of imbibed water (see p. 111) and ion concentrations within the fibre were calculated by the Donnan theory. Approximate agreement is obtained between this theoretical treatment and the experimental data.[40]

Experimental data on the basic dyeing of wool are even more meagre than data on direct dyeing. By analogy with acid dyeing, each cation should be bound by each free COO^- group, the cations of neutral salts competing for the sites. The foundation for a detailed study has been laid recently by Steinhardt and Zaiser,[41] who determined the titration curves at 0° C. of a series of the monovalent bases recorded in Chapter 5, p. 213. The displacements of the titration curves along the pH axis give direct measures of the affinities which are found to vary regularly with the length of the maximum dimensions of the molecules (but not with the molecular weight) of a series of organic cations. The affinity of the ions of the alkali metals, however, increases regularly both with molecular weight and size. Small organic ions (mol. wt. < 150) have an almost constant affinity, but a rapid increase with weight is observed above this critical value, when a shift of 1 pH unit per 45 units of weight is observed for asymmetrical ions.

The affinity of the symmetrical ion is always less than that of a corresponding asymmetrical ion. This critical size, below which the affinity remains small both for anions and cations, is similar to that for micelle formation in compounds containing these ions. These observations show the binding to be due to Van der Waals' forces which become operative only when long-range electrostatic forces increase the probability of a close approach. Steinhardt and Zaiser conclude that no specific chemical bonds are formed.[41]

It follows that the process of basic dyeing is exactly analogous to acid dyeing except that one dye cation is bound for each proton removed from the salt links by alkali. Neutral salts facilitate the absorption by their effect on the double layer, but the cations reduce the quantity of dye sorbed by competition. In practice, however, the few basic dyes which have been applied to wool (e.g. methylene blue and the Rhodamines) are usually applied in a neutral or weak acid solution to reduce alkali damage, and consequently rely on their very high affinities for absorption by a leading ion mechanism rather than by cation exchange. This mechanism is therefore comparable with neutral dyeing with acid dyes.

REFERENCES

[1] SPEAKMAN AND HIRST, *Trans. Faraday Soc.*, 1933, **29**, 148.
[2] MEYER, *Naturwiss.*, 1927, **15**, 129.
[3] STEINHARDT, FUGITT AND HARRIS, *J. Res. N.B.S.*, 1941, **26**, 293.
[4] STEINHARDT, FUGITT AND HARRIS, *ibid.*, 1942, **28**, 201.
[5] BEEK, *ibid.*, 1935, **14**, 217.
[6] SKINNER AND VICKERSTAFF, *J. Soc. Dy. Col.*, 1945, **61**, 193.
[7] GOODALL, *J. Soc. Dy. Col.*, 1939, **55**, 529.
[8] ALEXANDER AND KITCHENER, *Text. Res. J.*, 1950, **20**, 203.
[9] SOOKNE AND HARRIS, *J. Res. N.B.S.*, 1939, **23**, 471.
[10] SOOKNE AND HARRIS, *J. Res. N.B.S.*, 1941, **26**, 65.
[11] STEINHARDT, *J. Res. N.B.S.*, 1942, **28**, 191.
[12] SPEAKMAN AND CLEGG, *J. Text. Inst.*, 1941, **32**, T83.
[13] GILBERT, *Proc. Roy. Soc.*, 1944, **183A**, 167.
[14] LEMIN, (unpublished), quoted by Vickerstaff, *Physical Chemistry of Dyeing*, Oliver and Boyd, London, p. 343.
[15] KRESSMAN AND KITCHENER, *J. Chem. Soc.*, 1940, 1208.
[16] SCHRIM, *J. prakt. Chem.*, 1935, **144**, 69.
[17] VICKERSTAFF, *Physical Chemistry of Dyeing*, Oliver and Boyd, London, 1950, 305.
[18] PFEIFFER, *Organische Moleculverbindungen*, Stuttgart, 1932.
[19] ALEXANDER AND CHARMAN, *Text. Res. J.*, 1950, **20**, 761.
[20] STEINHARDT, FUGITT AND HARRIS, *J. Res. N.B.S.*, 1942, **29**, 417; STEINHARDT, *ibid.*, 425.
[21] HALE AND REICHENBERG, *Faraday. Soc. Discussions*, 1949, **7**, 79.
[22] BOYD, SCHUBERT AND ADAMSON, *J. Amer. Chem. Soc.*, 1947, **69**, 2818.
[23] DUNCAN AND LISTER, *J. Chem. Soc.*, 1949, 3285.
[24] GREGOR, *J. Colloid Sci.*, 1951, **20**, 245, 304, 325.
[25] PEPPER, REICHENBERG AND MACAULEY, *J. Chem. Soc.*, 1951, 493.
[26] GLUECKAUF, *Nature*, 1949, **163**, 414; GLUECKAUF AND DUNCAN, *A.E.R.E. Report* C/R 808, 1951.
[27] ELÖD, *Trans. Faraday Soc.*, 1933, **29**, 327.
[28] NEALE, *J. Colloid Sci.*, 1946, **1**, 371.
[29] LEMIN AND VICKERSTAFF, *J. Soc. Dy. Col.*, 1947, **63**, 405.
[30] PORAI-KOSCHITZ, *J. prakt. Chem.*, 1933, **137**, 179; GOODALL AND HOBDAY, *J. Soc. Dy. Col.*, 1939, **55**, 529.
[31] SPEAKMAN AND STOTT, *J. Soc. Dy. Col.*, 1934, **50**, 341.
[32] ASTBURY AND DAWSON, *J. Soc. Dy. Col.*, 1938, **54**, 6.

[33] ELLIOTT AND SPEAKMAN, *ibid.*, 1943, **59**, 124.
[34] SPEAKMAN AND ELLIOTT, *Fibrous Proteins Symp.*, Soc. Dy. Col., Bradford, 1946, 116.
[35] MARTIN, *ibid.*, p. 126; BARRER, *ibid.*, 125; CRISP, *ibid.*, 127.
[36] ROYER AND MARESH, *Amer. Dyestuff Reporter*, 1943, **32**, 181.
[37] HALSEY AND TAYLOR, *J. Chem. Phys.*, 1947, **15**, 624.
[38] WILLIS, WARWICKER, STANDING AND URQUHART, *Trans. Faraday Soc.*, 1945, **41**, 506.
[39] PETERS AND VICKERSTAFF, *Proc. Roy. Soc.*, 1948, **192A**, 292.
[40] MARSHALL AND PETERS, *Symposium on the Recent Advances in the Theory and Practice of Dyeing*, Soc. Dy. Col., Bradford, 1947, 82.
[41] STEINHARDT AND ZAISER, *J. Biol. Chem.*, 1950, **183**, 789.
[42] SMITH, *J. Biol. Chem.*, 1936, **113**, 473.

CHAPTER 8

The Disulphide Bond

THE proteins of the keratin group are characterized chemically by a high sulphur content which is present mainly in the amino acid cystine,[1] although a small quantity of methionine[2] is also present. Although the sulphur and cystine contents of wool are known to vary with fibre diameter or quality and small differences may even occur from sheep to sheep,[3] a fundamental problem of wool chemistry is the lack of correlation between the sulphur content of a given sample of wool and the content of its sulphur-containing amino acids, cystine and methionine. For some time it was accepted that the sulphur content of wool was significantly greater than that corresponding to its amino acid composition (see Table 8.1) and it has been proposed

TABLE 8.1

The sulphur distribution in root and tip wool[2]

Sulphur present as	Root (percentage)	Tip (percentage)
Disulphide	2·99	2·66
Thiol	0·14	0·15
Methionine	0·12	0·12
Sulphate	0·06	0·10
Br-oxidizable sulphur	0·00	0·02
	3·31	3·05
Total sulphur found	3·60	3·57

that the difference is due to elementary sulphur in the fibre.[3] It now seems more likely that the problem is one of analysis. The majority of methods used for determining the sulphur content of wool involve wet oxidation to sulphate ions and the determination of the latter gravimetrically as barium sulphate. The weight of the latter depends on a number of factors[4] and generally an arbitary procedure is employed, it being reported that deviations give high or low results.

Myers[5] has examined this method in detail and arrives at a value of 3·3 per cent. sulphur in 64^s wool. Earland[6] has reported a similar value using the Schöniger method employing oxidation in gaseous oxygen and the determination of sulphate ions volumetrically. The apparent cystine content of wool also varies with the analytical method employed, ion-exchange chromatography giving an appreciably lower value than Shinohara's method. Lewis, Robson and Tiler[7] have found a number of sulphur-containing compounds, in addition to methionine and cystine, in wool hydrolysates and in view of the fact that most of these are probably derived from cystine and are Shinohara-positive, the Shinohara method may well be as accurate as any at present available for determining cystine. The value of 12·3 per cent. obtained for 64^s quality wool[8] together with a value of 0·6 per cent. for the methionine content accounts for the 3·3 per cent sulphur present.

The chemical reactivity of keratin is attributed largely to the combined amino acid cystine, which may readily be oxidized, reduced or hydrolysed to give a variety of complex reaction products. As cystine is a diamino acid, it can combine with adjacent polypeptide chains to form the disulphide cross-link. This component of the keratin structure contributes largely to the physical and mechanical properties of the fibres, although Astbury and Woods[9] pointed out thirty years ago that it does not influence the crystal spacing of the three dimensional network. It is generally recognized that the cystine is incorporated in the wool macromolecule so as to cross-link two main polypeptide chains (Fig. 8.1a), although various other possibilities have been considered. Of these, all but the first have been dismissed at one time or another for a variety of reasons. Thus 1b is excluded as no free cysteine is obtained on disruption of the disulphide bond with a mild reducing agent.[10] This is also confirmed by experiments with peracetic acid (see p. 266). 1c has been excluded as the cystine content is not decreased by deamination, although this cannot be regarded as conclusive because of the method of cystine analysis. The structure 1d cannot be wholly eliminated on acid titration data as has been done by Patterson et al.[10] who claimed that all the free carboxyl groups can be accounted for as glutamic and aspartic acids. Even in the most accurate amino-acid determinations, however, considerable deviations are recorded (see Chapter 11). Thus, it is seen from Table 11.4 (p. 350) that even using the ion-exchange chromatographic technique, amino acid analyses can be regarded as reliable only to within about 10 per cent. It is therefore difficult to rule out a particular structure from the correlation of the dibasic acid content with acid titration data, and part of the cystine may well be accommodated in wool fibres as the structure 1*d*. The ring structure 1*e* can be excluded as no diketopiperazines have ever been found in wool hydrolysates, nor in the hydrolysate of any

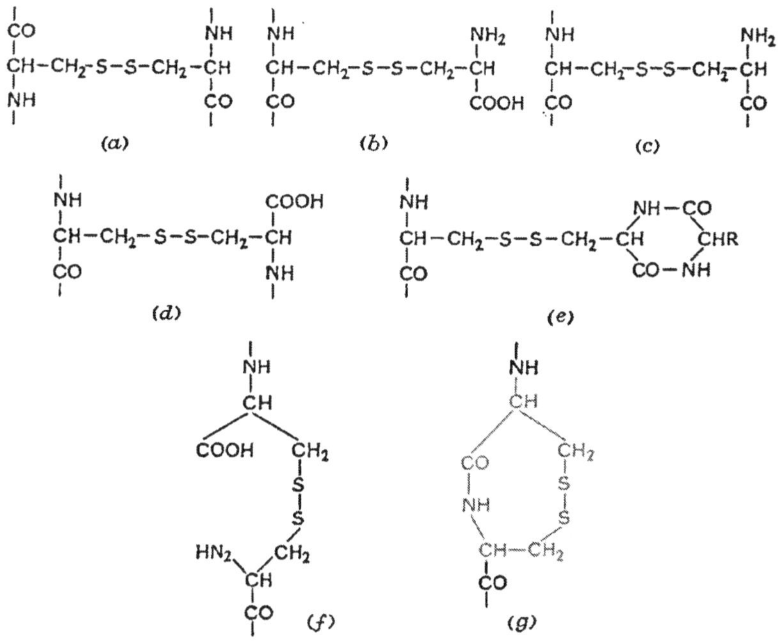

FIG. 8.1.—Possible ways of incorporating cystine into the polypeptide chain.

other protein although it is known that these rings are more stable than peptide bonds.

The suggestion has been made from time to time, that part at least of the disulphide sulphur is incorporated in the main polypeptide chain.[11] The first possible mode of linkage, 1*f*, has been rejected, mainly on acid-binding data although this is not conclusive as already mentioned. The second, 1*g*, however is unlikely on mechanical evidence. Thus 1*a* must be regarded as the only definite structure for the combined cystine. Owing to the varying reactivity, however, it is possible that part of the cystine is combined in a different way, and structures 1d and 1f cannot be excluded with certainty, and their presence was envisaged by Elöd and his co-workers[12] who found that part of the cystine appeared to have no influence on the physico-mechanical properties of the fibres. It is now generally recognized, however that the wool fibre consists of several phases with varying influence on the physical properties, although in each phase the cystine is probably incorporated as in 1a. The experimental work of Elöd et al. can therefore be explained on this basis, and actually provides powerful support for the structure of wool which is now normally accepted.

It should be pointed out that it is not essential for structure 1a to represent an inter-chain link. It is possible for the cystine residue to bridge a loop in the same chain to form an intra-chain linkage. Since this type of structure has been established by Sanger's classical work on insulin, this possibility cannot be rejected.

Owing to the high reactivity of the disulphide bond, disruption occurs in all but a few reactions, such as acid hydrolysis. In most of the treatments applied to wool commercially, the disulphide bond is attacked, although reaction usually proceeds with other groups in the wool macromolecule as well. Little progress was made in determining the role played by the disulphide bond in the many properties of wool until specific reagents for disrupting it were found. This problem was first solved by the use of reducing agents since, with the exception of the per-acids to be discussed in detail later (p. 266), all oxidizing agents are non-specific.

Reduction of the disulphide bond

It has already been mentioned that the disulphide bond may be reduced by a variety of reagents, but that reaction is not confined to this group alone in many cases. For this reason it is proposed to concentrate primarily on reactions which are specific for cystine, and non-specific reactions which have been studied in considerable detail. The reaction with thioglycollic acid was originally studied by Goddard and Michaelis[13] who showed that the disulphide bond is rapidly converted to thiol groups in alkaline solution. Strongly alkaline solutions were used as these authors thought that reduction occurred only above pH 10, owing to the insolubility of the wool in neutral or acid solutions of thioglycollate. However, model disulphide compounds are known to be reduced by thioglycollic acid over a wide pH range, so that this conclusion was soon questioned. Harris and his collaborators[10] showed that the reaction with thiols in neutral solution ($pH < 8$) was confined to the reduction of the disulphide bond, thus eliminating secondary reactions which undoubtedly occur in more alkaline media. The reduction in neutral or acid solution is, however, an equilibrium reaction, and consequently it is impossible to reduce all the disulphide bonds in one experiment

$$\text{CH·CH}_2\text{·S·S·CH}_2\text{·CH} \Big< + 2\text{CH}_2\text{SH·COOH} \rightleftharpoons 2 \Big> \text{CH·CH}_2\text{·SH} + (\text{S·CH}_2\text{·COOH})_2$$

Repeated treatments with a reducing agent are necessary to reduce the cystine content of the wool by more than 60 per cent., and it is possible to obtain a 90 per cent. reduction eventually. The amount

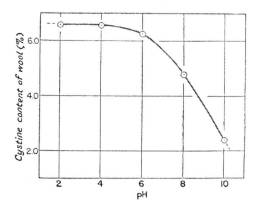

FIG. 8.2.—Cystine content of wool after reduction at various pH values for 20 hr. at 35° C. with 0·2 M thioglycollate solution.[10] (*Reproduced by permission.*)

of cystine reduced in one stage is found to be constant from pH 2–6, somewhat greater at pH 8, and markedly greater at pH 10 (Fig. 8.2). The increased reaction in the alkaline region is not due to destruction of cystine by alkali, but to reduction, since wool that has been reduced for short periods (2 hr.) can be re-oxidized to products containing nearly the original amount of cystine-sulphur. Longer exposures, (20 hr.) at pH 10, however, were found to give a material with an abnormally low cystine content after re-oxidation, and this was most probably due to the preferential extraction of cystine-rich material as a considerable proportion of the wool dissolves under these conditions. The increased reduction in alkaline solution may be a consequence of the ionization of the sulphydryl groups, which would increase swelling and facilitate penetration of the reagent, and might even change the mechanism of the reduction.

The reduction with thiols requires many hours to reach equilibrium, and the extent of the reaction is markedly affected by the concentration of thioglycollic acid in solution when the ratio of the quantity of reagent to weight of wool is maintained constant (see Fig. 8.3). In the experiments represented in Fig. 8.2, twenty times

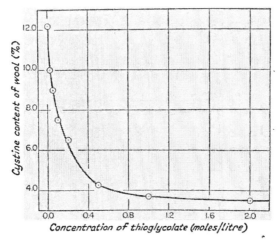

Fig. 8.3.—Cystine content of wool after reduction with thioglycollate solutions of various concentrations for 20 hr. at pH 4·5 and 35° C.[10] (*Reproduced by permission.*)

the reducing agent necessary to reduce all the disulphide groups of the wool was used. Temperature was found to affect the rate of reaction considerably, although the equilibrium was not significantly affected.

The use of many thiols for reducing the disulphide bond has been described in the patent literature, but there is no reason to suppose that these compounds react differently from thioglycollic acid.

The thiol groups produced by reduction are very labile and can easily be re-oxidized, either by prolonged exposure to the atmosphere or by treatment with mild oxidizing agents, e.g. hydrogen peroxide or potassium bromate. (This reversible process forms the basis of all the 'home permanent-waves' now on the market and is also used to impart permanent pleats to wool fabric.) Harris *et al.*[14] showed that the wet strength of a fibre, which had been reduced to less than one-tenth by reduction, could be restored to almost its original value by subsequent oxidation.

Instead of re-oxidizing the fibres in order to reform the disulphide bonds, other cross-links may be introduced by reaction of the reduced material with bi-functional compounds such as ethylene dibromide.[10, 15] In this way the disulphide bond has been replaced by a bond of the type S–R–S where R ranges from CH_2 to $(CH_2)_6$. Although the cross-links differ greatly in length, the properties of the fibre were found to remain largely unaffected, which supports the opinion, earlier expressed by Elöd,[11] that the disulphide bond cannot form part of the crystalline phase of the wool fibre.

Instead of reacting the reduced wool with bi-functional alkylating agents, Harris also used mono-functional reagents such as alkyl iodides, thus converting the thiol group to an S–R group which is stable and cannot be oxidized to form a new cross-link. In this way he was able to determine quantitively the influence of the disulphide cross-link on the mechanical properties of the fibre[14, 16] (see p. 61).

Of the other reducing agents which have been studied, sodium bisulphite and formaldehyde have received the most attention. The reaction of wool with bisulphite is of considerable industrial interest, as sulphur dioxide is used for bleaching wool, and in addition bisulphite is usually applied as an anti-chlor following treatments with halogens for rendering wool unshrinkable. If wool is treated with permanganate (either for bleaching or reducing shrinkage), bisulphite is again used to remove the manganese dioxide deposited on the fibres. The disulphide bond is probably not significantly attacked when dilute solutions of bisulphite are used to remove oxidizing agents left on the wool after processing,[17] but under more severe conditions, Elsworth and Phillips[18] have shown that 50 per cent. of the total cystine can be reduced by this reagent according to the equation:

$$
\begin{array}{c}
\mid \\ CO \\ \mid \\ CH-CH_2-S-S-CH_2-CH \\ \mid \\ NH \\ \mid
\end{array}
\begin{array}{c}
\mid \\ NH \\ \mid \\ \\ \mid \\ CO \\ \mid
\end{array}
+ NaHSO_3 \rightarrow
\begin{array}{c}
\mid \\ CO \\ \mid \\ CH-CH_2SH \\ \mid \\ NH \\ \mid
\end{array}
+ NaOSO_2 . S .
\begin{array}{c}
\mid \\ NH \\ \mid \\ CH_2-CH \\ \mid \\ CO \\ \mid
\end{array}
$$

It will be observed that the cystine is not completely reduced, but that half the sulphur forms S-cysteine sulphonate and only one half is converted to thiol groups. Simple disulphide compounds react in exactly the same way, and it is impossible to obtain complete reduction with bisulphite. Of the cysteine groups produced in the wool in this way, half are more labile than the remainder and revert to cystine on rinsing in cold water (p. 260). The disulphide bond can be converted completely into two cysteinyl sulphonate groups by a method based on the following oxidation-reduction cycle:

$$\text{R—S—S—R} + SO_3^{--} \longrightarrow \text{RS—SO}_3^- + \text{RS}^-$$
$$\uparrow \text{tetrathionate}$$
$$\text{or iodosobenzoate}$$

In this reaction, the thiol group is oxidized by sodium tetrathionate or iodosobenzoate to the disulphide form, which then reacts with the excess of sulphite.

Complete conversion into cysteinyl sulphonate is also achieved[18a] by addition of cupric ion. When wool is treated with a solution $0.02M$ in cuprammonium hydroxide, $0.5M$ in sodium sulphite and $8M$ in urea at pH 9·8–10·5 and at room temperature, up to 90 per cent. of the wool is dissolved in a few days.

$$\text{R—S—S—R} + 2Cu^{++} + 2SO_3^{--} \longrightarrow 2\text{RS—SO}_3^- + 2Cu^+$$

The cysteinyl sulphonate derivative of the wool is precipitated from the solution by removal of copper by dialysis against citrate or ethylenediamine-tetracetate at pH 7–9, followed by acidification. When copper is removed by dialysis against potassium cyanide, the cysteinyl sulphonate groups are converted into thiocyanate groups.[18b]

$$\text{RS—SO}_3^- + CN^- \longrightarrow \text{RSCN} + SO_3^{--}$$

The 50 per cent. fraction of the cystine remaining after the Elsworth–Phillips treatment can eventually be reduced by prolonged reaction at a high temperature to give, it is claimed, amino-acrylic acid and sodium thiosulphate. There is some doubt as to the precise nature of this reaction, which has never been observed with simple disulphides.

Stoves[19] has measured the decrease in strength of fibres after reaction with bisulphite solutions ranging from pH 3·6 to pH 11, and has concluded that maximum fission of the disulphide bond occurs at pH 4·5 and 11.

Sodium sulphide has long been known to react with wool fibres and to reduce them sufficiently to render them soluble in solutions known to be capable of breaking hydrogen bonds (see p. 362). Whilst the disulphide bond is undoubtedly reduced to thiol groups, general degradation of the peptide chains must also occur, probably because of the highly alkaline solutions which have to be applied. The extent of the degradation can be estimated directly, because the component of highest molecular weight found in wool keratin rendered soluble by treatment with sodium sulphide is 10,000, whereas it is thought that the actual weight is of the order of 70,000 (see p. 372).

Ethylene sulphide can reduce the disulphide bond in the same way as the thiols, but the reaction is followed by polymer formation within the fibre according to the following reaction:[20]

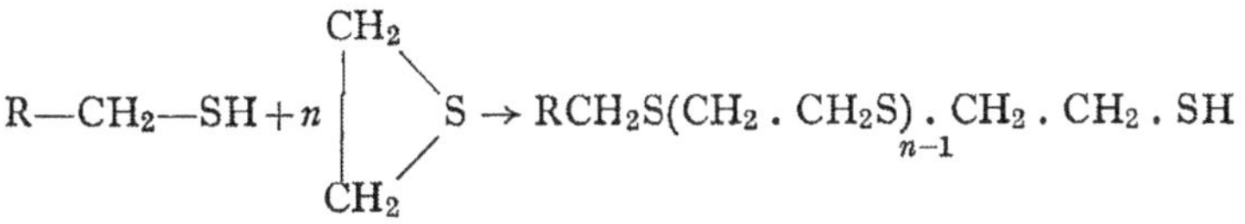

$$R-CH_2-SH + n \begin{matrix} CH_2 \\ | \\ CH_2 \end{matrix} \!\!\! \diagup\!\!\!\!\!\diagdown S \to RCH_2S(CH_2 . CH_2S)_{n-1} . CH_2 . CH_2 . SH$$

Consequently the number of free thiol groups remains very low throughout the treatment, which is reported to impart an unshrinkable finish to wool.[21]

Elöd and Zahn[22] have studied the reaction with wool of the powerful reducing agent, sodium sulphoxylate ($NaHSO_2$) which is widely used in industry. It is probably as specific as the thiols, although no detailed chemical analysis appears to have been performed. Neither has the reaction with the hydrosulphites, used commercially for stripping dyes, been investigated, although these reagents are certainly capable of reducing the disulphide bond.

The disulphide bond is more resistant to formaldehyde, and reduction does not occur at any pH in the cold.[23] Above 50° C. however part of the cystine is reduced to give combined thiazolidine 4-carboxylic acid, although the fate of half the cystine is unknown.

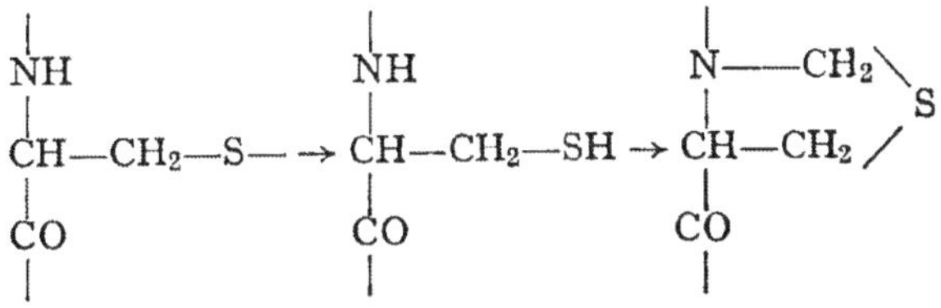

$$\begin{matrix} | \\ NH \\ | \\ CH-CH_2-S- \\ | \\ CO \\ | \end{matrix} \to \begin{matrix} | \\ NH \\ | \\ CH-CH_2-SH \\ | \\ CO \\ | \end{matrix} \to \begin{matrix} | \\ N\!\!-\!\!-\!\!-\!\!CH_2 \\ | \qquad\qquad \diagdown \\ CH-CH_2 \quad\; S \\ | \qquad\qquad \diagup \\ CO \\ | \end{matrix}$$

Middlebrook and Phillips[24] have shown that the cystine which eventually cyclizes in this way, is that which forms water-stable thiol and S-cysteinesulphonate groups with bisulphite. The fraction which forms water-labile groups with bisulphite does not normally react with formaldehyde, but at pH 10 is converted to lanthionine (see p. 254).

If the wool is previously treated with thioglycollic acid to convert 50 per cent. of the disulphide bonds to thiol groups, subsequent treatment with formaldehyde converts half the reduced cystine to thiazolidine 4-carboxylic acid and half to djenkolic acid:[24]

$$\begin{array}{c} | \\ \text{NH} \\ | \\ \text{CH}-\text{CH}_2-\text{S}-\text{CH}_2-\text{S}-\text{CH}_2-\text{CH} \\ | \\ \text{CO} \\ | \end{array} \qquad \begin{array}{c} | \\ \text{NH} \\ | \\ \\ | \\ \text{CO} \\ | \end{array}$$

The presence of this combined acid has been assumed in the reaction with formaldehyde alone, on the basis of indirect evidence,[25] but this must be rejected in the light of the direct chemical evidence quoted above. The reaction of formaldehyde is not confined to the cystine, and is extremely complex (see p. 327), so that changes in physical properties can provide little information about the actual reactive groups in the absence of chemical analysis.

Hydrolysis in neutral solution

As in the case of reduction, neutral or alkaline hydrolysis leads to the fission of the disulphide bond with the formation of a thiol and a –SOH group. Schöberl[26] investigated the action of boiling water on wool, and found that after twenty-one days, 75 per cent. of the sulphur had been removed as H_2S, accompanied by the evolution of an equivalent amount of ammonia. By means of very sensitive staining tests using nitroprusside, the presence of thiol groups was readily demonstrated.[27] Schöberl[28] was able to show that wool treated with water contained aldehyde groups, and on the basis of experiments with model substances he postulated the following series of reactions, in which the postulated unstable sulphenic acid decomposes to the aldehyde.

THE DISULPHIDE BOND

$$\begin{array}{c} \text{CH—CH}_2\text{—S—S—CH}_2\text{—CH} \\ \downarrow \text{H}_2\text{O} \\ \text{CH—CH}_2\text{—SOH} + \text{SH—CH}_2\text{—CH} \\ \downarrow \\ \text{CH—C}\!\!\begin{array}{c}\text{H}\\ \diagup \\ \diagdown \\ \text{O}\end{array}\!\! + \text{SH—CH}_2\text{—CH} \\ + \text{H}_2\text{S}. \end{array}$$

Table 8.2 shows the changes brought about in wool by water at 100° C. for different times. Although the disulphide bond is very reactive there are no direct analytical data to show that any hydrolysis occurs in steam in the short period necessary for setting a fibre, or in water at room temperature during the course of a few hours in which considerable creep occurs in stressed fibres (see Table 8.2). It is therefore unlikely that either of these phenomena can be explained in terms of disulphide bond breakdown (see p. 69).

TABLE 8.2

Changes in wool fibres produced by boiling water[29]

Time of treatment in days	Wool dissolving %	N content of residual wool %	Sulphur content %	Cystine content %	Acid uptake milli-equiv./g.
0	0	16·5	3·65	10·6	0·92
0·5	1·0	16·4	3·44	9·4	—
1·0	4·0	16·0	3·33	8·6	0·89
2·0	7·3	16·5	3·11	7·5	0·86
4·0	10·7	16·5	3·07	6·9	0·87
8·0	37·2	16·6	2·66	4·6	0·89

Elöd et al.[12] treated wool with water in the presence of finely divided silver, mercury and cadmium and their soluble salts, and found, in agreement with the earlier work of Speakman,[30] that the hydrolysis rate is greatly increased. Working in neutral solution, up to 50 per cent. of the sulphur may be removed in the form of metallic sulphide in 2–3 days at 80° C. The dry tensile strength and the crystallographic structure were completely unchanged after this extensive reaction. If the reaction with metals and water is prolonged in order to remove most of the sulphur, the wool is severely degraded, dissolving partially and losing its crystalline structure.

Wool is very resistant to dry heat, and it is found that the cystine content is decreased by twenty-four hours' heating only at temperatures exceeding 150° C. (see p. 301).

Reaction with alkalis

In the early stages of wool processing, when the raw wool is degreased and scoured, it is usually exposed to alkaline solutions, and consequently slight modification in the structure may occur. Reaction between alkali and wool has been studied at length by a number of investigators. Harris[31] found in 1935 that dilute solutions of caustic soda were able to attack 50 per cent. of the total cystine, and that little of the remaining 50 per cent. reacted even after very prolonged treatments. Sodium sulphide is immediately formed as a reaction product, which is in turn capable of reacting with the disulphide bond. Care must therefore be taken to ensure that the reaction with sodium sulphide does not interfere significantly with the primary reaction. Speakman and Whewell[32] have deduced from studies of supercontraction and permanent set that alkali disrupts the disulphide bonds in such a way that new cross-links are formed in a subsequent reaction. Various suggestions were advanced for the structure of this cross-link, and these workers proposed the C–S–C type of linkage. In 1941 Horn, Jones and Ringel[33] fully confirmed this hypothesis by isolating the hitherto unknown amino acid lanthionine, from alkali-treated wool. Lanthionine

$$\begin{array}{c} \text{NH}_2 \\ \diagdown \\ \text{CH}-\text{CH}_2-\text{S}-\text{CH}_2-\text{CH} \\ \diagup \qquad\qquad\qquad\qquad\qquad \diagdown \\ \text{HOOC} \qquad\qquad\qquad\qquad\qquad \text{COOH} \end{array} \qquad \begin{array}{c} \text{NH}_2 \\ \diagup \\ \\ \end{array}$$

was eventually synthesized by Schöberl[34] in 1947. Speakman[30] has also postulated that –S–NH linkages are formed in wool during alkaline treatment by the condensation of the sulphenic acid, obtained in the initial disulphide-bond hydrolysis, with side-chain amino groups from lysine. Phillips[35] and more recently Stoves[36] suggested that alkali-treated wool contains $-CH=N-$ linkages arising out of the condensation of a postulated aldehyde (produced by the disproportionation of the sulphenic acid) with amino groups. No evidence for the existence of either of these groups could, however, be found in careful analytical work by Phillips[37,38] and others[39] and it must be concluded that they are not formed. (cf. however, Speakman[39a]).

The early conclusion of Harris that only 50 per cent. of the cystine is normally reactive towards alkali was supported by Phillips and his co-workers,[37] who showed that this reaction proceeded even at pH 7 to yield lanthionine.[38] The much more resistant fraction may be attacked under more severe conditions, when however amino acrylic acid and not lanthionine is produced. Typical results showing the sulphur distribution on boiling in solutions at various alkalinities are recorded in Table 8.3. It will be observed that the two cystine fractions are not sharply distinguished and at high alkalinities they are attacked simultaneously.

During the past few years considerable progress has been made in elucidating the exact mechanism of lanthionine formation in wool by alkalis and it is now recognized that it proceeds by a bimolecular nucleophilic β-elimination reaction:[40]

$$\overset{|}{C}HCH_2SSCH_2\overset{|}{C}H + OH^- \rightarrow \overset{|}{C}HCH_2SS^- + \overset{|}{C}H_2 = \overset{|}{C} + H_2O$$

$$\downarrow H^+$$

$$\overset{|}{C}HCH_2SCH_2\overset{|}{C}H + S$$

Earlier workers such as Bergmann and Stather,[41] Nicolet[42] and Tarbell and Harnish[43] had proposed that the reactivity of many cystine derivatives depends upon the presence in the molecule of an activated hydrogen atom in a position beta to the disulphide bond. The above mechanism is supported by a number of facts. Thus, Bergmann and Stather[41] found that α-amino acrylic acid is produced on treating certain cystine peptides with alkali, Nicolet, Shinn and Saidel[44] were able to introduce a $-CH_2.S.CH_2$ linkage into silk by converting serine side groups to α-amino acrylic acid and reacting the latter with a thiol, Swan[40] found that αα'dimethylcystine is degraded only extremely slowly by alkali compared with free cystine and Earland and Raven[45] have shown that polyamides containing a

$$\overset{|}{\underset{|}{N}}CH_2SSCH_2\overset{|}{\underset{|}{N}}$$

cross-link do not undergo a lanthionine-type reaction with alkali.

TABLE 8.3

The sulphur distributions of wool before and after boiling in buffer solutions of pH 7–10 and the conversion of disulphide-sulphur into lanthionine-sulphur[37]

Duration of treatment (hr.)	Thiol- +disul- phide-S (%)	Br-oxidiz- able S + SO$_4$ as S (%)	Total-S (%)	Lanthio- nine-S (%)	Loss of disul- phide-S (%)
0	3·10	0·20	3·76	0·36	0·56*
At pH 7					
2	2·74	0·30	3·58	0·44	0·92
4	2·53	0·39	3·52	0·50	1·13
7	2·36	0·42	3·38	0·50	1·30
16	2·04	0·45	3·17	0·58	1·62
32	1·73	0·46	3·00	0·71	1·93
At pH 8					
1	2·28	0·35	3·47	0·72	1·38
2	2·08	0·42	3·42	0·82	1·48
4	1·72	0·39	3·18	0·97	1·94
6	1·61	0·38	3·02	0·93	2·05
9	1·36	0·35	2·77	0·96	2·30
13	1·17	0·28	2·55	1·00	2·49
At pH 9					
0·5	1·89	0·63	3·38	0·76	1·77
1	1·57	0·48	2·92	0·77	2·09
2	1·17	0·42	2·64	0·95	2·49
4	0·97	0·31	2·44	1·06	2·69
At pH 10					
0·5	0·90	0·45	2·44	0·99	2·76
1	0·67	0·45	2·27	1·05	2·99
2	0·57	0·40	2·14	1·07	3·09

* Due to weathering or other causes.

THE DISULPHIDE BOND 257

In addition to the above mechanism, Zahn, Kunitz and Hildebrand[46] have suggested that lanthionine in wool may also be produced by direct nucleophilic attack by the $-CH_2S^-$ group on the $-C-S-S$ bond of the combined cystine.

$$\begin{array}{c} S-S-CH_2-W \\ | \\ W-CH_2 + -S-CH_2-W \\ \uparrow\_\_\_\_\_| \end{array} \rightarrow \begin{array}{c} -S-S-CH_2-W \\ + \\ W-CH_2-S-CH_2-W \end{array}$$

The above mechanism is dependent on the small number of naturally occurring thiol groups in wool (see p. 317) playing an important role in the reaction.

Zahn and Osterloh[47] have found that the addition of organic polar solvents, such as acetone and dioxan, to solutions of weak alkalis enables lanthionine formation in wool to be brought about far more effectively than in their absence, but no satisfactory explanation for this had been advanced.

For a considerable time all attempts to convert cystine itself or simple derivatives to lanthionine failed, and it was assumed that the disulphide bond of wool reacted in a manner entirely different from a disulphide bond in simple compounds. Simultaneously, however, Schöberl and Wagner[48] and Swan[49] showed that this was not so. The former workers converted cystine to lanthionine with alkali, whilst Swan converted a number of cystine derivatives such as biscarbobenzoxy-L-cystine and its dihydrazide to lanthionine with aqueous solutions of sodium carbonate or cyanide. The differences in reactivity appear, therefore, to be rather of degree than of sharp demarcation.

Solutions of strong alkali in alcohols are effective in preventing felting. The use of alcohol as a solvent for this purpose has been described by Freney and Lipson,[50] and of a mixture of butyl alcohol and white spirit by Wood.[51] In this way the degradative reaction is confined to the surface of the unswollen fibres, thus reducing damage. Nothing is known of the chemical reaction taking place under these conditions, but it would appear that lanthionine is not formed since it is necessary to break disulphide bonds without forming new cross-links to cause the degradation necessary for the production of non-felting. It would appear that the process is in principle similar to an earlier one[52] utilizing concentrated aqueous solutions of strong alkalis such as caustic soda of at least 35 per cent. concentration or potassium hydroxide greater than 50 per cent. concentration. These solutions contain definite hydrates and the alkali is virtually present in non-aqueous solution.

It must be realized that alkali solutions react rapidly with many other groups of wool (see p. 291), and for this reason should be considered as general degradative reagents, although the disulphide and acid amide groups are probably the first to react.

Reaction with potassium cyanide. The reaction of wool with solutions of potassium cyanide provides an example of a qualitative difference in the reaction of the disulphide bond in simple molecules and in wool. Mauthner[53] showed that cystine reacts with potassium cyanide in aqueous solution to give one mole of cysteine and one mole of α-amino rhodanine propionic acid.

$$\underset{NH_2}{\overset{HOOC}{>}}CH.CH_2.S.S.CH_2.CH\underset{NH_2}{\overset{HOOC}{<}} + KCN \rightarrow$$

$$\rightarrow \underset{NH_2}{\overset{HOOC}{>}}CH.CH_2.SCN + \underset{NH_2}{\overset{HOOC}{>}}CH.CH_2.SH$$

By continually oxidizing the cysteine formed to cystine, Schöberl[54] was able to convert the whole of the sulphur to thiocyanate.

In wool, on the other hand, it was shown by Cuthbertson and Phillips[37] that the disulphide bonds are almost quantitatively converted to lanthionine by treatments with aqueous solutions of cyanide, e.g. boiling in 0·65 per cent. for 30 min. or treatment with 1 per cent. KCN for sixteen hours at 66° C. The reaction is thought to proceed in the following way and the thiocyanate formed was quantitatively detected in the solution:

$$>CH.CH_2SSCH_2.CH< + KCN \rightarrow >CH.CH_2SK + NCSCH_2CH<$$

$$\underset{-KCNS}{\longrightarrow} >CH.CH_2SH + H_2C=\underset{|}{\overset{|}{C}} \rightarrow >CH.CH_2.S.CH_2.CH<$$

Swan[55] has, however, suggested the possibility of lanthionine formation by SCN displacement without removal of beta hydrogen atoms:

$$\overset{|}{\underset{|}{C}}HCH_2S^- + NCSCH_2\overset{|}{\underset{|}{C}}H \rightarrow \overset{|}{\underset{|}{C}}HCH_2SCH_2\overset{|}{\underset{|}{C}}H + CNS^-$$

Zahn et al.[46] have also proposed this mechanism and Earland and Raven[56] have shown that a lanthionine type of reaction occurs when N-mercaptomethylpolyhexamethyleneadipamide disulphide is reacted with solutions of cyanide, where clearly β-elimination is not possible

$$\overset{|}{\underset{|}{N}}CH_2SSCH_2\overset{|}{\underset{|}{N}} + CN^- \longrightarrow \overset{|}{\underset{|}{N}}CH_2SCH_2\overset{|}{\underset{|}{N}} + CNS^-$$

The reactions by which lanthionine is produced by alkali only and by cyanide differ both quantitatively and qualitatively. In the action due to hydroxyl ions the mode of combination of the sulphur split off has not been completely determined, but is thought to be largely in the form of the element, whereas with cyanide it is eliminated as thiocyanate. Further, all the disulphide sulphur can be converted to lanthionine by cyanide and only 50 per cent. by alkali.

Classification of the cystine in wool into four fractions differing in reactivity with alkali and reducing agents. The detailed work of Phillips[57] and his co-workers on the reaction of alkali, sodium bisulphite, formaldehyde and thioglycollic acid with wool showed that the cystine appeared to be divided into two main fractions (A+B) and (C+D), each of which appeared to be sub-divided again into the four fractions indicated. The evidence on which this division is based is summarized in Table 8.4. The fractions were determined primarily by the different products formed and not by relative accessability or reactivity since excess reagent was always used.

After considering the possible reasons for the existence of these different fractions, Phillips has rejected different ways of incorporating the cystine bond which were considered on page 245, and favours the suggestion that the variations are due to different amino acids combining with the cystine (i.e. variation in side-chain environment). Several facts indicate that the more reactive fraction (A+B) is near

TABLE 8.4

The subdivision of the combined cystine of wool into four subfractions[57]

Wool: Cystine-S, 3·6%	
Cystine fraction (A+B), 1·8%	Cystine fraction (C+D), 1·8%
(i) NaHSO$_3$	
Subfraction A (0·8%) gives water-labile −SH and NaO.SO$_2$SCH$_2$- groups; subfraction B (1·0%) gives water-stable- SH and NaO.SO$_2$.SCH$_2$-groups.	Subfraction C (0·8%) is inert. Subfraction D (1·0%) decomposes in hot bisulphite and yields combined aminoacrylic acid.
(ii) Alkalis	
Converted into lanthionine. Subfraction A changes more rapidly than subfraction B.	Both fractions give aminoacrylic acid: subfraction C slowly.
(iii) HCHO	
Subfraction A does not react: subfraction B gives combined thiazolidine-4-carboxylic acid.	No evidence of any reaction with either subfraction.
(iv) HS.CH$_2$.COOH	
Subfractions A and B are reduced at pH 4·5. With HCHO subfraction A yields djenkolic acid: subfraction B yields thiazolidine-4-carboxylic acid. Subfraction A reacts with methylene dibromide: subfraction B does not react.	Subfractions C and D reduced much more slowly at pH 4·5. Reaction with HCHO and methylene dibromide in the reduced state is unknown.

or surrounded by polar side chains, while fraction (C+D) is associated with non-polar side chains. Thus, it has been clearly demonstrated that esterification of the carboxyl groups, and deamination, influence the reactivity of only one of the cystine fractions. Differences in crystallinity or arrangement of the molecular chains within the fibre are insufficient to explain the existence of four subfractions, since accessibility to the reagent appears to play no part.

In addition to the evidence provided by modifying the polar groups in wool, experiments with model substances also indicate that it is side-chain environment which produces the four subfractions. For instance, Schöberl[26] found that cystine itself and the various cystine peptides which he examined, hydrolysed in water to give a thiol group and an aldehyde group. On the other hand, Bergmann and Stather[41] found that dialanyl cystine anhydride hydrolysed to give an SH group and combined α-amino acrylic acid, as Phillips has postulated for the reaction with wool. The reduction of cystine peptides with bisulphite also varies from peptide to peptide. Fox[58] found that cystinyl diglycine and diglycyl cystine were both reduced completely by bisulphite at pH 4·8, whereas some of the larger peptides were not reduced until more alkaline solutions of bisulphite were used. None of these examples closely parallels the differences in reactivity of combined cystine in wool found by Phillips, but they do illustrate that side-chain environment greatly influences the reactivity of the SS bond.

Blackburn[59] has pointed out that the validity of Phillips' work on the reactivity of wool towards formaldehyde depends on the assumption that the reaction products, thiazolidine-4-carboxylic acid and djenkolic acid, are stable to the acid hydrolysis conditions of 5N hydrochloric acid employed by Phillips prior to demonstrating the presence of these substances. Blackburn has, however, shown that pure djenkolic acid on hydrolysis yields an appreciable amount of thiazolidine-4-carboxylic acid, and the latter on hydrolysis yields djenkolic acid. Under these circumstances it is obviously very difficult to decide whether djenkolic acid or thiazolidine-4-carboxylic acid were originally present in a formaldehyde-treated protein that has been hydrolysed with acid. Further, Blackburn and Lee[60] have shown by direct analysis that for every cystine residue attacked during the alkaline treatment of wool, one residue of lanthionine is produced, and reactions other than lanthionine formation do not take place to any appreciable extent. These workers cannot, therefore, substantiate the formation of aminoacrylic acid by the action of alkali on the cystine sub-fraction (C+D) of Phillips. Although the painstaking work of Phillips and his co-workers, based on detailed chemical analysis, cannot be discarded at this stage, it seems probable that a reappraisal of the conclusions of Phillips will be necessary in the future.

To explain the observed mechanical properties, such as permanent set and supercontraction of fibres after different chemical treatments, Speakman[61] postulated that the reactivity of the SS bond is greatly increased when the fibre is stressed. No chemical evidence has been advanced in support of this hypothesis. On the contrary Phillips[57] found that the reactions with bisulphite, with alkali, and with formaldehyde were unaffected on stretching fibres by 70 per cent.; nor is the division into four subfractions modified. A more probable explanation of the effects observed by Speakman is that on stretching there is a change in the nature of hydrogen bonding within the fibre, a factor which has been often neglected.

Reaction with peroxides

Hydrogen peroxide. One of the most widely used methods for bleaching is the immersion of wool in stabilized alkaline hydrogen peroxide solutions for several hours either at or slightly above room temperature. The reaction of this reagent with wool protein is slow and with care it is usually possible to remove the pigment before substantial attack on the protein has occurred.[62] To keep damage to a minimum it is usually desirable to impregnate the wool with the hydrogen peroxide and then to store it at an elevated temperature rather than to leave it in the actual peroxide bath.[63] Damage can be reduced further by applying the peroxide from a non-swelling organic solvent, when the reaction is more likely to be confined to the surface, although this condition can never be entirely achieved since peroxide molecules can diffuse into the interior of a wool fibre even when this is not swollen.[63] Except in alkaline solution the reaction with H_2O_2 at room temperature is very slow. When wool is immersed in hydrogen peroxide, some is initially sorbed by the amino and imino groups without reaction.[64] The amount thus taken up depends on the concentration of peroxide in solution and the reaction is completely reversible; for instance 10 mg. of hydrogen peroxide per gram of wool are sorbed from a 3 per cent. solution. On rinsing with water all this peroxide can be removed in the wash liquor. This sorbed peroxide, which appears to be very non-reactive, is similar to the stable crystalline addition compound between urea and H_2O_2 and can be recovered after many hours. When bleaching wool, therefore, it is essential, after the reaction is complete, to rinse the wool with water so as to remove the sorbed peroxide, which otherwise remains in the wool and may cause skin irritation.

The cystine group is the first to be attacked by this reagent and Breine and Baudisch[65] found that sulphuric acid was a reaction product. Smith and Harris[62] have investigated the action of hydrogen peroxide on wool extensively and found that all the cystine could be oxidized, but that the rate of reduction in cystine content varied with the concentration of the reagent, temperature, the time of treatment and the pH (see Figs. 8.4 and 8.5; Table 8.5). The last factor is particularly important and the degradation is greatly increased at a pH above 7. For this reason it is desirable in practice to accelerate the rate of bleaching of wool preferably by increasing the temperature and not by increasing the pH of the solution. Smith and Harris[62]

TABLE 8.5

The effect on wool of 2-volume hydrogen peroxide solutions at temperatures varying from 23° to 80° C. for 3 hours and the alkali solubility of the treated wool[62]

Temperature of H_2O_2 solutions ° C	Cystine content after treatment with H_2O_2 percentage	Alkali solubility percentage
23	11·2	9·7
35	11·3	13·6
50	10·8	19·4
65	9·8	31·6
80	7·4	100·0

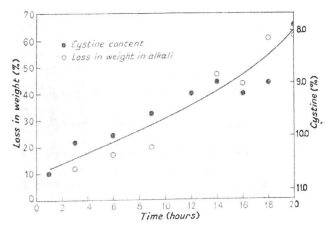

FIG. 8.4.—Effect on wool of 2 vol. hydrogen peroxide solution at 50° C. for different lengths of time.[62]

considered that cystine analyses of wool oxidized with peroxide were of doubtful value, since they showed clearly that one of the reaction products was the intermediate disulphoxide. This decomposes partially to cysteic acid and reverts partially to cystine during

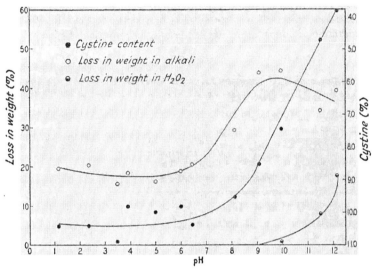

FIG. 8.5.—Effect on wool of 2 vol. hydrogen peroxide solution of varying pH at 50° C. for 3 hr.[62]

the extensive hydrolysis necessary for these analyses. Besides this disulphoxide, Consden and Gordon[66] have shown that cysteic acid is also formed and, as already mentioned, some of the sulphur is split off as sulphate. The attack of hydrogen peroxide on many other groups was indicated by Smith and Harris,[67] who found that the acid-combining power of the wool decreased with increasing oxidation by peroxide. They at first attributed this to preferential reaction with the amino groups but direct analysis by Rutherford and Harris[68] indicated that these do not participate in the reaction. Acid-binding data, however, are very difficult to interpret when reactions of such a general nature are being considered. Elöd, Nowotny and Zahn[69] showed that hydrogen peroxide also reacted surprisingly rapidly with peptide bonds, since both silk (containing negligible cystine) and wool dissolve completely in a solution of 3 per cent. hydrogen peroxide in three days at 60° C. The proteins were complete degraded, and no high molecular weight peptides could be isolated from the reaction liquor. These authors found a close relation between the change in mechanical properties and cystine content of wool which had been allowed to react with hydrogen peroxide. However, they showed this to be fortuitous since the reaction is non-specific and simultaneous peptide and SS bond breakdown occurred. Similarly, conclusions drawn by Stoves[70] from purely mechanical measurements, concerning the state of the disulphide bond in wool after reaction with peroxide are erroneous since no account was taken of the attack on the peptide bond.

The reaction between wool and hydrogen peroxide is powerfully catalysed by heavy metal ions such as copper or nickel. The effect has not been studied in detail but forms the basis of an anti-shrink process.[71] It is not known if the catalysed and uncatalysed oxidation proceed by the same chemical reaction although there is no doubt that the SS bond is very readily oxidized in the presence of heavy metals, and a hundredfold increase in rate for this reaction has been observed.[72] Sulphate is also produced in the reaction of wool with ozone, although the reaction proceeds only under certain conditions.[73]

Photo-chemical oxidation

Harris and Smith[74] have shown that ultra-violet light can disrupt the disulphide bond of dry wool, so that eventually all the cystine is oxidized. The resulting degradation in the wool, determined by the alkali solubility, measurements and estimation of liberated ammonia and sulphate, is directly related to the decrease in cystine content.[75] Although the overall rate of the reaction is greatly increased by increasing the water content or the acidity, these factors do not influence the first stage in the reaction which Harris and Smith believe to be a direct activation of the disulphide bond. This probably leads to the transient formation of ions, followed by hydrolysis (which was also shown to be a photo-chemical process), leading to the evolution of hydrogen sulphide from the unstable sulphenic acid.

$$\begin{aligned}
&\diagdown\!CH\cdot CH_2\cdot S\cdot S\cdot CH_2\cdot CH\diagup \rightarrow \diagup\!CH\cdot CH_2\cdot S^+ +{}^-S\cdot CH_2\cdot CH\diagup \\
&\xrightarrow{+H_2O} \diagdown\!CH\cdot CH_2\cdot SH + HOS\cdot CH_2\cdot CH\diagup
\end{aligned}$$

Crawshaw and Speakman,[76] on the other hand, favour homolytic fission of the disulphide bond, rather than a heterolytic fission.

The hydrogen sulphide evolved is eventually oxidized to sulphate. Although all the cystine can be destroyed photo-chemically, the total reaction cannot be represented exclusively by the above mechanism, as the disappearance of cystine and loss of total sulphur do not coincide and some cysteic acid is definitely formed. It seems likely by analogy with other photo-chemical reactions, that hydroxyl radicals play a part, and in this respect the reaction may be similar to the reaction with peroxides. In view of the improved methods of protein analysis which have been developed since the work of Harris and colleagues, a re-examination of this reaction is desirable. The tips of wool fibres are known to have a low cystine content due to weathering, and it has been shown that cysteic acid is present.[77] As weathering involves the prolonged interaction of light and water with wool, the processes which occur should be similar to the photo-chemical reaction.

Oxidation with per-acids

Organic per-acids do not react in the same way as hydrogen peroxide with wool. Toennies and Homiller[78] showed that performic acid, although a very powerful oxidizing agent, does not oxidize amino acids in general, but reacts only with tryptophan, methionine and cystine, the last compound being oxidized quantitatively to cysteic acid. Sanger[79] used this reagent to split the SS bonds in insulin without affecting any of the other amino-acid residues, and in this way obtained two polypeptide chains which were originally bound together by SS bridges. Performic acid, however, is not very stable and undergoes decomposition to carbon dioxide and water. Alexander, and co-workers found that stable dilute aqueous solutions of the higher per-acids can be prepared, and investigated the action of peracetic and the higher alkyl per-acids[80,81] on wool. Working in dilute aqueous solutions it was considered that all the disulphide groups were readily accessible and could be oxidized completely, the time of reaction depending on the concentration, temperature of the reagent (see Table 8.6) and the diameter of the fibres. Recent work has, however, thrown doubts on this conclusion. It was shown that peracetic acid was as specific as performic acid and did not oxidize any amino acids other than cystine, methionine and tryptophan. Experiments[58] with a number of peptides showed that peracetic acid does not attack the peptide bond, and does not bring

TABLE 8.6

Oxidation of disulphide bond in wool by aqueous solutions of peracetic acid at 22° C.[81]

Time	Percentage cystine oxidized	
	1·6% peracetic acid	30% peracetic acid
1 min.	—	29·7
5 mins.	19·8	38·6
30 mins.	46·3	—
1 hr.	71·1	81·8
2 hrs.	75·0	100·0
4 hrs.	77·0	—
24 hrs.	92·0	—

about main-chain degradation, at least in the time necessary for complete reaction with the disulphide bond. During the treatment with aqueous solutions of these per-acids, none of the wool substance went into the solution. Blackburn and Lowther[82] oxidized wool with performic acid dissolved in anhydrous formic acid and confirmed that the disulphide bond was readily oxidized but a substantial part of the wool substance dissolved during the reaction.

Products of reaction

After acid hydrolysis, all the cystine oxidized by the peracetic acid could be determined as cysteic acid.[83] This and the observation that no loss of sulphur occurred suggested the following mechanism:

$$\overset{|}{C}H-CH_2-S-S-CH_2-\overset{|}{C}H + 5O + H_2O \rightarrow 2\overset{|}{C}H-CH_2-SO_3H$$

However, attempts to detect sulphonic acid groups in the wool were unsuccessful. Thus, no ion-exchange of the expected sulphonic acid groups could be detected, although this is quantitative in the case of synthetic resins containing strong acidic groups. Alexander, Fox and Hudson[83] suggested that on oxidation of wool with peracetic acid, intra-molecular cyclization occurs to produce a cyclic sulphocarboxylic acid-imide, which on subsequent treatment with alkali, hydrolyses to a sulphonamide as shown in Fig. 8.6.

Although the evidence for these reactions was indirect, it is known that saccharin, which has a ring structure very similar to that of the cyclic sulphocarboxylic acid-imide, behaves in a similar manner under these conditions.[84]

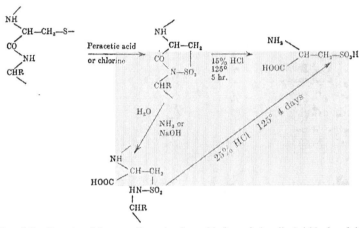

Fig. 8.6.—Postulated intermediates in the oxidation of the disulphide bond by peracetic acid.

Earland and Knight[85] in a re-examination of the problem attempted to detect the presence of sulphonic acid groups in oxidized wool by preparing the ammonium salt and analysing it for an increased nitrogen content. These workers agreed with the findings of Alexander et al.[83] that sulphonic acid groups are not detected in oxidized wool by chemical methods. The work of Weston[86] on the other hand, who examined the infra-red absorption spectra of wool oxidized with peracetic acid, can leave little doubt that in point of fact sulphonic acid groups are present, since wool treated with peracetic acid gives a spectrum containing peaks at 1180 cm^{-1} and 1040 cm^{-1}, characteristic of the sulphonic group, whereas the sulphonamide peak at 1300–1350 cm^{-1} is entirely absent. Further, the infra-red spectra of oxidized wool and cysteic acid are very similar (see Fig. 8.7).

It must be concluded that the sulphonic acid groups present in oxidized wool are masked in such a way that they cannot be detected chemically. Earland and Knight[85] suggested that this is due to the formation of strong salt links $-SO_3^-\ldots\,^+H_3N-$. Woodin[87] has pointed out that an arginine-bisulphate interaction is such a stable linkage that it would be difficult to distinguish it from a covalent

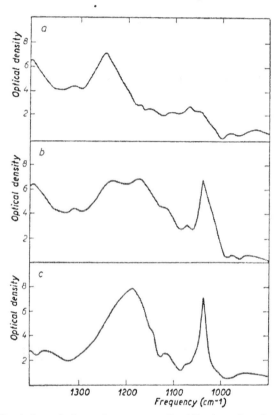

FIG. 8.7.—The infra-red absorption spectra of oxidized wool and cysteic acid.[88] (a) virgin wool; (b) peracetic acid-treated wool; (c) L-cysteic acid.

bond. It is relevant that the infra-red absorption spectrum favours the sulphonic acid group being present as the ionized form, rather than as $-SO_3H$. Since O'Donnell and Thompson have shown[88] that oxidized wool behaves as a carboxylic acid ion-exchange resin it seems most likely that the salt-link concept is correct.

Perphosphomolybdic and perphosphotungstic acids

Although their reaction with wool has not been examined as

extensively as in the case of the aliphatic peracids, the only investigation reported[89] shows that they behave in a very similar manner.

Perbenzoic acid

The reaction of this compound with wool has not been studied, but Lavine, Toennies and Wagner[90] obtained a product from the reaction with cystine which had taken up the maximum of four atoms of oxygen per mole, and which on hydrolysis was converted to a sulphonic and sulphinic acid. The structure was therefore given as a sulphone (see equation):

$$RS-SR \rightarrow R-\underset{\underset{O}{\|}}{\overset{\overset{O}{\|}}{S}}-\underset{\underset{O}{\|}}{\overset{\overset{O}{\|}}{S}}-R \rightarrow RSO_3H + RSO_2H$$

Caro's acid. Toennies[91] showed that this acid oxidized cystine to cysteic acid although it was possible to stop the reaction at intermediate stages. A preliminary examination of the effect of Caro's acid on wool showed that although the disulphide bond was oxidized as effectively as with peracetic acid, the reaction was not confined to this compound and loss of tyrosine also results.[94] Its superior activity compared with peracetic acid as an anti-shrink agent can be attributed to these additional reactions.[81, 93]

Pervanadic and perchromic acids. A cursory examination of the reaction of these acids with wool[81] showed that in their behaviour they resembled hydrogen peroxide and not the reactive per-acids considered above. They react only very slowly with wool and it would appear that the reaction is far from specific.

Permanganate. Acid solutions of potassium permanganate were recognized as effective agents for rendering wool unshrinkable.[92] This reagent only produces non-felting from acid solution, and Alexander, Carter and Hudson showed that solutions with pH greater than 3 were ineffective for this purpose.[17] A detailed study of the reaction between wool and permanganate at pH 2 and 9·2 was made by Alexander, Hudson and Fox[80] who found that acid solutions rapidly react with up to 30 per cent. of the cystine in the wool. It was not possible to exceed this value and if more permanganate was used this merely reacted with other groups in the wool leaving intact 70 per cent. of the cystine. Alkaline permanganate oxidized very little of the cystine until severe degradation of the fibre had occurred (see Table 8.7). The products of the oxidation of the cystine in wool by acid permanganate solution were shown[83] to be cysteic acid and sulphate in equal proportions. After the severe treatment with alkaline permanganate necessary to oxidize some of the cystine, it is found that lanthionine is a reaction product. The reaction of

TABLE 8.7

Cystine in wool oxidised by potassium permanganate at 25° C.[80]

pH	$KMnO_4$ reduced (g./100 g. of wool)	Cystine oxidized (percentage of that originally present)
2·0	6·25	18·6
2·0	12·50	21·8
2·0	16·70	30·7
2·0	25·00	29·6
9·2	6·25	3·0
9·2	25·00	17·7

both acid and alkaline permanganate with cystine and its peptides is entirely different, as both reagents readily oxidize the SS bond in the simple compounds and produce cysteic acid quantitatively, although in the presence of excess permanganate general degradation leading to the formation of sulphate occurs.[58] It is particularly surprising that alkaline permanganate should react preferentially with other groups in the wool and that acid permanganate, after having oxidized 30 per cent. of the cystine, should also react with other groups and not with the remaining cystine, in view of the fact that in experiments[17] with model substances only SS and phenolic groups are oxidized at all readily in the cold. (See p. 274).

Reactions with halogens

Halogens or derivatives with oxidizing properties are widely used industrially for rendering wool unshrinkable. As a result of the usually accepted hypothesis of Speakman *et al.*[93] that disruption of the disulphide bond is necessary for unshrinkability produced chemically, it has been assumed for many years that these reagents oxidize this bond.

Until the advent of paper chromatography, no chemical investigation had been made, although the formation of various compounds (e.g. ·SCl) had been postulated and oxidation of SH compounds with chlorine has been found to give sulphur chlorides. Consden, Gordon and Martin[77] originally showed that cysteic acid is produced on oxidizing wool with chlorine and bromine in acid solution. It had been known since 1903 that bromine oxidized free cystine to this acid, and it may be assumed that chlorine reacts similarly. Chlorine reacts with all the amino acids except glycine, rapidly at pH 2 and more slowly at pH 10.[95] Deamination followed by complete decomposition takes place, although oxidation of the disulphide bond occurs preferentially. With a stoichiometric proportion of chlorine or hypochlorite at any pH, cystine peptides are converted quantitatively to the corresponding combined cysteic acids.

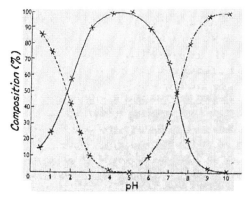

Fig. 8.8.—Variation in composition of a solution of 0·05 mol. of chlorine dissolved in water at different hydrogen ion concentrations.[80]
Cl_2(- - - - -); HOCl(———); OCl^-(—·—).

It will be realized that the constitution of an aqueous chlorine solution is highly dependent on pH, chlorine, hypochlorous acid and hypochlorite ions existing in equilibrium at a given acidity or alkalinity (Fig. 8.8). Solutions of chlorine and hypochlorous acid can oxidize all the cystine if sufficient reagent be employed,[80] whereas hypochlorite solutions at pH 10 can remove a maximum of only 25 per cent. of the cystine, like acid solutions of permanganate (see Fig. 8.9). This 25 per cent. cystine fraction is well characterized, as it corresponds to the cystine which reacts and is the first to be oxidized by chlorine in acid solution.

It therefore appears at first sight that oxidizing agents in the form of anions can only react with a 25 per cent. fraction of the cystine, whereas unionized molecules (peracetic acid, hydrogen peroxide, chlorine, hypochlorous acid and chlorine peroxide) can remove all the cystine. This apparent relation however requires further investigation and may well prove to be fortuitous.

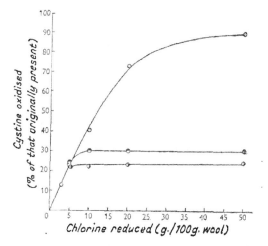

FIG. 8.9.—Removal of cystine in wool oxidized at pH 2 and 10 with different quantities of chlorine.
O, chlorination at pH 2 and 25° C. ◐, chlorination at pH 10 and 25° C.; ◉ chlorination at pH 10 and 80° C[80]

The cause of the non-reactivity of 75 per cent. of the combined cystine towards certain oxidizing agents has been considered.[80] Thus explanations depending on morphological differences, electrical repulsion and crystallinity of the fibre have been rejected. Moreover, it is unlikely that side-chain environment can be responsible, because all the peptides of widely varying amino-acid composition which have been examined[58, 96] could be oxidized readily by chlorine and hypochlorite ions. There is no relation between these fractions and the subfractions of Phillips already considered (see p. 259). Thus on removing half of the cystine in wool (i.e. Phillips's fraction AB) with alkali, the remaining cystine reacts in the same way as untreated wool towards chlorine and hypochlorite. Esterification and deamination affect the reactivity of the Phillips fractions but not that of the fractions discussed here, which also proves their difference.

It has been pointed out that both permanganate and hypochlorite are non-specific for oxidizing disulphide bonds in keratin.[97] Thus in the case of sodium hypochlorite only 20 per cent. of the reaction may be confined to oxidizing this group. Furthermore, the oxidized protein goes rapidly into solution. Under these conditions it is hardly surprising that the wool after oxidation always contains unattacked cystine, since two competing reactions of approximately equal speed occur, followed by solution of the reaction products. Although the experimental work of Alexander *et al.*[80] has been confirmed[97] its relationship to the structure of keratin has not been interpreted unequivocally.

Products of reaction. The products of the reaction of chlorine and hypochlorous acid with the disulphide bond appear to be the same as those obtained with peracetic acid, although the reaction is not nearly so specific and many other groups are also oxidized (see p. 268). Moreover, after extensive oxidation all the cystine is no longer present as cysteic acid, and the sulphur content of the fibre as a whole decreases[83] (see Table 8.8).

TABLE 8.8

Disulphide sulphur, cysteic acid and total sulphur analyses of wool treated with various concentrations of chlorine at pH 2, 4, 8 at 25° C.[83]

pH	Chlorine reduced (g./100 g. of wool)	g./100 g. wool		
		Disulphide sulphur	Cysteic acid	Total sulphur
2·0	Nil	2·66	—	3·50
	5	2·05	0·6	3·57
	10	1·57	1·3	3·35
	20	0·72	1·9	3·40
	50	0·27	1·4	2·50
4·0	20	0·98	1·7	3·54
	50	0·16	1·7	2·80
8·0	20	2·17	0·1	3·33
	50	1·40	0·4	2·69

Further evidence for the occurrence of side reactions with cystine is provided by the slight increase in cystine content of chlorinated wool after an alkali treatment (see Table 8.9), which can be explained by analogy with the results of Lavine.[98] After chlorination, the wool behaves as an oxidizing agent and the residual oxidizing power in wool after chlorination correspond to approximately 1 per cent. of the total amount of oxidizing agent reacted.[99] It is unlikely to be sorbed chlorine since the oxidizing power is not lost on washing the wool, but may be a labile sulphoxide or a chloramine. The presence of this residual oxidizing capacity makes it necessary in practice to treat the wool after chlorination with a reducing agent, usually bisulphite.

TABLE 8.9

The cystine content of wool oxidized by chlorine solution at pH 2 before and after extraction with 3N ammonia[83]

Chlorine reduced (g./100 g. of wool)	Cystine content (%)	
	Before extraction	After extraction
Nil	10·00	9·65
13	5·32	6·60
35	2·02	2·95
50	0·39	2·20

In addition to the presence of intermediate oxidation compounds, there is also evidence[83] for the formation of a stable cross-link on chlorination. Thus it can be seen from Fig. 11.1, on p.368 that wool, in which all the SS bonds have been oxidized by peracetic acid, dissolves (with the exception of the cortical cell membranes) completely in dilute alkali, whereas complete oxidation with chlorine only renders 70 per cent. of the wool soluble in dilute ammonia. Moreover, the chlorinated wool does not supercontract unlike the wool oxidized with peracetic acid, and it is therefore suggested that new cross-links are formed on acid chlorination which prevent supercontraction and increase the resistance to alkali.

Compounds containing SCl or SO_2Cl groups, though formed by treatment of disulphides and sulphydryl compounds with chlorine, are not present in wool since no combined chlorine could be found in chlorinated wool. If they are formed at all they can only be present as intermediates which decompose with elimination of hydrochloric acid.

The cystine fraction (i.e. 25 per cent.) capable of reacting with chlorine in alkaline solution is converted rapidly in the cold to lanthionine[83] as in the reaction with alkaline permanganate. The alkalinity of the solution would not under the conditions of the oxidation experiments bring about any significant conversion of cystine to lanthionine, and the hypochlorite and permanganate ions must therefore play a part in this reaction. The state of combination of the ejected sulphur atom is unknown and has not been identified either in the present case or when lanthionine is formed by alkalis.

Formation of new cross-links by hypochlorite is supported by the observation that above pH 8·5 chlorine is ineffective in reducing shrinkage (see p. 282), and by the load–extension data given in Table 8.10. Although chlorination at pH 2 greatly reduces the elasticity of the fibre, reaction at pH 10 only effects a small decrease, even when the same proportion of cystine has reacted at the two different pH values. Furthermore, a peracetic-acid treatment sufficient to oxidize all the disulphide bonds does not render wool which had been chlorinated at pH 10 soluble in dilute ammonia.

The reaction of bromine with wool has not been investigated in detail, although it has been used for rendering wool non-felting and

TABLE 8.10

The reduction in work (R.W.) obtained on stretching the fibres to a constant extension, after treatment with chlorine solutions at pH 2 and 10 at room temperature[83]

pH	Chlorine reduced (g./100 g. of wool)	Percentage cystine oxidized (percentage of that originally present)	R.W. at pH 7
2·0	4	16·5	21·0
	6	25·5	35·0
10·0	4	20·0	3·3
	6	22·2	4·6
	8	—	10·4
	10	22·4	14·4

can therefore be assumed to sever the SS bond. No indications have been obtained to suggest that this reagent should not behave like chlorine, and Consden, Gordon and Martin[77] found that cysteic acid is produced on hydrolysis of brominated wool. Fluorine reacts rapidly with wool and is known to render this unshrinkable.[73] In the early stages of the reaction, only gaseous sulphur fluorides (SF_6 and S_2F_2) are given off by the wool, and no hydrogen fluoride or carbon fluorides could be detected, although some of these compounds were given off in severe treatments. Because of the extremely reactive nature of fluorine, the reaction had to be examined in a gaseous system, and a very detailed study could not be made. It was found, however, that a small amount of combined fluorine remained in the wool. In view of the elimination of SF_6, it seems unlikely that cysteic acid is produced, as it cannot be found in the initial reaction. Wool which had been exposed to fluorine gas for about 30 minutes contained some cysteic acid in its acid hydrolysate, but no quantitative study has been made. Hudson and Alexander[73] considered that fluorine reacted only with the SS bond at first since no hydrogen fluoride was eliminated. This is, however, inconclusive as it is possible that limited amounts of hydrogen fluoride formed could have been retained by the wool by strong hydrogen bonding or in combination with the amino groups. In view of the fact that fluorine is so effective an anti-shrink agent it now seems probable that it cannot be as specific in its reaction with the SS bond as was at first thought (see p. 283).

Iodine, unlike the more reactive halogens, does not oxidize the SS bond, and Blackburn and Phillips[100] were unable to detect any reduction in the cystine content of wool after very prolonged treatment with cold aqueous solutions of iodine in potassium iodide. Haller[101] and Hove[102] found, however, that iodine both combined with and was absorbed by wool in large quantities and an uptake of up to 16 per cent. of iodine has been observed. The total quantity taken up varies with the concentration of iodine in solution and decreases with increasing pH; the iodine so combined could be divided into three parts:[100]

    (i) combined iodine which could not be removed by washing, and which is thought to have substituted in the benzene nucleus of tyrosine;
   (ii) absorbed iodine which can be washed out as free iodine;
  (iii) anionic iodine which can be washed out as the iodide ion.

It is postulated[100] that the absorbed and anionic iodine are held as a quaternary ammonium complex with the free NH_2 groups of the amino acids lysine and arginine, and the following reaction mechanism has been postulated which, it should be pointed out, is highly tentative and lacks direct experimental proof:

$$R.COO^-.^+NH_3.R$$
$$\downarrow KI+I_2$$
$$R.COOK \ldots I_3^- NH_3^+ . R$$

Reaction with chloramines. In order to reduce the speed of the chlorination of wool (see p. 164), nitrogen compounds have been added[103] to the solution which under the acid conditions employed produce chloramines:

$$R.NH_2 + Cl_2 \rightleftharpoons RNHCl + HCl$$

Other compounds such as chloramine T, dichloracetamide, chlorphthalimide[104] and chlorsulphamic acid[105] have also been used to render wool unshrinkable. Alexander, Carter and Earland[106] showed that in their general reactivity with wool they are similar to solutions of chlorine but there are distinct differences. It was shown[106] that although all the chloramines reacted rapidly with wool, only those which had a strongly electro-negative group adjacent to the reacting NCl group were capable of rendering it unshrinkable. Thus, alkyl chloramines were completely ineffective, chlorurea had a limited effectiveness and chlorsulphamic acid was highly effective. The quantity of chloramine required to reduce shrinkage was considerably less in the presence of high concentrations of chloride ions; no such effect was observed in treatments with solutions of chlorine. Extensive treatment with chlorurea and chlorsulphamic acid rendered wool completely soluble in dilute ammonia and produced powerful supercontraction of the fibres; a similar treatment with chlorine water does not bring about either of these changes. A further significant difference between the chloramines and chlorine is that after reaction with the former, wool does not retain any residual oxidizing capacity. From Table 8.11 it is seen that chlorsulphamic acid oxidizes the SS bond as effectively as acid solutions of chlorine both in the absence and presence of chloride ions. The attack on the tyrosine in wool, however, is greatly enhanced by the presence of chloride ions in solution,

TABLE 8.11

Loss of cystine and tyrosine after treatment with chlorsulphamic acid[106]

Weight of available chlorine (%)	Tyrosine oxidized* (%)		Cystine oxidized† (%)	
	No NaCl	10% NaCl	No NaCl	10% NaCl
3·0	5·50	10·3	22·8	16·3
5·0	6·56	18·9	24·6	22·4
10·0	14·54	35·1	53·6	45·4
20·0	46·10	68·4	77·5	83·1
50·0	—	—	97·0	96·3

* The tyrosine content of the untreated wool was 5·5%.
† The cystine content of the untreated wool was 10·4%, and loss of cystine was calculated from this value.

and in their absence chlorsulphamic acid is much less effective in attacking this group than chlorine. It will be shown that the influence of salt on the non-shrink action can be interpreted in terms of the attack on the tyrosine (p. 283). The products of oxidation with chlorsulphamic acid have not been investigated in detail although cysteic acid has been found in the hydrolysate of treated wools.

Although it was known[106] that the assumption that chloramines react with wool via hydrolysis to free chlorine which effects a simple chlorination of the wool was untenable, the precise mechanism of these reactions remained obscure. Later work[107] has shown that different chloramines do not necessarily react with wool by the same mechanism. Thus chlorurea, which is a neutral molecule, diffuses into the fibre where the chlorine produced by the equilibrium

$$NH_2 . CO . NHCl + H^+ + Cl^- \rightleftharpoons NH_2 . CO . NH_2 + Cl_2$$

reacts with the keratin. This explains why the properties of the wool after reaction, e.g. a nearly complete lack of surface degradation, differ from those of wool which has been treated with chlorine in solution, where obviously reaction proceeds from the outside to the interior of the fibre. Sulphamic acid, on the other hand, is a strong acid and may be titrated with wool. Kinetic studies[107] indicate that the rate-determining step in the reaction between chlorsulphamic acid and wool is the ion-exchange process that determines the rate of access of the chlorsulphamic acid to the wool. This rate is reduced by increasing the concentration of other anions, including chloride ions, in the solution.

Bromamines

Compounds containing the >NBr grouping are, in general, neither as stable nor as readily prepared as the corresponding chlorine derivatives. N-Bromoacetamide and N-bromosuccinimide are, however, well defined stable bromamines and their reactions with wool have been studied in some detail.[108] Reaction with wool occurs in acid solution, and to produce non-felting wool it is unnecessary for the corresponding anion to be present as is the case of the analogous chlorine compounds. These reactions are essentially brominations, since considerable amounts of free bromine can be detected in solutions during reaction, and they probably proceed as follows:

$$W + >NBr + H_2O = WO + >NH + H^+ + Br^-$$
$$>NBr + H^+ + Br^- = >NH + Br_2$$
$$W + Br_2 + H_2O = WO + 2H^+ + 2Br^-$$

where W = wool.

The precise mechanism of the initial reaction is obscure, since under the reaction conditions used the bromamines did not undergo any appreciable hydrolysis.

Chlorine peroxide

In a detailed investigation on the action of chlorine peroxide with amino acids and proteins, Schmidt and Braunsdorf[109] found that this reagent confines its attack to cystine and tyrosine. These workers did not study the more recently isolated amino acids, tryptophan and methionine, which can also be oxidized by chlorine peroxide.[94] As they are, however, present in wool only to a very limited extent they need not be considered here. Das and Speakman[110] examined the action of chlorine peroxide on wool in considerable detail and found that it was capable of oxidizing all the SS bonds, bringing about supercontraction and rendering wool non-felting. The cystine was not oxidized exclusively to cysteic acid but some sulphate was also formed and this shows that chlorine peroxide is much less specific than peracetic acid. This point is also indicated by the colour changes which take place on treating wool with this reagent since the wool turns yellow, brown and eventually white again. Qualitative analyses by paper chromatography showed that tyrosine was oxidized but no analytical data giving the relative rates of reaction with these two centres in the wool molecule are available. During the reaction the wool partly dissolves and the remainder is readily soluble in dilute alkali. High molecular weight fractions[111] could be obtained but these were of lower molecular weight than those obtained with peracetic-acid-treated wool (p. 372), and it must therefore be assumed that chlorine peroxide has broken the main chain, possibly at the point where tyrosine is incorporated.

Acid solutions of sodium chlorite[112] have been used for producing unshrinkable wool, but sodium chlorate does not react with wool except at high temperatures or under very acid conditions when it is known to dissociate, in part at least, into free chlorine.

Bromic acid (i.e. bromate in acid solution) renders wool non-felting and has been shown to oxidize the disulphide bond[113] but no reaction appears to take place in neutral solution. Iodates oxidize cystine[114] and were shown to render wool non-felting[115] and it seems probable that the disulphide bond in wool is oxidized. Neither of these reagents is specific as tyrosine groups are also oxidized.

Activated reactions

It has been shown that the course of the reactions between wool and chlorites[116] or bromates[117] is greatly influenced by the addition of reducing agents such as formaldehyde, sulphides and thiocyanates. In general the wool is rendered more resistant to felting. The reactions are very complex, free halogens, halogen oxyacids and free radicals being possible intermediates.

Relationship between disulphide bond breakdown and the production of non-felting wool

In Chapter 2 the different ways of bringing about non-felting have been discussed. The method which has received the most detailed study, and which is used in all the industrial non-shrink processes, consists of degrading the surface of the fibre which causes a loss of the directional friction effect (D.F.E.) without seriously weakening the fibre. As already seen, it is possible to prevent fibres from felting by modifying their elastic properties, and Whewell[118] showed that fibres treated with hydrogen peroxide lost all tendency to felt as a result of the general weakening brought about by this reagent, although they still possessed a considerable D.F.E. It is therefore clear that all degradative reagents are capable of bringing about a measure of non-felting if the reaction is extensive. Speakman *et al*.[93] pointed out in 1938 that all reagents which are capable of bringing about non-felting by surface degradation without extensive modification of the elastic properties were capable of reacting with the SS bond in wool. They postulate that all reagents capable of reacting with the SS bond, and in particular those which reacted without the formation of new cross-links, should bring about non-felting by surface degradation.

Direct support for this concept was provided by the experiments[119] with wool in which the disulphide bonds had been replaced by more stable cross-links such as $-C.S.C-$; $-C.S.(CH_2)_n.S.C-$ and $-C.S.Hg.S.C-$. Such wools could no longer be rendered non-felting by treatment with chlorine, alcoholic solutions of potassium hydroxide, suphuryl chloride, chlorsulphamic acid, permanganate and fluorine. In addition it is found that chlorine between pH 2–8·5 renders wool completely unshrinkable, but in more alkaline solutions this is not the case because of the formation of lanthionine.[83] It was concluded[119] that these reagents, though capable of severing the SS bond, cannot disrupt the more stable cross-links. Sufficient lanthionine is formed already when wool is dyed at a pH greater than 7 to reduce the effectiveness of these oxidizing agents in reducing felting. Thus if wool is dyed at pH 8, twice as much chlorine is required to prevent felting than with an acid-dyed sample.

The theory relating disulphide-bond breakdown to the production of non-felting has proved most useful. It has been responsible for the very great development in non-shrink processes during the last two decades and has provided the impetus for some of the chemical studies of the reactivity of the disulphide bond which have been reported in this chapter. This simple theory has however to be modified to be in harmony with later findings.

Alexander, Carter and Earland[81] showed that peracetic acid was unable to bring about non-felting by surface degradation, and it has been well known for a long time that reducing agents such as bisulphite and thioglycollic acid cannot produce unshrinkable wool.

All these reagents sever the SS bond, but react specifically in that they do not attack other groupings in the wool molecule. Actually a certain measure of unshrinkability can be obtained with aqueous solutions of peracetic acid but only after severe degradation which results in loss of elasticity. Admittedly, peracetic acid from organic solvents is able in this way to render wool non-felting by completely solubilizing the cuticle of the wool, since a fibre without scales has no D.F.E. Under similar conditions hydrogen peroxide, which also oxidizes the disulphide bond, fails to bring about unshrinkability since it cannot effect dissolution of the scales. Dilute acid solutions of chlorine and 1·6 per cent. peracetic acid solution react with wool at roughly the same rate and oxidize 14 per cent. of the cystine in 2–3 minutes at 20° C. Yet this amount of reaction with chlorine (3 per cent. on the weight of wool) completely removes the D.F.E. of the fibres, whereas the peracetic acid hardly reduces it. That some cuticular factor is involved is indicated by the findings of Farnworth[119a] that alcoholic pre-extraction or the use of alcoholic solutions allows wool to be rendered unshrinkable by, e.g., bisulphites.

It is difficult to avoid the conclusion that although attack on the disulphide bond is essential, another point in the wool structure must also be attacked. All the effective non-felting agents such as chlorine, permanganate, chlorine peroxide and chloramines are of course non-specific in their reaction and attack many other groups besides the SS bond. Moreover, the effectiveness of combined treatments of two reagents in rendering wool non-felting also suggests that another group besides the SS group must be attacked. Thus, neither bisulphite nor alkaline solutions of permanganate by themselves render wool non-felting, but when applied successively give an effective non-shrink finish.[121] Similarly a mild treatment with peracetic acid followed by one with alkaline permanganate renders wool unshrinkable while a much more severe treatment by one of these reagents alone does not do so.[115] On the basis of these observations Alexander, Carter and Earland[81] have postulated that to obtain the degree of surface degradation necessary to remove the D.F.E. the SS bond must be severed and the main chain broken. It seems probable that most of the common anti-shrink agents attack the tyrosine residue at the main chain for the following reasons.

(1) Although chlorine is capable of oxidizing all the amino acids of wool it reacts preferentially with tyrosine and cystine and in wool oxidizes the other amino acids only very slowly.[95]

(2) Permanganate at pH 2 reacts rapidly with cystine and tyrosine and at room temperature not detectably with other individual amino acids.[120] In wool, acid permanganate does attack amino-acid residues other than cystine and tyrosine but only very slowly.

(3) Schmidt and Braunsdorf[109] showed that chlorine dioxide oxidized only tyrosine and cystine and it appears to be specific for these amino acids even when they are combined in proteins, although no study comparable with that made with chlorine, permanganate and peracetic acid has been carried out with this reagent.

(4) Chloramides and chlorsulphonamides render wool unshrinkable much more readily[122] in the presence of a high concentration of chloride ions, the presence of which was shown from analytical data not to influence the reaction with the disulphide bond but to promote the reaction with phenolic groups such as the tyrosine residue in wool (see p. 279). The necessity for oxidizing a group other than the disulphide bond was demonstrated unambiguously with chlorsulphamic acid since, using the equivalent of 3 per cent. of chlorine on the weight of wool, the rate of reaction and the quantity of cystine oxidized were approximately the same in the presence and absence of salt and yet only in the former case was the wool rendered non-felting. The only apparent difference was that in the presence of salt the tyrosine content of the wool was reduced by 10·3 per cent. as compared with only 5·5 per cent. in the absence of salt (see Table 8.11).

Since publication of the modification of Speakman's theory[93] relating disulphide breakdown to non-felting, proposed by Alexander, Carter and Earland, subsequent experimental work produced by different workers is contradictory as to the modification's validity. It has been shown[123] that a mixture of peracetic acid and hypochlorite is an excellent non-felting agent. Since peracetic acid alone is nearly ineffective for anti-felting and hypochlorite readily oxidizes tyrosine,[95] this must be regarded as strong evidence in favour of the modified theory. McPhee has found that many oxidizing agents, in addition to those described by Alexander et al.,[81] can attack the disulphide bond extensively without causing shrinkproofing.[124] McPhee has also examined the action of permanganate in neutral saturated sodium chloride solution on wool[124] and has found that the loss of cystine is the same for permanganate treatments in sodium chloride or in water, yet the former, but not the latter, shrinkproofs wool. This is analogeous to the position with chlorsulphamic acid which Alexander et al. have put forward in support of their theory.

Analysis of treated wools for loss of tyrosine, however, suggests that it is unnecessary to attack this residue to produce unshrinkability. Diehl and Zahn[125] have analysed wool after reaction with chlorine under various conditions of pH, alone and in the presence of permanganate. Their main conclusions were that in the production of non-felting wool a reduction in cystine content is not invariably accompanied by a reduction in tyrosine, although the tryptophan content is always greatly reduced. McPhee[124] has also reported that shrinkproofing effects may be obtained without detectable change in amino acids other than cystine, or simultaneous peptide bond hydrolysis. Although the modified theory is capable of predicting reagents or mixtures of reagents, which in aqueous solution are capable of rendering wool non-felting, and there is much experimental evidence in its support, it must be concluded that the exact nature of the chemical changes leading to shrinkproofing is unknown.

REFERENCES

[1] BAILEY, *Biochem. J.*, 1939, **31**, 1396.
[2] CUTHBERTSON AND PHILLIPS, *Biochem. J.*, 1945, **39**, 7.
[3] SIMMONDS, *Proc. Int. Wool Text. Res. Conf.*, Australia, 1955, C65.
[4] GLYNN, *Text. Res. J.*, 1958, **28**, 744.
[5] MYERS, *J. Text. Inst.*, 1959, **50**, T494.
[6] EARLAND, *Text. Res. J.*, 1961, **31**, 492.
[7] LEWIS, ROBSON AND TILER, *Second Quinquennial Wool Text. Res. Conf.*, (*J. Text. Inst.*), Harrogate, 1960, T653.
[8] CORFIELD AND ROBSON, *Proc. Int. Wool Text. Res. Conf.*, Australia, 1955, C79.
[9] ASTBURY AND WOODS, *Phil. Trans. Roy. Soc.*, 1932, **232A**, 333.
[10] PATTERSON, GEIGER, MIZELL AND HARRIS, *J. Res. N.B.S.*, 1941, **27**, 89.
[11] ELÖD, *Kolloid Z.*, 1941, **96**, 284.
[12] ELÖD, NOWOTNY AND ZAHN, *Kolloid Z.*, 1940, **93**, 2.
[13] GODDARD AND MICHAELIS, *J. Biol. Chem.*, 1934, **106**, 605; 1935, **112**, 361.
[14] HARRIS, MIZELL AND FOURT, *Ind. Eng. Chem.*, 1942, **34**, 833.
[15] GEIGER, KOBAYASHI AND HARRIS, *J. Res. N.B.S.*, 1942, **29**, 381.
[16] HARRIS AND BROWN, *Symp. Fibrous Proteins*, 1946, p. 203 (publ., Soc. Dyers & Col, Bradford).
[17] ALEXANDER, CARTER AND HUDSON, *J. Soc. Dy. Col.*, 1949, **65**, 152.
[18] ELSWORTH AND PHILLIPS, *Biochem. J.*, 1938, **32**, 837; 1941, **35**, 135.
[18a] SWAN, *Aust. J. Chem.*, 1961, **14**, 69.
[18b] SWAN in 'Sulphur in Proteins'. Academic Press, 1959, p. 11.
[19] STOVES, *Trans. Faraday Soc.*, 1942, **38**, 261.
[20] BLACKBURN AND PHILLIPS, *J. Soc. Dy. Col.*, 1945, **61**, 203.
[21] BARR AND SPEAKMAN, *ibid.*, 1944, **60**, 238.
[22] ELÖD AND ZAHN, *Textil Praxis*, 1949, p. 27.
[23] MIDDLEBROOK AND PHILLIPS, *Biochem. J.*, 1942, **36**, 294.
[24] MIDDLEBROOK AND PHILLIPS, *ibid.*, 1947, **41**, 218.
[25] STOVES, *Trans. Faraday Soc.*, 1943, **39**, 294.
[26] SCHÖBERL, *Collegium*, 1936, p. 412.

[27] CROWDER AND HARRIS, *J. Res. N.B.S.*, 1943, **30**, 47.
[28] SCHÖBERL, *Angew. Chem.*, 1940, **53**, 227.
[29] ZAHN, *Melliand Textilber.*, 1950, **31**, 481.
[30] SPEAKMAN, *Nature*, 1933, **132**, 930.
[31] HARRIS, *J. Res. N.B.S.*, 1935, **15**, 63.
[32] SPEAKMAN AND WHEWELL, *J. Soc. Dy. Col.*, 1936. **52**, 380.
[33] HORN, JONES AND RINGEL, *J. Biol. Chem.*, 1941, **138**, 141.
[34] SCHÖBERL, *Ber.*, 1947, **80**, 379.
[35] PHILLIPS, *Nature*, 1936, **138**, 121.
[36] STOVES, *Trans. Faraday Soc.*, 1942, **38**, 254.
[37] CUTHBERTSON AND PHILLIPS, *Biochem. J.*, 1945, **39**, 7.
[38] LINDLEY AND PHILLIPS, *ibid.*, 1945, **39**, 17.
[39] COCKBURN, DRUCKER AND LINDLEY, *ibid.*, 1948, **43**, 438.
[39a] SPEAKMAN AND ASQUITH, *Proc. Int. Wool. Text Res. Conf. Aust.*, 1955, C.302.
[40] SWAN, *Nature*, 1957, **179**, 965.
[41] BERGMANN AND STATHER, *Hoppe-Seyler Z.*, 1926, **152**, 189.
[42] NICOLET, *J. Amer. Chem. Soc.*, 1931, **53**, 3066.
[43] TARBELL AND HARNISH, *Chem. Rev.*, 1951, **49**, 1.
[44] NICOLET, SHINN AND SAIDEL, *J. Biol. Chem.*, 1942, **142**, 609.
[45] EARLAND AND RAVEN, *Nature*, 1961, **191**, 384.
[46] ZAHN, KUNITZ AND HILDEBRAND, *Second Quinquennial Wool Text. Res. Conf.* (*J. Text. Inst.*), Harrogate, 1960, T740.
[47] ZAHN AND OSTERLOH, *Proc. Int. Wool Text. Res. Conf.*, Australia, 1955, C-18.
[48] SCHÖBERL AND WAGNER, *ibid.*, C-11.
[49] SWAN, *ibid.*, C-25.
[50] FRENEY AND LIPSON, *Nature*, 1940, **145**, 25.
[51] WOOD, British Patent 603,494.
[52] BLEACHERS ASSOCIATION, British Patent 546,529.
[53] MAUTHNER, *Hoppe-Seyler Z.*, 1912, **78**, 32.
[54] SCHÖBERL AND HAMM, *Biochem. Z.*, 1948, **318**, 331.
[55] SWAN, *Current Trends in Heterocyclic Chemistry*, Butterworths, London, 1958, p. 65.
[56] EARLAND AND RAVEN, *J. Amer. Chem. Soc.*, 1958, **80**, 3430.
[57] PHILLIPS, *Symp. Fibrous Proteins*, 1946, p. 39 (publ. Soc. Dy. Col., Bradford).
[58] Fox, Ph.D. thesis, 1951 (London Univ.).
[59] BLACKBURN, *J. Soc. Dy. Col.*, 1957, **73**, 506.
[60] BLACKBURN AND LEE, *Biochim. Biophys. Acta*, 1956, **19**, 505.
[61] SPEAKMAN, *J. Soc. Dy. Col.*, 1936, **52**, 423.
[62] SMITH AND HARRIS, *J. Res. N.B.S.*, 1936, 16, 301; 1936, 16, 309; 1936, 17. 97; 1936, **17**, 577; 1937, **18**, 623.
[63] ALEXANDER AND EARLAND, British Patents 637,150, 637,140 (1950).
[64] ALEXANDER, CARTER AND EARLAND, *Biochem. J.*, 1950, **52**, 159.
[65] BREINE AND BAUDISCH, *Hoppe-Seyler Z.*, 1907, **52**, 159.
[66] CONSDEN AND GORDON, *Biochem. J.*, 1950, **46**, 8.
[67] SMITH AND HARRIS, *J. Res. N.B.S.*, 1937, **19**, 81.
[68] RUTHERFORD AND HARRIS, *J. Res. N.B.S.*, 1938, **20**, 559.
[69] ELÖD, NOWOTNY AND ZAHN, *Melliand Textilber.*, 1942, **23**, 313.
[70] STOVES, *Trans. Faraday Soc.*, 1942, **38**, 501.
[71] CUNLIFFE, SHARP AND ASHWORTH, British Patent 614,966.
[72] ANDREWS, *J. Biol. Chem.*, 1933, **102**, 253.
[73] HUDSON AND ALEXANDER, *Sym. Fibrous Proteins*, 1946, p. 193 (publ. Soc. Dy. Col., Bradford).
[74] HARRIS AND SMITH, *J. Res. N.B.S.*, 1936, **17**, 97; 1938, **20**, 563.
[75] HARRIS AND SMITH, *ibid.*, 1936, **17**, 577.
[76] CRAWSHAW AND SPEAKMAN, *J. Soc. Dy. Col.*, 1954, **70**, 81.
[77] CONSDEN, GORDON AND MARTIN, *Biochem. J.*, 1946, **40**, 580.
[78] TOENNIES AND HOMILLER, *J. Amer. Chem. Soc.*, 1942, **64**, 3054.
[79] SANGER, *Cold Spring Harbor Symposia*, 1947, **12**, 237.
[80] ALEXANDER, HUDSON AND FOX, *Biochem. J.*, 1950, **46**, 27.
[81] ALEXANDER, CARTER AND EARLAND, *J. Soc. Dy. Col.*, 1951, **67**, 23.

[82] BLACKBURN AND LOWTHER, *Biochem. J.*, 1951, **49**, 554.
[83] ALEXANDER, FOX AND HUDSON, *ibid.*, 1951, **49**, 129.
[84] MCLELLAND, WARREN AND JACKSON, *J. Chem. Soc.*, 1929, p. 1582.
[85] EARLAND AND KNIGHT, *Biochim. Biophys. Acta*, 1955, **17**, 457.
[86] WESTON, *ibid.*, 1955, **17**, 462.
[87] WOODIN, *Discussions Faraday Soc.*, 1953, No. 13, p. 281
[88] O'DONNELL AND THOMPSON, *Aust. J. Biol. Sci.*, 1959, **12**, 490.
[89] EARLAND, MACRAE, WESTON AND STATHAM, *Text. Res. J.*, 1955, **25**, 963.
[90] LAVINE, TOENNIES AND WAGNER, *J. Amer. Chem. Soc.*, 1934, 56, 242.
[91] TOENNIES, *ibid.*, 56, 2198.
[92] KAMMERER, British Patent 5,612 (1907).
[93] SPEAKMAN, NILSSEN AND ELLIOTT, *Nature*, 1938, **142**, 1035.
[94] ALEXANDER, unpublished.
[95] ALEXANDER AND GOUGH, *Biochem. J.*, 1951, **48**, 504.
[96] ALEXANDER AND FOX, *Proc. Int. Wool Text. Res. Conf.*, Australia, 1955, C-35
[97] EARLAND AND RAVEN, *Biochim. Biophys. Acta*, 1958, **27**, 41.
[98] LAVINE, *J. Biol. Chem.*, 1936, **113**, 583.
[99] ALEXANDER, GOUGH AND HUDSON, *Trans. Faraday Soc.*, 1949, **45**, 1112.
[100] BLACKBURN AND PHILLIPS, *J. Soc. Dy. Col.*, 1945, **61**, 100.
[101] HALLER, *Hebr. Chim. Acta*, 1930, **13**, 620.
[102] HOVE, *Angew. Chem.*, 1934, **47**, 756.
[103] French Patent 812,191; U.S.A. Patent 2,144,824.
[104] German Patent 647,566.
[105] U.S.A. Patent 2,427,097.
[106] ALEXANDER, CARTER AND EARLAND, *J. Soc. Dy. Col.*, 1951, **67**, 17.
[107] EARLAND, *J. Soc. Dy. Col.*, 1955, **71**, 89.
[108] EARLAND AND RAVEN, *Text. Res. J.*, 1954, **24**, 108.
[109] SCHMIDT AND BRAUNSDORF, *Ber.*, 1922, **55**, 1529.
[110] DAS AND SPEAKMAN, *J. Soc. Dy. Col.*, 1950, **66**, 583.
[111] SPEAKMAN, *J. Text. Inst.*, 1941, **32**, T83.
[112] OVERBEKE AND MAZINQUE, *Bull. Inst. Textile*, France, 1950, **18**, 11.
[113] LAVINE, Private communication.
[114] WILLIAMS AND WOODS, *J. Amer. Chem. Soc.*, 1937, **59**, 1408.
[115] HUDSON, unpublished.
[116] EARLAND AND JOHNSON, *Text. Res. J.*, 1952, **22**, 591.
[117] STATHAM, *Text. Res. J.*, 1957, **27**, 41.
[118] WHEWELL, CHARLESWORTH AND KITCHEN, *J. Text. Inst.*, 1949, **40**, 769.
[119] FARNWORTH, NEISH AND SPEAKMAN, *J. Soc. Dy. Col.*, 1949, **65**, 447.
[119a] FARNWORTH, *J. Soc. Dyers Col.*, 1961, **77**, 483.
[120] ALEXANDER, *J. Soc. Dy. Col.*, 1950, **66**, 349.
[121] RAYNES AND STEVENSON, British Patent 638,580.
[122] ALEXANDER, CARTER AND EARLAND, *J. Soc. Dy. Col.*, 1950, **66**, 538.
[123] DAVIDSON AND PRESTON, *J. Text. Inst.*, 1956, **47**, p.685.
[124] MCPHEE, *Text. Res. J.*, 1960, **30**, 329.
[125] DIEHL AND ZAHN, Private communication.

CHAPTER 9

Chemical Reactivity

(EXCLUDING THE DISULPHIDE BOND)

Acid hydrolysis

THE peptide links which give rise to the protein chain are very stable, although they can be hydrolysed by acids, alkalis and enzymes. With one exception, all the side chains of the amino acids in wool are substantially stable to acids. Some of the amino acids are decomposed by acids in the presence of carbohydrates, but no such interference arises with wool. In boiling $5N$–HCl, tryptophan is totally destroyed, whilst cystine, serine, and threonine are decomposed to analytically significant extents. Acid hydrolysis, however, is the most satisfactory method of converting wool fibres quantitatively into their individual amino acids.

Acid amide groups are attacked first by acid; this is followed by general hydrolysis of the main peptide chain when certain amino acid residues in the main chain are more easily split off than others.[1] Determination of the order of the amino acids in the chain by studying the products of partial hydrolysis have until recently been unsuccessful in the main owing to the exceptional experimental difficulties involved. The development of more delicate analytical methods for the separation and detection of the constituent amino acids and peptides in aqueous mixtures has simplified these studies considerably. Even so the resolution of the highly complex mixtures is extremely tedious, and in spite of the outstanding pioneer work of Martin, and his collaborators, relatively little progress in determining the order of the amino acids in the chains has been made (see p. 355). It is impossible to stop the hydrolysis at a stage corresponding to the formation of peptides of a given size, and under even the mildest conditions free amino acids are present in addition to lower peptides and polypeptides. Some polypeptide groups are relatively unstable, giving rise to the appropriate amino acid in the early stages of the hydrolysis. In particular aspartic and glutamic acids, and serine are readily extracted by partial hydrolysis while most of the protein is still in the polypeptide form.[1, 2]

Zahn and Meienhofer [3, 4] have shown that simple peptides, termed wool gelatins, can be isolated from wool merely by cold water extraction. These contain chiefly aspartic and glutamic acids, serine, glycine and alanine residues. Material of nearly identical composition has been isolated from cold dilute peracetic acid that has been used to oxidize wool.[5] Although protein in nature, in view of the ease with which they can be extracted from wool, it would appear to be better to regard these substances as extraneous material like the wool waxes, rather than an integral part of the keratin structure. These water-soluble peptides are probably material that has escaped complete keratinization.

The disintegration of the main chain produced by acids can be detected readily by measuring the change in the mechanical properties of the fibres. Many authors have commented on the degradative action of acids and it has been found that even the conditions used in acid dyeing lead to limited main-chain breakdown. This is not appreciated sufficiently and few workers realize that the acid dyeing process may be much more degradative to the wool than subsequent

TABLE 9.1

Changes in wool produced by N . HCl *at* 80° C.[6]

Time of treatment (hr.):	0	1	2	4	8
Nitrogen content (%)	16·5	15·4	16·0	15·1	14·8
NH_3–Nitrogen content (%)	1·00	0·77	0·52	0·35	0·29
Acid-combining power millim./g.	0·82	0·88	0·95	1·03	1·12
Cystine content	11·2	12·1	12·9	12·5	12·4
Fibre dissolved (%)	—	0·3	3·6	18·1	52·6
Strength expressed as (%) dry	100	83	75	51	4
Strength expressed as (%) wet	100	78	49	10	5
Proportion of peptide bonds hydrolysed (%)	0·00	0·92	2·58	4·78	35·70

finishing treatments which are carefully controlled in order to minimize damage, although Elöd, Nowotny and Zahn[6] have published an excellent paper on the effects of acids on the properties of wool.

TABLE 9.2

Supercontraction of fibres after treatment with acids and acidic salts at 100° C.[7]

Treatment	Time (hr.) of treatment	Contraction (%)
3% HCl	1·0	13·5
10% H_2SO_4	0·5	13·6
Conc. acetic acid	2·0	9·6
Conc. formic acid	2·0	20·8
1% Thorium chloride	5·0	10·0
0·3% Silver sulphate	5·0	21·4

The extensive weakening of fibres which results from the breakdown of a relatively small number of peptide bonds can be seen from the results in Fig. 9.1. The percentage of peptide bonds broken was determined from the increase in amino-nitrogen after treatment with acid. Tables 9.1 and 9.2 show the effect on the various mechanical and physical properties of the fibre after different treatments with acids.[6,7] On prolonged treatments, the swelling anisotropy which is so marked in native wool fibres (see p. 87) disappears. At the same time the X-ray configuration changes, and the fibre gives an X-ray diagram typical of dis-oriented β-keratin. Almost all acids cause super-contraction in the absence of any disulphide-bond breakdown (see Table 9.2). This effect was first discovered by Harrison[8] who found zinc chloride to be an effective supercontracting agent.

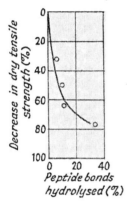

Fig. 9.1.—Relationship between *dry* tensile strength of wool fibres and peptide bond hydrolysis by 0·1N hydrochloric acid.

As expected, the most striking change in chemical properties of the fibres after acid treatment is the increase in acid- and alkali-combining power (Table 9.1). The increase is partly due to the amino and carboxyl groups liberated by the breakdown of the peptide bond particularly after prolonged treatments. The rapid initial increase in the acid-combining power, however, is due to the hydrolysis of the amides of glutamic and aspartic acid with the liberation of ammonia.

Approximately 45 per cent. of the acidic amino acids are combined with ammonia in native wool, and as was shown by Steinhardt[9] these amides are hydrolysed by acid in preference to the peptide bonds. Steinhardt and Fugitt[10] found that the hydrolysis of the peptide bond and more particularly the amide bond was increased at a given acid concentration by the presence of neutral salts. At constant ionic strength it was found that acids of high anion affinity (see p. 222) produced a greater amount of hydrolysis. Thus, by adding 0·1 N sodium sulphate to hydrochloric acid, the rate of deamidation was increased tenfold. In this way it proved possible to deamidate wool completely with only a very small amount of peptide hydrolysis. This effect has been interpreted by Peters and Speakman[11] by the Donnan membrane concept in that more free acid is present inside the fibre in the presence of neutral salts.

An important point to note from Table 9.1 is that the cystine content of the wool is unaffected by treatment with acids. Sanger[12] showed that the peptide bond adjacent to a cysteic acid group is very reactive and that cysteic acid peptides are hydrolysed under mild conditions. It seems probable, therefore, that wool which has been oxidized will be more liable to acid degradation than untreated wool. In support of this suggestion it is found that acid dyeing, after an oxidative non-shrink treatment, produces considerable damage.

As inferred previously, little is known about the optimum conditions for the complete hydrolysis of proteins, and widely varying conditions have been used by different investigators. An important paper by Gordon, Martin and Synge[13] stresses that the stability of the peptides varies greatly, and that no general rules can be formulated. With some proteins strong mineral acids do not appear to be sufficient for complete breakdown, and Miller and Du Vigneaud[14] have used a mixture of 20 per cent. hydrochloric and 50 per cent. formic acid for insulin. Phillips[15] has found that wool can be completely hydrolysed by refluxing with 10 N. hydrochloric acid for 4 hours and has used these conditions in all his work. To prevent aerial oxidation, however, the method of Sanford and Humoller[16] is to be preferred. These workers treat 0·5 g. of wool with 10 ml. of 5 N. hydrochloric acid in a sealed tube at 125° C. for five hours, and the authors have found these conditions to be highly satisfactory.

Alkaline hydrolysis

Alkali is more destructive and less selective than acid, so that smaller peptides are formed on alkaline hydrolysis of the peptide bonds. Some of the amino acids are decomposed, particularly cystine, and according to Warner and Cannan[17] arginine, histidine and serine are also destroyed. Tryptophan on the other hand is decomposed by acids but is stable in mild alkalis: strong alkalis decompose it to some extent and, for analytical purposes, it is preferable to use weaker alkalis such as strontium hydroxide.[17a] Solutions of barium hydroxide at the boil are sometimes used for alkaline hydrolysis of wool but, generally, complete alkaline hydrolysis has only been investigated to a limited extent.[18]

The main object of such studies has been to correlate the quantity of wool dissolving in alkali with the damage done to the fibres in a previous reaction or with the extent of cross-linking. Harris and

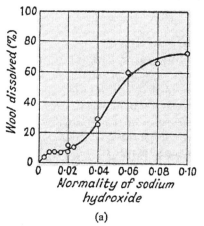

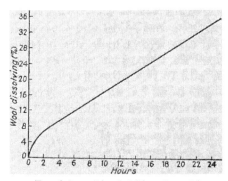

FIG. 9.2.— (b) solubility of wool.[6]
(a) Influence of concentration of sodium hydroxide applied for 1 hr. at 100° C.[20]
(b) Influence of time of treatment, 0·065N sodium hydroxide at 65° C.[19]

Smith[19] used a treatment with 0·10N sodium hydroxide at 65° C. for one hour as a standard test, and Zahn[20] has shown that similar results are obtained by using 0·02 N sodium hydroxide for the same time at 100° C., i.e. at a much more convenient temperature. As would be expected, the amount of protein dissolving depends on the time of treatment, temperature and concentration of the alkali (Fig. 9.2).

The alkali solubility increases when the wool is degraded either by peptide-bond breakdown or rupture of the disulphide bonds (see Table 9.3) and is a most useful method for assessing the damage done

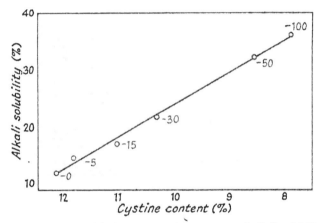

Fig. 9.3.—Relationship between cystine content and alkali solubility of wool irradiated with ultra-violet light.[19] Figures on the curves represent the duration of exposure in hours. (*Reproduced by permission.*)

in processing (e.g. effect of irradiation with ultra-violet light, see Fig. 9.3). Moreover, disorientation of the macromolecular structure resulting from the breaking of hydrogen bonds only (e.g. with lithium bromide or phenol) also greatly increases the alkali solubility although no covalent bonds have been broken.

In addition the test has been used effectively to determine the amount of cross-linking in the fibre, for Harris has shown that the alkali solubility decreases greatly when the disulphide bond is replaced by a more stable link (see Chapter 8). It is important to realize that alkali itself gives rise to such a link in wool, so that alkali solubility cannot be used to assess damage due to peptide-bond hydrolysis occurring in alkaline treatments. Elöd *et al.*[6] originally thought that

TABLE 9.3

Alkali solubility of wool after different pretreatments related to industrial processes

Treatment	Alkali solubility (%)	Related industrial process
None	10·5	—
0·1 N. hydrochloric acid, 65° C., 24 hr.	20·0	Acid dyeing
0·1 N. hydrochloric acid, 65° C., 240 hr.	80·0	—
Sulphuric acid, 125° C., 1 min.	19·4	carbonizing
2 vol. hydrogen peroxide, 50° C., 1 hr.	15·5	bleaching
2 vol. hydrogen peroxide, 18 hr. at pH 9	28·0	bleaching
Chlorine (3% on weight of wool) pH 1·5	20·3	rendering non-felting
Chlorine (3% on weight of wool) pH 9·0	17·3	rendering non-felting
1% potassium permanganate, 35° C., 1 min.	17·5	rendering non-felting
1% sodium bisulphite, 100° C., 1 hr.	15·2	clearing
10% sodium hydrosulphite, 50° C., 2 hr.	39·0	stripping

cystine played a minor rôle in the properties of the fibre, because of the lack of correlation between cystine content and strength of alkali-treated wool. This disparity is now known to be due to lanthionine formation.

In addition to cases of lanthionine formation, the alkali solubility also decreases when entirely new cross-links are introduced by treatment with one of the cross-linking agents discussed in Chapter 10.

Although the quantity of wool becoming soluble in alkali depends on the reaction conditions (Fig. 9.2), it seems that the first 10 per cent. can be extracted much more readily than the remainder. In addition to the indications from the alkali test, it is known that the alkali solubility of the wool is greatly reduced by treatment with the enzymes from pancreatin[19] which remove about 10 per cent. of the wool.

CHEMICAL REACTIVITY 295

The chemical basis of the alkali-solubility test is therefore rather indefinite, so that discretion must be used in employing it. Nevertheless it has been specified by the U.S. Quartermaster as a rapid means of detecting damage, and has been used widely for this purpose.

In contrast to the action of acids, even prolonged treatments with alkali cause little weakening of dry fibres although the cystine content has been reduced substantially. The fibres become weakened only after substantial quantities of the wool substance have gone into solution (Fig. 9.4).

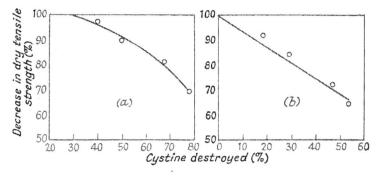

Fig. 9.4.—Relationship between *dry* tensile strength of wool fibres and reduction in cystine content by sodium carbonate (curve *a*) and hydrogen peroxide (curve *b*).[6]

It is surprising that even after very severe alkali treatments which dissolve substantial quantities of the protein, the fibre retains its α configuration, and a change to disoriented β keratin is brought about much more readily by acid than by alkali. Elöd and Zahn[21] made a detailed examination of the X-ray patterns of an alkali-treated wool sample, the cystine content of which had been reduced by 75 per cent. They were unable to find any change even in the longest spacings measurable, proving conclusively that the disulphide bond plays no part in the crystalline structure of the wool, in agreement with the conclusion of Astbury and Woods.[22] These workers found that fibres when stressed in caustic soda, for example in a 1 per cent. solution for 15 minutes at 20° C., contracted in length by 10 per cent. although the α pattern was retained. Zahn[23] found that fibres even contract in 1·0 N. sodium hydroxide without previous stressing, but no contraction occurs in more dilute solutions, even at higher temperatures. Fibres supercontracted in this way retain their α configuration but on immersion in boiling water contract further and are changed into a disoriented β structure. In this respect the fibres resemble those treated with peracetic acid very closely (see p. 81). Zahn attributes this behaviour to hydrogen-bond breakdown and claims that only concentrated solutions are effective. It seems more likely that dilute alkali solutions cause cross-linking because of lanthionine formation, whereas more concentrated solutions cause complete disulphide-bond breakdown. This alternative explanation is supported by the similarity between fibres treated with concentrated sodium hydroxide and peracetic acid. It is interesting to note that very concentrated solutions of sodium hydroxide no longer behave as strong alkalis since a hydrate is formed and the solution contains few free hydroxyl ions. Wool can be treated in these solutions without severe degradation and a rapid treatment with concentrated sodium hydroxide has been used for bleaching. If however, a weaker solution, e.g. 10 per cent. is used, the wool completely dissolves. A similar hydrate of potassium hydroxide is formed but at much higher concentrations. (See p. 258.)

Enzymatic hydrolysis

Block and Vickery[24] suggested that one criterion for distinguishing keratin from other proteins is its resistance to proteolytic enzymes. Since this suggestion was put forward a number of detailed investigations have been made in this field. It is now known that although whole wool can be disintegrated by enzymes, in general not more than 10 per cent. of the wool is digested even when the fibre is mechanically or chemically broken up or has its cuticle removed.

Of all the enzymes pancreatin has been examined in most detail and Meunier and others[25] were able to disintegrate wool into its morphological components (i.e. cuticle and spindle cells) by prolonged action of this enzyme on wool which had been partly degraded in alkali. Elöd and Zahn[26] showed that chemical damage was not necessary as finely sectioned fibres could also be disintegrated by this enzyme into their morphological components and the preparations so obtained were suitable for electron-microscopy. Trypsin[27] has been used in exactly the same manner as pancreatin for breaking up wool fibre, and in particular the classical work of Woods on the behaviour of spindle cells was carried out in this way. In these enzymatic treatments, when the wool has not been previously degraded, more than 90 per cent. of the wool remains in the solid state although disintegrated into cortical cells and scales. The properties of these morphological components have been investigated mainly after isolation in this way. The fibre is much more resistant to the action of pepsin, and Geiger et al.[28] were able to disintegrate wool into its components with this enzyme only after partial reduction of the disulphide bond. When all the disulphide bonds had been reduced, the bulk of the fibre could then be digested by pepsin and became soluble. From this it was concluded that the SS bonds prevent attack by enzymes and that their reduction is a prerequisite for enzymatic digestion.

In their detailed study of the action of pancreatin on wool, Elöd and Zahn[29] disproved this conclusion at least for this enzyme. They examined the influence of different pretreatments on the extent of digestion by pancreatin which, as has already been mentioned, can only digest about 10 per cent. of untreated wool. From Table 9.4 it is seen that extensive SS-bond breakdown alone, which may lead to very high alkali solubilities, does not lead to increased

TABLE 9.4

Relationship between chemical degradation of wool fibre and quantity digested by pancreatin[29]

Pretreatment	X-ray diagram of fibres after treatment	Diminution in cystine content %	Percentage dissolved by pancreatin in 3 days
None	α	0	8
5% thioglycollic acid; 42° C., 20 hr.	α	50	22
0·1 N. sodium hydroxide, 50° C., 4 hr.	α	67	23
0·1 N. hydrochloric acid; 100° C., 23 hr.	disorientated β	0	80
Pure formamide, 122° C., 3 hr.	disorientated β	25	87

digestion in pancreatin. At the same time when X-ray analysis shows that the structure is changed to a disorientated β configuration the fibres become highly susceptible to attack by the enzyme. Thus pretreatment in acid or in hydrogen-bond-breaking solvents without any SS-bond breakdown makes it possible for the enzyme to digest the fibre almost completely, whereas a fibre in which 50 per cent. of the disulphide bonds had been reduced is not attacked so extensively. When all the disulphide bonds have been broken, however, the fibre can be converted into a disorientated β form by very mild treatments such as boiling in water for a few minutes, when the fibre becomes extensively soluble in pancreatin solutions. In Fig. 9.5 the rate of

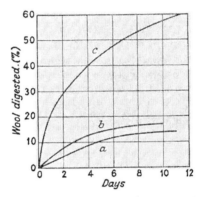

FIG. 9.5.—Relationship between quantity of wool dissolved by pancreatin and crystalline structure.[29]
(a) Untreated wool.
(b) 40 per cent. of cystine reduced by thioglycollic acid (α keratin).
(c) Above followed by treatment with 0·01N hydrochloric acid for 30 min. at 100° C. (disoriented β keratin).

enzymatic digestion of a reduced wool in the α form and a reduced wool in the disorientated β form is compared and brings out clearly the importance of molecular configuration. Even when a fibre is in the disorientated β form only 90 per cent. of it can be digested and the residue left behind is in the form of a membrane with a few adhering scales. (See p. 11.)

Papain has been studied in great detail by Schöberl[30] and by workers at the Wool Industries Research Association.[31] In common with all other enzymes it does not attack wool extensively unless this has received some pretreatment. Schöberl found that cyanide was very effective for this purpose and worked out the optimum conditions for the attack of this enzyme on the treated wool. At pH 10 the reaction appears to be at its maximum, and 50 per cent. of the wool could be dissolved, the remainder being left behind as spindle cells.

It seems probable that drastic pretreatment with cyanide changes the fibres into a disorientated β form. Trotman[32] showed that an unshrinkable finish could be obtained by treating wool with enzymes under mild conditions, and this led the workers at the Wool Industries Research Association to carry out their detailed investigation of the interaction of papain with wool since this proved to be the most effective enzyme for this purpose. In the presence of bisulphite, papain at pH 7 is able to dissolve up to 10 per cent. of the wool and by taking certain precautions its attack can be confined to the surface of the fibres to give a very permanent anti-shrink finish. In an effort to determine the mode of action of the enzyme, Blackburn[33] studied the peptides formed by papain under these conditions. A considerable number of different peptides was found but no definite conclusions could be drawn from this work.

In nature, the problem of a complete digestion by proteolytic enzymes has been solved in an ideal manner by the larvae of the clothes moth. Linderstrøm-Lang and Duspiva[34] have studied this digestion in detail and in particular the behaviour of the proteolytic enzymes in the digestive tract of the larvae. The digestive juices were found to have the surprisingly high pH of 9·9 and contained a substance capable of reducing disulphide bonds which was essential for the action of the proteolytic enzymes present. These workers conclude that the digestive juices of the larvae can digest keratin not because of any special enzymes but because of their reducing power which enables the ordinary enzymes to attack keratin. (see also Waterhouse *et al*.[34a]). In view of the work of Elöd and Zahn[29] it would be valuable to know whether the wool is disorientated during the high pH reduction.

Utilizing these results, Harris and his colleagues[35] were able to develop a treatment for rendering wool resistant to the attack of moths. They showed that if wool is first reduced and the SS bond rebuilt with alkyl dihalides so that the disulphide bond was changed into a $-S-(CH_2)_n-S-$ cross-link (see p. 339), wool could no longer be attacked by moths.

Detailed studies of the action of fungi on wool have been carried out by Siu and his colleagues[36] in the Quartermaster's Laboratories of the U.S. Army. They were able to show that *Microsporum gypseum* brought about hydrolysis of the disulphide bond and stressed that disulphide-bond breakdown is not essential for the attack by the complex enzyme system of this fungus and that the rate of digestion of a number of fibrillar proteins depended on their degree of crystallinity, a conclusion which is in complete agreement with the findings of Elöd and Zahn.[29]

The influence of heat

As already mentioned (p. 253) the rate of reaction of water with wool at 100° C. is slow, and only 20 per cent. of the cystine is lost in 24 hours. Above 100° C., however, the degradation caused by water is much more pronounced. The hydrothermal degradation of wool has been studied by Stirm and Rouette[37] and, more recently, by Lennox.[37a] Schmidt[38] showed that wool supercontracts on heating to a high temperature which according to Elöd and Zahn[39] is due to the presence of moisture. In the absence of water, only very small contractions occur (see Tables 9.5 and 9.6). Heating wool in the presence of water brings about a change in the crystal structure, and an

TABLE 9.5
Supercontraction of wool fibres produced by heating in sealed tubes with water for 3 hours at different temperatures[41]

Temperature °C.	Contraction (%)
100	0·0
115	3·5
120	22
125	31
130	33
135	38
140	38
160	53

X-ray picture of disorientated β keratin is observed after 10 minutes at 130° C. or 8 days at 100° C.

Rutherford and Harris[40] have shown that dry heat is much less damaging, which is clearly demonstrated by the data obtained by Zahn[41] in Table 9.6. It was observed that hair retains most of its chemical and physical properties after being heated for 24 hours at 120° C., and that serious modifications only occur when the temperature exceeds 150° C. In addition, in the dry state conversion of the α to disorientated β keratin only occurs at temperatures above 180° C.

According to Mecham and Olcott[42] who worked with horn keratin,

TABLE 9.6

The effect of dry heat for 24 hours on horse hair[41]

Temperature (°C.)	Colour	Contraction during heating (%)	Percentage Contraction in		Acid binding milliequiv/ 100g.	Alkali solubility (%)	Cystine content (%)
			5% NaHSO$_3$	20% Phenol			
120	white	—	24·5	35·5	98	18·0	13·5
140	white	—	24·6	35·7	—	—	—
150	white	—	22·4	31·3	—	—	—
160	yellowish	1·7	22·6	26·0	81	22·7	12·8
170	yellow	2·1	21·8	24·5	78	—	10·8
180	orange	2·4	18·5	18·5	75	22·2	8·6
188	light brown	2·4	12·2	15·0	67	24·4	7·8
210	dark brown	2·6	12·5	15·0	67	40·4	—
	black	—	—	—	55	99·0	6·8

general degradation only occurs above 180° C., and at a temperature above 100° C. new cross-links are produced between amino and carboxyl groups. Lundgren *et al.*[43] also came to the conclusion that amide cross-links can form by heating at 165° C. on the basis of the increased strength of their synthetic protein fibres. The decreased supercontraction above 160° C. (*cf.* Table 9.5) supports this contention, although it must be acknowledged that such interpretations from indirect evidence cannot be considered as conclusive, owing to the complexity of the reactions which are promoted by heating.

The strong resistance of wool to dry heat is also stressed by McCleary and Royer,[44] who conclude that heating dry wool to 150° C. for 1 hour brings about no degradation so long as care is taken to ensure that the fibre is neutral. As a result of these studies they conclude that finishing treatments involving the polymerization of synthetic resins in wool at high temperatures bring about a negligible amount of degradation. More prolonged heating, on the other hand, can bring about chemical modification of wool. Thus, Mazingue and Van Overbèke[45] found that heating for 7 hours at 160° C. reduced the side-chain amino groups by 30 per cent.

Ionizing radiations

In contradistinction to ultra-violet radiation (see p. 265), high energy radiation has little effect on wool. O'Connell and Walden[46] found that, unless the time of irradiation were prolonged, gamma radiation from cobalt-60 neither materially degraded the fibre nor produced new cross-linkages. The damage produced was considerably less than that caused by ultra-violet radiation.

The reactivity of tyrosine

The phenolic residue introduced into the wool structure by the amino acid tyrosine is a centre of high reactivity and probably influences the chemical properties of the fibre significantly. It has been suggested[47] that two tyrosine residues may unite to form a phenolic ether cross-link between peptide chains, but it is very doubtful whether such links exist in wool since this becomes largely soluble on disruption of the disulphide bonds only (p. 368). For this reason it is unlikely that any other covalent cross-links are present in the molecule, although the phenolic hydroxyl group is capable of forming powerful hydrogen bonds. The insolubility of the cortical cell membranes (see p. 368) in solvents which can break hydrogen bonds, after all the disulphide bonds have been oxidized, may, however, be due to ether linkages as the membrane becomes soluble after reaction with chlorine or chlorine peroxide, both of which react with tyrosine.[48] In the absence of more direct experimental data however, this suggestion can only be regarded as hypothetical.

Accessibility. All the tyrosine in wool (and in silk) can be methylated by diazomethane,[49] and it is probable that most, if not all of the tyrosine can also be acetylated by acetic anhydride. 85 per cent. of the tyrosine of wool can be converted to the dinitrobenzene ether by reaction of the phenolic group with 2:4 dinitrofluorobenzene.[50] Zahn and Kohler[51] showed that N. nitric acid is capable of nitrating all the tyrosine in wool in 24 hours at 70° C. when apparently the cystine is not oxidized. The nitrated wool shows new X-ray reflections at 40 and 83 Å and it is suggested that tyrosine repeats in the crystalline phase at these distances. Although inconclusive, this is the first evidence of a definite stereo-chemical order detectable by X-rays in protein fibres, and if confirmed will be of the greatest value in the further elucidation of the molecular arrangement in the peptide chains.

Treatment of globular proteins such as egg albumin with the Folin phenol reagent[52] and with ketene[53] revealed that some of the hydroxyl groups of tyrosine are inaccessible to these reagents. In addition, ultraviolet spectra of proteins show that ionization of tyrosine affects the position of its absorption band. Although the pK is at 10·2, a large proportion of the tyrosine gives the unchanged spectrum even at

pH 12.[54, 55] Cramer and Neuberger[54] in explaining the lack of ionization concluded that the tyrosine was inaccessible to the small hydroxyl ion, but a more probable explanation would seem to be that the hydroxyl groups take part in hydrogen bonding thus increasing the pK of combined tyrosine beyond that of the uncombined amino acid.[56] Although no similar spectroscopic investigation has been carried out with wool fibres, the soluble high molecular weight derivative, α keratose, exhibits exactly the same effect as egg albumin,[55] Further, there is no indication from alkali absorption data that the tyrosine in wool does in fact combine with alkali (i.e. ionize) at pH 12 when the phenolic group in tyrosine and tyrosine peptides are fully ionized since the amount of the alkali combined can be accounted for in the titration of the basic groups of lysine, arginine and histidine (see p. 212).

The Cramer–Neuberger effect (i.e. increase of the pK value of tyrosine) does not occur in denatured proteins or in α keratose, which has been treated with phenol, and the tyrosine appears to ionize at the expected pH. No alkali combination data is available for wool which has been contracted in phenol but by analogy it is to be expected that such wool would show increased alkaline combination at pH 12 due to ionization of the tyrosine.

Chemical reactivity of tyrosine. Mirsky and Anson[57] consider that the tyrosine residue may be responsible for the reducing action of proteins, but this has only been clearly demonstrated in the case of iodine when mono- and di-iodo tyrosine were identified in a number of proteins. Blackburn and Phillips[58] consider that the covalently bound iodine in wool treated with solutions of iodine in potassium iodide is present as di-iodo tyrosine although no direct chemical evidence is given. Richards and Speakman[59] have succeeded in converting 96 per cent. of the tyrosine residues in wool to 3:5–di–iodotyrosine using an alcoholic solution of iodine. The oxidation of the amino acid tyrosine itself with alkaline potassium permanganate gives rise to a mixture of oxalic, acetic and para-hydroxy-benzoic acids,[60] whereas the main product on oxidation with alkaline hypochlorite is 4-hydroxy-phenyl-acetaldehyde.[61] Oxidation products of the tyrosine in wool or other proteins have not been isolated, but indirect evidence suggests that an indole is formed on oxidizing a variety of proteins with potassium permanganate or ceric sulphate.[62] The formation of a quinone on oxidation of the combined tyrosine with permanganate[63] and chlorine peroxide[64] has also been postulated.

After oxidation with permanganate, wool assumes a yellow tinge, the intensity of which varies with the extent of treatment with alkali, but which disappears when the wool is acidified.[63] In this way the wool behaves as an indicator, and the yellow colour is quite distinct from that produced by heat, alkali or sodium hypochlorite, which cannot be discharged by acid. Of all the amino acids present in wool, tyrosine is oxidized in aqueous solution by acid or alkaline permanganate to a yellow colour which shows an indicator-like behaviour, so that this yellow colour in the wool may be identified with the oxidation product of the tyrosine. The product is thought to be a quinone by analogy with the similar oxidation products of para-substituted aromatic compounds, e.g. p-hydroxy- and p-amino-benzoic acid. Oxidation with chlorine peroxide causes a much deeper colour, and only part of it acts as a reversible indicator. Das and Speakman,[64] who do not appear to have observed the indicator-like behaviour, suggest the formation of an ortho-quinone although by analogy with the permanganate reaction a para-quinonoid structure is more likely. The whole question, however, requires further investigation particularly by direct chemical analysis. Tyrosine is not attacked by peracetic acid either as the free amino acid or when incorporated in the wool structure.[48] Certain inorganic per-acids, however, are less specific and the tyrosine in wool is oxidized by permonosulphuric acid (Caro's acid).

Earland, Stell and Wiseman[65] examined the action of potassium nitroso-disulphonate on a number of proteins including wool and silk. This reagent is very effective in oxidizing phenols to quinones, reactions which in many cases occur quantitatively.[66]

$$HO\text{-}C_6H_4\text{-} + 2ON(SO_3K)_2$$
$$\rightarrow O=C_6H_4=O + HON(SO_3K)_2 + HN(SO_3K)_2$$

$$HO\text{-}C_6H_4\text{-}OH + 2ON(SO_3K)_2 \rightarrow O=C_6H_4=O + 2HON(SO_3K)_2$$

The colouration produced with wool was nearly identical with that produced by chlorine dioxide and there can be little doubt it is due to oxidation of tyrosine residues.

The division of tyrosine into two fractions of varying reactivity. The combined tyrosine, like the combined cystine, is not all equally reactive, and some reagents, e.g. sodium hypochlorite and potassium permanganate, can only react with 30 per cent. of that present in wool.[48] The more powerful oxidizing agents, chlorine and chlorine peroxide, can eventually remove all the tyrosine (see Fig. 9.6). Alkaline permanganate solutions can react with all the tyrosine and in this respect the tyrosine behaves quite differently from the cystine in

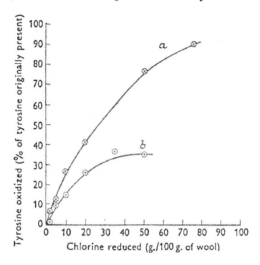

FIG. 9.6.—Diminution in tyrosine content of wool oxidized with different quantities of chlorine at pH 2 (curve a) and pH 10 (curve b).[48]

wool, which is very resistant to attack by alkaline permanganate (see p. 270). It is important to bear this difference in mind when attempting to correlate the factors responsible for the non-reactivity of these two amino acids with certain oxidizing agents. When mixtures of cystine and tyrosine in solution are oxidized it is found that permanganate oxidizes the tyrosine preferentially while chlorine reacts much more readily with the cystine.[67] In wool no such preferential action could be observed and both amino acids were oxidized to the same extent by different quantities of chlorine and acid permanganate. This provides further support for the suggestion that the chemical reactions are rapid compared with the rate of diffusion of reactant through the fibre (see p. 163).

As in the case of cystine, the inability of 70 per cent. of the tyrosine to react with hypochlorite and acid permanganate cannot be due to (1) morphological causes, since powdered horn exhibits exactly the same behaviour, (2) differences between crystalline and amorphous regions, since supercontracted fibres contain the same quantity of non-reactive tyrosine, (3) the environment of polar groups, which are responsible for the Phillips sub-fractions, since the proportion of the two fractions is not changed on esterification or acetylation, (4) the order of the amino acids since different tyrosine peptides (excepting cystine-tyrosine peptides) are equally readily oxidized by chlorine and hypochlorite.[67]

It has been found[48] that chlorine and hypochlorite are equally effective in oxidizing the tyrosine in simple di- or tri-peptides except those containing cystine.[67] In the proteins, silk and α keratose, chlorine reacts much more readily with the tyrosine than hypochlorite, but if a sufficient quantity of the latter is applied all the tyrosine can be oxidized.[48]

It will be shown later (see p. 391) that the wool fibre can be considered as a two-phase system made up of micelles of high molecular weight embedded in a cement of low molecular weight material. In such a model it can be supposed that chlorine would oxidize the tyrosine in both phases, whereas hypochlorite would oxidize preferentially only the tyrosine in the low molecular weight fraction. Excess hypochlorite would then react with other amino acids in the low molecular weight fraction rather than attack the tyrosine in the less reactive high molecular weight fraction. The 70 per cent. of tyrosine which is not oxidized by hypochlorite and permanganate is therefore not sterically inaccessible but is present in a phase of the wool which is less reactive as a whole. Chlorine at pH 2 and hypochlorite at pH 10 react equally readily with the 30 per cent. fraction of low molecular weight. It must be pointed out that this general explanation is based on very indirect evidence only, and must be regarded as a tentative suggestion pending more quantitative analytical work. The importance of the discovery in recent years of the varying reactivities of two constituent amino acids, cystine and tyrosine, cannot be estimated, but may open a new approach to the difficult problem of investigating the molecular structure of wool proteins.

Oxidation by chloramides. Chloramines and chloramides oxidize the disulphide bond in the same way as equivalent quantities of chlorine.[68] They react much less readily however with phenolic groups both in wool (see p. 278) and in simple compounds. The presence of a high concentration of chloride ions greatly increases the reactivity of the chloro compounds with phenols bringing it to the same order as that with the disulphide bond (see Table 9.7). The effect of chloride ions is specific, as sodium nitrate, sulphate and fluoride do not promote the tyrosine reaction.

TABLE 9.7

Influence of sodium chloride on reactivity with chloramines[68]
(Half-time of reaction in min. at 18° C.)

Substance	N-Monochlorourea		Chlorosulphamic acid		Chlorine at pH 1·5
	No NaCl	10% NaCl	No NaCl	10% NaCl	
Cysteine	1·0	<0·5	<0·5	<0·5	<0·5
Cystine	2·0	<0·5	<0·5	<0·5	1·0
pp′ Diamino diphenyl disulphide	3·0	<0·5	<0·5	<0·5	<0·5
Phenol	20·0	1·0	84·0	<0·5	<0·5
Tyrosine	70·0	3·0	∞	3·0	2·0
Silk	>100·0	10·0	80·0	8·0	4·0
Glycine	∞	70·0	∞	∞	∞
a-Alanine	∞	40·0	90·0	60·0	8·0
Polyglycine	∞	∞	∞	∞	∞
Polyalanine	∞	∞	∞	∞	40·0
Nylon	∞	∞	∞	∞	6·0

Reactivity of the carboxyl groups

The free carboxyl groups in the peptide chain are provided by the side chains of the two acidic amino acids, glutamic and aspartic acids, and also the terminal groups of the main polypeptide chains. All the acidic side chains of these two amino acids do not, however, exist as free carboxyl groups, but some are present as amide groups, e.g.

$$\begin{array}{c} | \\ CO \\ | \\ CH-(CH_2)_2-CONH_2 \\ | \\ NH \\ | \end{array}$$

On acid or alkaline hydrolysis, ammonia is released by the wool due to the breakdown of these amide links, and from the quantity of amide nitrogen determined in this way it is possible to determine the proportion of carboxyl groups which are blocked. The amide link is more labile than the imino link of the main peptide chain and it is possible to bring about deamidation by treatment with acid under conditions where main-chain hydrolysis is negligible.[10] Middlebrook[69] considers that all the aspartic acid in wool is present as the amide and that the free carboxyl groups arise solely from the glutamic acid present, some of which, however, must also be present as amide. When studying the acid combination of wool in which the free carboxyl groups are back titrated (see p. 182) it is important to bear in mind that some amide hydrolysis occurs at high acidities and this must be corrected for when calculating the true maximum acid-combining power.

Esterification of the carboxyl group. In a series of patents, Schlack[70] described the alkylation of wool by a large number of compounds including epoxides, alkylhalides, β-chlorethylamines, dimethyl sulphate and alkyl or aryl sulphonic acids. The affinity of wool for dyes is increased by reaction with these compounds and it can therefore be assumed that all of them esterify the free carboxyl groups, although they probably react also with the amino groups in the wool. As acid dyes are held as gegen-ions to the charged amino groups in the wool and the ionized carboxyl groups compete with the dye ions for the cationic groups, it follows that the affinity of wool for acid dyestuffs is increased by blocking the carboxyl groups and preventing these from ionizing.[71]

Rutherford, Patterson and Harris [49] demonstrated that diazomethane esterifies the free carboxyl groups in proteins as well as alkylating the phenolic group of tyrosine. Similarly Blackburn, Carter and Phillips,[72] in a detailed study of the methylation of wool, showed that methyl sulphate, methyl iodide and methyl bromide at ordinary temperatures were capable of introducing methyl groups into wool. With methyl sulphate a maximum of 20 per cent. of methyl groups could be introduced, although only a small proportion of the free carboxyl groups in the wool were blocked. Alexander found[73] that after an extensive treatment of 30 hours, only 25 per cent. of the total free carboxyl groups were esterified by this reagent (corresponding to 0·3 per cent. of methyl groups). This is not surprising since methyl sulphate in the pH range used (pH 3–7) does not behave as an esterifying agent and does not normally methylate organic acids. Under these conditions some methyl sulphate reacts with the amino groups and the remainder is considered by Blackburn *et al.*[72] to methylate the peptide chain which they believe is capable of enolizing in the following manner:

$$-HN.CH.CO.NH.CH.CO- \rightleftharpoons HN.C=C.NH.CH.CO-$$
$$||||$$
$$R_1R_2R_1OHR_2$$

or

$$-HNCH.CO.NH.CH.CO- \rightleftharpoons HN.CH.C=N.CH.CO-$$
$$||||$$
$$R_1R_2R_1OHR_2$$

This hypothesis is supported by the fact that sulphate is bound by the wool and that a large part of the methyl groups introduced (i.e. those combined with the peptide link) is very labile and removed by treatment with steam which would not hydrolyse a carboxylic ester. Suspensions of methyl iodide and methyl bromide in water behave in a similar way to methyl sulphate, and besides esterifying some of the carboxyl groups also appear to methylate some amino and peptide groups. In view of the side reactions, these alkylating agents can not be used as specific reagents for esterifying the carboxyl groups in wool. The methyl halides react with amino and peptide groups to a more limited extent than dimethyl sulphate or methyl p-toluidine sulphonate.

Anhydrous methanol in the presence of traces of a strong mineral acid is a very satisfactory esterifying agent,[74] and reacts with the free carboxyl groups in proteins without side reactions or main-chain degradation. Blackburn and Lindley[75] used this method to introduce 0·8 per cent. of methyl groups into wool corresponding to an esterification of 60 per cent. of the total free carboxyl groups. This reaction has also been studied in detail by Alexander, Carter, Earland and Ford[71] who used a large range of alcohols to determine the accessibility of wool to molecules of different sizes. The reaction was found to proceed satisfactorily at elevated temperatures, and was complete in six hours instead of six days by working at 65° instead of 25° C. The absence of main-chain degradation under anhydrous conditions was illustrated[71] by the very low alkali solubility and the unchanged elasticity of the treated wool. Also, re-determination of the total acid-combining power after saponification of the esterified wool showed that no new carboxyl groups were formed, which proves that no hydrolysis of peptide bonds or amides had occurred. If more than

3 per cent. of water is present in the solvent, however, damage can be detected by all the methods mentioned. The rate of reaction is increased by increasing the concentration of acid catalyst and is decreased in the presence of water. In Fig. 9.7 the rate of esterification under comparable conditions using three different alcohols is shown, and it is seen that in every case all the carboxyl groups which can be esterified had reacted in 4–6 hours.

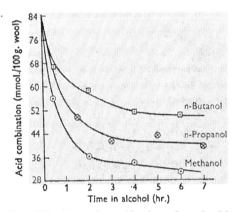

Fig. 9.7.—Rate of esterification of wool with anhydrous alcohols containing 0·1N hydrochloric acid; methanol at 65° C., n-propanol at 97° C., n-butanol at 100° C.[71]

The maximum number of carboxyl groups which can be esterified by different alcohols is shown in Table 9.8 from which it is seen that all the carboxyl groups are not accessible even to the smallest molecule, methyl alcohol. The number of groups reacting decreased successively with increase in the molecular weight of the alcohol until with the higher alcohols such as benzyl or cetyl a small percentage only can be esterified. The accessibility of wool to these two molecules is greatly increased when the fibre is swollen with water, and benzyl and cetyl iodide suspended in a swelling solvent are capable of esterifying up to 40 per cent. of the carboxyl groups. In the presence of small quantities of water the number of carboxyl groups which the alcohols esterify is increased, but this number decreases if more than 5 per cent. of water is present owing to the reverse reaction:

$$\text{Wool—COOR} + H_2O \rightleftharpoons \text{Wool—COOH} + R.OH$$

The esters formed are fairly stable to acid although hydrolysed by

TABLE 9.8

Proportion of the carboxyl groups in wool accessible to molecules of varying sizes[71]

Alcohol containing 0·1 N. HCl*	Acid-combining value of wool (m. equiv./100 g. wool)	Proportion of carboxyl groups available for esterification
Methanol	26	69
Ethanol	36	56
n-propanol	38	54
Isopropanol	70	15
n-butanol	50	39
Isobutanol	65	21
Sec.-butanol	70	15
n-amyl alcohol	65	21
Ethylene glycol	40	51
Glycerol	44	46
1 : 5-pentanediol	65	21
Benzyl alcohol	72	12
Cetyl alcohol in xylene	72	12
Iodides:		
Benzyl†	51	38
Cetyl‡	50	39

* Reacted for 6 hours at 100° C. or at boiling point of alcohol if below 100° C.

† pH 8 phosphate buffer containing 5 per cent. benzyl iodide in suspension 2 hours at 95° C.

‡ 5 per cent. solution in isopropyl alcohol, 2 hours, 70° C.

conditions necessary for peptide-bond hydrolysis. They are readily saponified by alkali; even the alkalinity of ordinary tap-water (pH 7·5–8) is sufficient to cause slow hydrolysis and 30 per cent. of the esters formed are hydrolysed by running tap-water in 54 hours and 90 per cent. in 135 hours. Normal ammonia saponifies the esters completely in one hour at 60° C. and 50 per cent. saponification may be achieved in one hour at 25° C. Soap solution is slightly less effective than ammonia, but nevertheless brings about rapid saponification. There appears to be very little difference in the rate of saponification of the esters of different alcohols. This is surprising as simple esters hydrolyse at considerably different rates in alkali.

Fraenkel–Conrat[76] claimed that soluble proteins could be esterified by epoxides such as propylene oxide, glycidol and epichlorhydrin if these were present in high concentration. This reaction is not confined to the carboxyl groups but reaction may occur with amino and phenolic hydroxyl groups according to the following scheme:

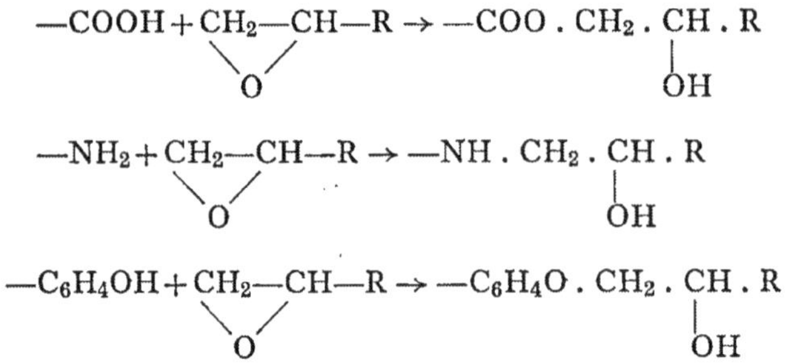

$$-COOH + CH_2 \underset{O}{-\!\!\!\diagdown\!\!\!\diagup\!\!\!-} CH-R \rightarrow -COO \cdot CH_2 \cdot \underset{OH}{CH} \cdot R$$

$$-NH_2 + CH_2 \underset{O}{-\!\!\!\diagdown\!\!\!\diagup\!\!\!-} CH-R \rightarrow -NH \cdot CH_2 \cdot \underset{OH}{CH} \cdot R$$

$$-C_6H_4OH + CH_2 \underset{O}{-\!\!\!\diagdown\!\!\!\diagup\!\!\!-} CH-R \rightarrow -C_6H_4O \cdot CH_2 \cdot \underset{OH}{CH} \cdot R$$

When studying the process for increasing the affinity of wool for dyes described by Schlack,[70] Alexander et al.[71] found that epichlorhydrin readily esterified the carboxyl groups in wool but no other epoxide examined was capable of reacting with more than 10 per cent. of the carboxyl groups when applied under identical conditions. The maximum number of carboxyl groups which could be esterified with epichlorhydrin was 50 per cent. The number of phenolic hydroxyl or amino groups which have reacted has not been determined, but judging from the weight increase of 10 per cent. which can be obtained by extensive reaction with epichlorhydrin, a large number of these groups probably react. The failure of the other epoxides to bring about esterification is surprising, especially since Fraenkel-Conrat claimed[76] to have obtained this reaction. However, Fraenkel-Conrat's technique for detecting esterification is open to severe criticism and this is in fact admitted by him in a later paper.[74] Alexander and co-workers[77] have suggested that at the isoelectric point the salt link reacts not as $-COO^- {}^+NH_3-$ but behaves chemically as $-COOH \ldots H_2N-$. Since epoxides react preferentially with ionized carboxyl groups, the almost complete absence of reaction at pH 7 would be explained. This theory also explains why reaction with amino groups occurs under these conditions since epoxides will react only with unionized amines.

The group introduced by epichlorhydrin on esterification is very labile and treatment with mild alkali, e.g. pH 9 buffer at room

CHEMICAL REACTIVITY 313

temperature, leads to the formation of an epoxide ring as follows:[71]

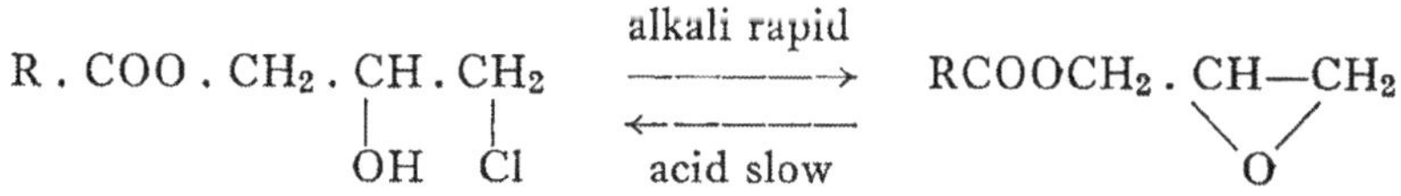

$$R \cdot COO \cdot CH_2 \cdot \underset{OH}{CH} \cdot \underset{Cl}{CH_2} \underset{\text{acid slow}}{\overset{\text{alkali rapid}}{\rightleftarrows}} RCOOCH_2 \cdot \underset{O}{CH - CH_2}$$

Reactivity of the amino groups

The amino groups at the end of the peptide chains and the side groups from the basic amino acids, lysine and arginine, provide the basic groups in wool. The amino groups can be determined by a method evolved by Sanger[78] in which the protein is treated with a solution of 2 : 4-dinitrofluorobenzene which reacts with both the terminal amino groups and those of lysine according to the following scheme:

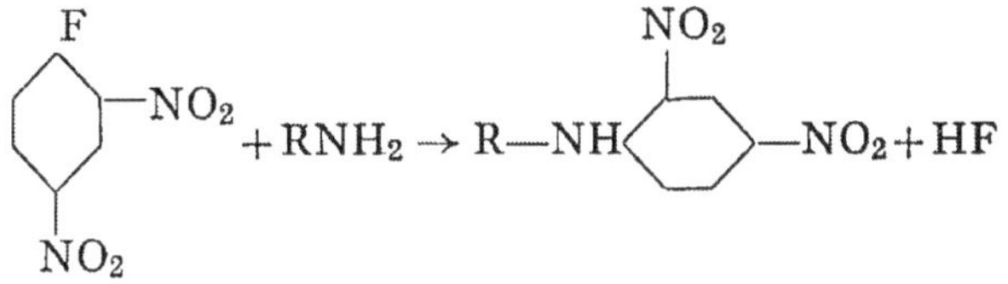

$$\underset{NO_2}{\underset{|}{\overset{F}{\underset{|}{\bigcirc}}}}-NO_2 + RNH_2 \rightarrow R-NH-\bigcirc\underset{NO_2}{\overset{NO_2}{-}}-NO_2 + HF$$

A coloured derivative is formed which is not hydrolysed by acid under the conditions of normal protein hydrolysis. The amino acids carrying free amino groups can therefore be isolated as the dinitrobenzene derivatives in which form they can be readily estimated. In this way Middlebrook[79] and Blackburn[33] independently showed that seven different amino acids give rise to terminal amino groups (see p. 351). The large number of different amino acids which form the terminal groups reflects the heterogeneous nature of wool and shows clearly that wool cannot be considered in any way as a single chemical species. Unfortunately this method cannot be used to detect the terminal amino groups of cystine which can provide either one or two of these per molecule, viz.,

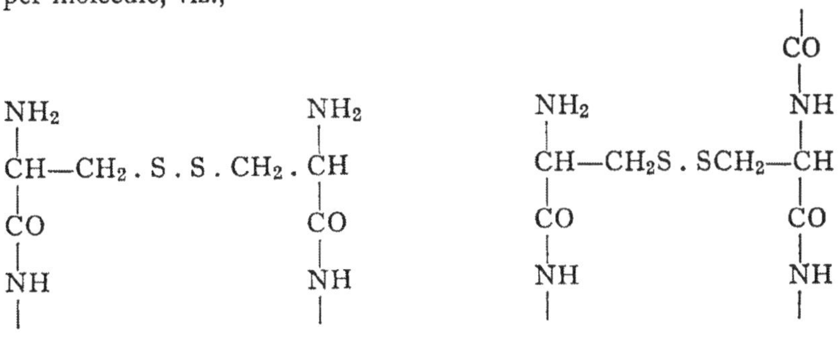

$$\begin{array}{cccc} NH_2 & NH_2 & NH_2 & \overset{|}{\underset{|}{CO}} \\ | & | & | & NH \\ CH-CH_2 \cdot S \cdot S \cdot CH_2 \cdot CH & & CH-CH_2S \cdot SCH_2-CH \\ | & | & | & | \\ CO & CO & CO & CO \\ | & | & | & | \\ NH & NH & NH & NH \\ | & | & | & | \end{array}$$

since Sanger[80] found that the bis-dinitrobenzene derivative of cystine undergoes a disulphide interchange reaction[81] during hydrolysis in the presence of free cystine.

$$\begin{array}{c}\text{DNP-NH-CH(COOH)-CH}_2\text{-S-S-CH}_2\text{-CH(NO}_2\text{-DNP-NH)-COOH} \\ + \\ \text{H}_2\text{N-CH(COOH)-CH}_2\text{-S-S-CH}_2\text{-CH(NH}_2\text{)-COOH} \end{array} \rightleftarrows 2\ \text{DNP-NH-CH(COOH)-CH}_2\text{-S-S-CH}_2\text{-CH(NH}_2\text{)-COOH}$$

The amino group of lysine which reacts with dinitrofluorobenzene can be also estimated in this way. The pK of the substituted amino group, both in lysine and the terminal amino acids, is not greatly altered and reaction with dinitrofluorobenzene does not influence the titration curve of wool. Van Slyke[82] showed that there

CHEMICAL REACTIVITY 315

was a very small proportion (0·2 per cent.) of hydroxylysine

$NH_2 . CH(COOH) . CH(OH) . CH_2 . CH_2 . CH_2 . NH_2$

present in wool but this does not react with dinitrofluorobenzene. Van Slyke's finding was confirmed by Kersten and Zahn[83a], who found $0·20 \pm 0·02$ per cent. by the Moore and Stein technique. However, there are other workers who consider that hydroxylysine is completely absent from wool. Middlebrook[79] concludes that the amino groups of the side chains must be combined in the wool in a way at present unknown and only become free on hydrolysis.

The basic groups in wool can be removed effectively either by acetylation, which converts the amino group to an amide group which does not ionize except in very acid solution, or by deamination with nitrous acid.

$Wool—CH_2—NH_2 + (CH_3CO)_2O \rightarrow$

$\rightarrow Wool—CH_2—NHCOCH_3 + CH_3COOH$

$Wool—CH_2—NH_2 + HNO_2 \rightarrow Wool—CH_2OH + N_2 + H_2O$

Elliott and Speakman[83] showed that the acetylating mixture normally used for treating cellulose (a mixture of acetic acid with catalytic quantities of sulphuric acid and pyridine) was capable of acetylating some of the amino groups in wool, but that sulphonic acid groups were simultaneously introduced. Lindley and Phillips[84] found that acetic anhydride alone acetylated all the terminal groups, and those from lysine and arginine. Unfortunately the treatment produced considerable discoloration and degradation accompanied by a decrease in cystine content. When the reagent was used at lower temperatures much less damage was done, but the acetylation was incomplete. The most effective acetylating process is that developed by I. G. Farben[85] and used by them commercially for producing wool with decreased dye affinity. The reagent, consisting of:

> 1·3 kg. acetic anhydride
> 3·7 kg. glacial acetic acid
> 0·12 kg. sulphuric acid monohydrate
> 0·07 kg. dimethylaniline

is applied to dry wool at the lowest possible liquor ratio for 45–60 minutes at 56°–58° C. The temperature conditions are very critical and at lower temperature acetylation is incomplete, and already at

60° C. the wool is badly discoloured and weakened. Under the exact conditions of treatment, however, the wool is perfectly white, no loss in cystine occurs[86] and more than 90 per cent. of the amino groups are blocked. Acetylation by this method is undoubtedly the ideal way of removing the basic groups from wool. Earland and Williamson[87] have shown that 8·6 per cent. of sulphate (as SO_4) is introduced into wool by this method, which corresponds very closely with the free amino groups present and the increase in weight on treatment. It therefore seems most probable that this dye resist process is basically one of sulphonation rather than acetylation.

Deamination with nitrous acid is the oldest method of removing the basic groups in wool and this method was used by Speakman and his collaborators in their classical researches on the "salt link" (see p. 213). This reagent was shown by Van Slyke[88] to react rapidly and completely with the terminal amino groups and the free amino groups of lysine under the conditions of his experiments which involve very efficient agitation. Arginine provides the largest number of free basic groups, which are reported to react rather slowly with nitrous acid,[89] and are not estimated by the Van Slyke test. Speakman[90] and others[91] found that most of the basic groups in wool could be removed by nitrous acid after a very prolonged reaction period (e.g. 6 days). Under these conditions, however, considerable oxidation of the disulphide bond occurs and in addition Cockburn, Drucker and Lindley[92] have shown that in these long reaction periods the tyrosine was completely removed. The basic groups remaining after the severe deamination by nitrous acid are probably due to unreacted arginine, but the mechano-elastic method used by Speakman for determining free amino groups does not allow distinction between the different types of amino groups. Deamination is therefore a highly unsatisfactory method of deactivating amino groups, and it is to be hoped that future workers will employ other methods for this purpose.

Elliott and Speakman[83] found that they could block a large number of the amino groups in wool with glyoxal-bis-sodium bisulphite and they believe that the following reaction occurs with some of the arginine side chains of wool:

$$RC\underset{NH_2}{\overset{NH}{\diagup\!\!\!\diagdown}} \rightarrow RC\underset{N=CH}{\overset{N-CHSO_3^-Na^+}{\diagup\!\!\!\diagdown}}$$

When wool is oxidized with permanganate the manganese dioxide formed is powerfully held and can only be removed by treatment with reducing agents. Indirect evidence indicates that the oxide may be held on the amino groups.[93]

Reactivity of the thiol groups

In general, methods of cystine analysis do not distinguish between cystine and its precursor in the reduced state, cysteine. Thus, the Shinohara method determines cystine colorimetrically after reduction to cysteine, and prior to separation on an ion-exchange column using the method of Moore and Stein, the cystine is often oxidized to cysteic acid. Further, even if no deliberate attempt is made to modify the amino acids, the hydrolysis conditions alone can oxidize cysteine to cystine. It is therefore essential that the presence of thiol groups be demonstrated in intact wool if their existence is to be established with any degree of certainty. Burley[94, 95, 96] has utilized organomercuri chlorides and N-ethyl maleimide to detect and quantitatively determine thiol groups since these substances react with intact wool as follows:

$$W\text{—}SH + ClHgX \rightarrow W\text{—}S\text{—}HgX + HCl$$

$$W\text{—}SH + \begin{matrix} CH\text{—}CO \\ \| \\ CH\text{—}CO \end{matrix} \!\!\!\! \diagdown\!\!\!\diagup \,NCH_2CH_3 \rightarrow \begin{matrix} W\text{—}S\text{—}CH\text{—}CO \\ | \\ CH_2\text{—}CO \end{matrix} \!\!\!\! \diagdown\!\!\!\diagup \,NCH_2CH_3$$

If a coloured organomercury halide, such as 1-(4-chloromercuri-phenylazo)-naphthol-2, be used, the reaction with wool can be followed quantitatively using a photometer. The reaction with N-ethyl-maleimide can be followed similarly from its absorption in the ultra-violet. Zuber, Traumann and Zahn[97] by reacting wool with 2:4-dinitrofluorobenzene at pH 5 and 60° C. have converted the cysteine to the S-D.N.P. derivative, which after hydrolysis is determined spectroscopically at 330 mμ. Their results agree well with those of Burley and the more recent observations of Maclaren et al.[97a] and show that the cysteine content of virgin wool is about 0·4 per cent.

Reaction with concentrated acids

Harris, Mease and Rutherford[98] found that when wool is treated with 80 per cent. sulphuric acid, sulphur is introduced into the fibres and the gain in weight can be explained by sulphonation. These authors believe that the reaction with the amino groups occurs according to the following scheme:

$$W \cdot NH_2 + H_2SO_4 \rightarrow W \cdot NHSO_3H + H_2O$$

Speakman[99] indicated that tyrosine also took part in this reaction and in a detailed study of the reaction Lustig and Kondritzer[100] found that the sulphuric acid-treated wool had a reduced content of serine, tryptophan, phenylalanine, arginine (see Table 9.9) and tyrosine and the reaction is therefore obviously more complex than at first assumed by Harris et al.

TABLE 9.9

Sulphur and amino-acid distribution before and after sulphuric-acid treatment of hair[100]

	Normal (%)	H_2SO_4 treated (%)
Ash	0·42	0·86
N content	15·50	14·05
Total S	5·24	8·41
SO_4-S	—	3·51
Organic S	5·24	4·90
Cystine S	4·46	4·08
Organic non-cystine S	0·76	0·82
N: Organic S	2·96	2·87

Amino-acid nitrogen distribution as percentage of total nitrogen

Humin	0·64	1·26
Cystine	12·66	12·56
Arginine	19·31	14·11
Histidine	1·59	1·67
Threonine	5·29	5·31
Serine	6·30	5·70
Tryptophan	0·37	0·26
Tyrosine	1·45	0·95
Phenylalanine	1·99	1·73

Lemin and Vickerstaff,[101] who studied the acid- and base-combining properties of wool that had been carbonized (i.e., heated in presence of sulphuric acid in order to remove vegetable matter), concluded that the amino groups were involved in the reaction. Reitz et al.[102], however, showed that sulphuric acid at 0° brought about mainly sulphonation of the aliphatic hydroxyl groups of serine in a number of proteins, including wool. With chlorsulphonic acid in pyridine, most of the amino, amide, and aliphatic and phenolic hydroxyl groups were sulphonated.[103] Carbonized wool has been shown[103a] to contain, in addition to the O-sulphuric esters of serine and threonine, increased contents of serine and threonine N-terminal groups: this increase is due to a rearrangement, followed by fission, of the peptide chain at serine and threonine sites. This reaction is partly reversed during the neutralization stage of the carbonization process.

The reaction of wool with sulphuric acid, as judged by the changes in tensile strength and alkali-solubility, depends on the water content of the acid-wool system and it appears that sulphuric acid attacks wool by at least two different mechanisms.[103b]

Ferrel, Olcott and Fraenkel-Conrat[104] showed that the hydroxyl groups of serine in proteins could be effectively phosphorylated with phosphorus pentoxide in phosphoric acid, when reaction is confined almost entirely to the serine. It seems likely that phosphate groups can be introduced in this way into wool although this does not as yet appear to have been done.

Reactions with miscellaneous reagents

The following substances probably combine with wool by the reactions indicated. It must, however, be borne in mind that most of these reagents are capable of polymerization, and since polymers formed *in situ* are very difficult to remove from wool, unequivocal proof of chemical reaction is difficult to obtain.

Acrylonitrile

Bikales, Black and Rapoport[105] by reacting wool with acrylonitrile and alkali have introduced 5 per cent. by weight of cyanoethyl groups into the fibre.

$$W-OH + CH_2=CHCN \rightarrow W-OCH_2CH_2CN$$
$$W-NH_2 + CH_2=CHCN \rightarrow W-NHCH_2CH_2CN$$

Unsaturated aldehydes

The action on wool of acraldehyde, α-methylacraldehyde, crotonaldehyde and cinnamaldehyde has been studied in some detail by McPhee and Lipson.[106] Of these compounds acraldehyde reacted most readily, combining with the free sulphydryl groups and reacting under alkaline conditions with the lysine:

$$W-CH_2-SH + CH_2=CH.CHO \rightarrow W-CH_2.S.CH_2CH_2CHO$$
$$W-CH_2-NH_2 + CH_2=CH.CHO \rightarrow$$
$$\rightarrow W-CH_2.NHCHOH.CH=CH_2$$

There was no evidence from physical measurements to suggest the presence of new cross-links.

Isocyanates

Reaction of wool with monofunctional isocyanates has been reported by Farnworth[107] and Moore.[108] According to Farnworth, who studied the reaction using phenylisocyanate dissolved in anhydrous pyridine at 70° C., reaction occurs definitely with amino, carboxyl and thiol groups.

$$WNH_2 + C_6H_5NCO \rightarrow WNHCONHC_6H_5$$
$$WCOOH + C_6H_5NCO \rightarrow WCOOCONHC_6H_5$$
$$WSH + C_6H_5NCO \rightarrow WSOCNHC_6H_5$$

and probably with phenolic, amide, guanidino and glyoxaline groups.

Propiolactone

Lundgren and co-workers[109, 110] have studied the reaction of β-propiolactone with wool, both from solution in carbon tetrachloride and by direct application of the vapour. It is postulated that the following reactions occur:

$$W-CH_2OH + \underset{\underset{O\!-\!-\!CO}{|\quad\quad|}}{CH_2-CH_2} \rightarrow W-CH_2OCH_2CH_2COOH$$

$$W-NH_2 + \underset{\underset{O\!-\!-\!CO}{|\quad\quad|}}{CH_2-CH_2} \rightarrow W-NHCOCH_2CH_2OH$$

CHEMICAL REACTIVITY 321

$$W-C=CH \quad CH_2-CH_2 \rightarrow W-C=CH$$
$$\begin{array}{ccc} | & | & | \\ N & NH + O\!-\!-\!CO & N \quad NCOCH_2CH_2OH \\ \diagdown\!CH\!\diagup & & \diagdown\!CH\!\diagup \end{array}$$

$$W-COOH + CH_2-CH_2 \rightarrow W-COOCH_2CH_2COOH$$
$$\begin{array}{c} | \quad\quad | \\ O\!-\!-\!CO \end{array}$$

The claim of these workers that reaction with propiolactone increases the felting power of wool has not been confirmed by Fearnley and Speakman.[111]

REFERENCES

[1] BULL, *Cold Spring Harbor Sym. Quant. Biol.*, 1949, **14**, 1.
 PARTRIDGE AND DAVIS, *Nature*, 1950, **165**, 62.
[2] LEACH, *Proc. Int. Wool Text. Res. Conf.*, Australia, 1955, C-181.
[3] ZAHN AND MEIENHOFER, *ibid.*, C-62.
[4] ZAHN AND MEIENHOFER, *Text. Res. J.*, 1955, **25**, 738.
[5] EARLAND AND KNIGHT, *Biochim. Biophys. Acta*, 1956, **22**, 405.
[6] ELÖD, NOWOTNY AND ZAHN, *Kolloid Z.*, 1940, **93**, 50.
[7] *Ibid.*, 1942, **100**, 297.
[8] HARRISON, *J. Text. Inst.*, 1937, **28**, T110.
[9] STEINHARDT, *J. Biol. Chem.*, 1941, **141**, 995.
[10] STEINHARDT AND FUGITT, *J. Res. N.B.S.*, 1942, **29**, 315.
[11] PETERS AND SPEAKMAN, *J. Soc. Dyers and Col.*, 1949, **65**, 63.
[12] SANGER, *Biochem. J.*, 1949, **45**, 563.
[13] GORDON, MARTIN AND SYNGE, *Biochem. J.*, 1941, **35**, 1369.
[14] MILLER AND DU VIGNEAUD, *J. Biol. Chem.*, 1937, **118**, 101.
[15] PHILLIPS, *Symp. Fibrous Protein*, 1946, 39 (publ. Soc. Dyers and Col., Bradford).
[16] SANFORD AND HUMOLLER, *Analyt. Chem.*, 1947, **19**, 404.
[17] WARNER AND CANNAN, *J. Biol. Chem.*, 1942, **142**, 725.
[17a] CORFIELD AND ROBSON, *Biochem. J.*, 1955, **59**, 62.
[18] RUTHERFORD, PATTERSON AND HARRIS, *J. Res. N.B.S.*, 1940, **25**, 451.
[19] HARRIS AND SMITH, *ibid.*, 1936, **17**, 557.
 HARRIS, *Amer. Dyestuff Rep.*, 1935, **24**, 306.
[20] ZAHN, *Textil Praxis*, 1949, **2**, 70.
[21] ELÖD AND ZAHN. *Kolloid Z.*, 1942, **100**, 293.
 KRATKY AND SEKORA, *Z. elektrochem.*, 1942, **48**, 409.
[22] ASTBURY AND WOODS, *Proc. Roy. Soc.*, 1933, **232A**, 333.
[23] ZAHN, *Kolloid Z.*, 1950, **117**, 102.
[24] BLOCK AND VICKERY, *J. Biol. Chem.*, 1930, **86**, 107.
[25] MEUNIER, CHAMBARD AND COMTE, *Compt. rend.*, 1927, **184**, 1208.
 FROMAGEOT AND PORCHARD, *ibid.*, 1931, **193**, 738.
[26] ELÖD AND ZAHN, *Melliand Textilber.*, 1943, **24**, 157, 245.
[27] BURGESS, *J. Text. Inst.*, 1934, **25**, T289.
 WALDSCHMITZ-LEITZ AND SCHUCKMANN, *Ber. dtsch. chem. Ges.*, 1929, **62**, 1891.
[28] GEIGER, PATTERSON, MIZELL AND HARRIS, *J. Res. N.B.S.*, 1941, **27**, 459.
[29] ELÖD AND ZAHN, *Melliand Textilber.*, 1946, **27**, 68.
[30] SCHÖBERL AND HAMM, *Biochem. J.*, 1948, **318**, 331.
[31] MIDDLEBROOK AND PHILLIPS, *J. Soc. Dy. Col.*, 1941, **57**, 137.
[32] TROTMAN, *The bleaching, dyeing and chemical technology of textile fibres*, Griffin, 1925, 196.

[33] BLACKBURN, *Biochem. J.*, 1950, **47**, 443.
[34] LINDERSTRØM-LANG AND DUSPIVA, *Z. physiol. Chem.*, 1935, **237**, 131; 1936, **241**, 177.
[34a] WATERHOUSE, *Aust. J. Sci. Res.*, 1952, B5, 444.
[35] GEIGER, KOBAYASHI AND HARRIS, *J. Res. N.B.S.*, 1942, **29**, 38.
[36] STAHL, MCQUE, MANDELS AND SIU, *Arch. Biochem.*, 1949, **20**, 422. *Text. Res. J.*, 1950, **20**, 481.
[37] STIRM AND ROUETTE, *Melliand Textilber.*, 1935, **16**, 4.
[37a] LENNOX, *Second Quinquennial Wool Text. Res. Conf.* (*J. Text. Inst.*), Harrogate, 1960, T1193.
[38] SCHMIDT, *Z. wiss. Biol. B. Zellf. u. mikrosckop. Anatom.*, 1932, **15**, 188.
[39] ZAHN, *Naturw.*, 1943, **31**, 137; ELÖD, NOWOTNY AND ZAHN, *Melliand Textilber.*, 1950, **31**, 481.
[40] RUTHERFORD AND HARRIS, *J. Res. N.B.S.*, 1939, **23**, 597.
[41] ZAHN, *Textil Praxis*, 1950, **1**, 7; *Melliand Textilber.*, 1950, **31**, 481.
[42] MECHAM AND OLCOTT, *Ind. Eng. Chem.*, 1947, **39**, 1023.
[43] LUNDGREN, HIGH, LINDQUIST AND WARD, *J. Polym. Sci.*, 1946, **1**, 22.
[44] MCCLEARY AND ROYER, *Text. Res. J.*, 1949, **19**, 457.
[45] MAZINGUE AND VAN OVERBÈKE, *Bull. Inst. Text. France*, 1956, **62**, 27.
[46] O'CONNELL AND WALDEN, *Text. Res. J.*, 1957, **27**, 516.
[47] LUNDGREN, *Ad. in Protein Chem.*, 1949, **5**, 316.
[48] ALEXANDER AND GOUGH, *Biochem. J.*, 1951, **48**, 504.
[49] RUTHERFORD, PATTERSON AND HARRIS, *J. Res. N.B.S.*, 1940, **25**, 457.
[50] FRITZE AND ZAHN, *Proc. Int. Wool Text. Res. Conf.* Australia, 1955, **C-119**.
[51] ZAHN AND KOHLER, *Z. Naturf.*, 1950, **56**, 137.
[52] HERRIOT AND NORTHRUP, *J. Gen. Physiol.*, 1936, **19**, 283.
[53] ROVERY AND DESNUELLE, *Biochim. Biophys. Acta*, 1948, 2, 5, 14.
[54] CRAMER AND NEUBERGER, *Biochem. J.*, 1943, **37**, 302.
[55] BEAVEN, HOLIDAY AND JOPE, *Faraday Soc. Discussion*, 1950, **9**, 406.
[56] ALEXANDER, *Kolloid Z.*, 1951, **122**, 8.
[57] MIRSKY AND ANSON, *J. gen. Physiol.*, 1936, **19**, 451.
[58] BLACKBURN AND PHILLIPS, *J. Soc. Dy. & Col.*, 1945, **61**, 100.
[59] RICHARDS AND SPEAKMAN, *J. Soc. Dy. Col.*, 1955, **71**, 537.
[60] DENIS, *J. biol. Chem.*, 1912, **10**, 74.
[61] LANGHELD, *Ber. dtsch. chem. Ges.*, 1909, **42**, 2371.
[62] DRAKE AND SMYTHE, *Arch. Biochem.*, 1944, **4**, 255.
[63] ALEXANDER, CARTER AND HUDSON, *J. Soc. Dy. Col.*, 1949, **65**, 152.
[64] DAS AND SPEAKMAN, *ibid.*, 1950, **66**, 583.
[65] EARLAND, STELL AND WISEMAN, *Second Quinquennial Wool Text. Res. Conf.*, (*J. Text. Inst.*), Harrogate, 1960, T817.
[66] TEUBER, AND RAU, *Ber.*, 1953, **86**, 1036.
[67] Fox, Ph.D. Thesis (London Univ), 1951.
[68] ALEXANDER, CARTER AND EARLAND, *J. Soc. Dy. Col.*, 1951, **67**, 17.
[69] MIDDLEBROOK, *Biochem. J.*, 1949, **44**, 17.
[70] SCHLACK, U.S. Patents 2,131,121 (1938); 2,238,949 (1941); 2,261,294 (1941); 2,291,061 (1942).
[71] ALEXANDER, CARTER, EARLAND AND FORD, *Biochem. J.*, 1951, **48**, 629; ALEXANDER, *Melliand Textilber.*, 1953, **34**, 756.
[72] BLACKBURN, CARTER AND PHILLIPS, *Biochem. J.*, 1941, **35**, 627.
[73] ALEXANDER (unpublished).
[74] FRAENKEL-CONRAT AND OLCOTT, *J. biol. chem.*, 1945, **161**, 259.
[75] BLACKBURN AND LINDLEY, *J. Soc. Dy. Col.*, 1948, **64**, 505.
[76] FRAENKEL-CONRAT, *J. biol. chem.*, 1944, **154**, 227.
[77] ALEXANDER, FOX, STACEY AND SMITH, *Biochem. J.*, 1952, **52**, 177.
[78] SANGER, *Biochem. J.*, 1945, **39**, 507.
[79] MIDDLEBROOK, *Nature*, 1949, **164**, 501; *Biochim. Biophys. Acta*, 1951, **4**, 547.
[80] SANGER, *Nature*, 1953, **171**, 1025.
[81] MCALLAN, CULLUM, DEAN AND FIDLER, *J. Amer. Chem. Soc.*, 1951, **73**, 3627.
[82] VAN SLYKE, HILLER AND MACFAYDEN, *J. Biol. Chem.*, 1941, **141**, 687. CHIBNALL, *Proc. Roy. Soc.*, 1942, **131A**, 136.

[83] ELLIOTT AND SPEAKMAN, *J. Soc. Dy. Col.*, 1943, **59**, 185.
[83a] KERSTEN AND ZAHN, *Nature*, 1959, **184**, 1490.
[84] LINDLEY AND PHILLIPS, *Biochem. J.*, 1947, **41**, 34.
[85] Quoted by ALEXANDER AND WHEWELL, *British Intelligence Objective Subcommittee Report* No. 1472, H.M. Stationery Office, 1947.
[86] ALEXANDER, HUDSON AND FOX, *Biochem. J.*, 1950, **46**, 27.
[87] EARLAND AND WILLIAMSON, Unpublished.
[88] VAN SLYKE, *J. Biol. Chem.*, 1911, **9**, 185.
[89] KANAGY AND HARRIS, *J. Res. N.B.S.*, 1935, **14**, 563.
[90] SPEAKMAN, *J. Soc. Dy. Col.*, 1936, **52**, 335.
[91] RUTHERFORD, HARRIS AND SMITH, *J. Res. N.B.S.*, 1937, **19**, 467.
[92] COCKBURN, DRUCKER AND LINDLEY, *Biochem. J.*, 1948, **43**, 438.
[93] ALEXANDER AND HUDSON, *J. Phys. Chem.*, 1949, **53**, 733.
[94] BURLEY, *Nature*, 1954, **174**, 1019.
[95] BURLEY, *Text. Res. J.*, 1956, **26**, 332.
[96] BURLEY AND HORDEN, *ibid.*, 1957, **27**, 615.
[97] ZUBER, TRAUMANN AND ZAHN, *Proc. Int. Wool Text. Res. Conf.*, Australia, 1955, C-127.
[97a] MACLAREN, LEACH AND SWAN, *Second Quinquennial Wool Text. Res. Conf.* (*J. Text. Inst.*), Harrogate, 1960, T665.
[98] HARRIS, MEASE AND RUTHERFORD, *J. Res. N.B.S.*, 1937, **18**, 343.
[99] SPEAKMAN, *J. Text. Inst.*, 1941, **32**, T83.
[100] LUSTIG AND KONDRITZER, *Arch. Biochem.*, 1945, **8**, 51.
[101] LEMIN AND VICKERSTAFF, *Symp. Fibrous Proteins* (Soc. Dy. Col., Bradford), 1940, 129.
[102] REITZ, FERREL, FRAENKEL-CONRAT AND OLCOTT, *J. Amer. Chem. Soc.*, 1946, **68**, 1026.
[103] *Ibid.*, 1946, **68**, 1031.
[103a] HILLE AND ZAHN, *Second Quinquennial Wool Text. Res. Conf.* (*J. Text. Inst.*), Harrogate, 1960, T1162.
[103b] MIZELL, DAVIS AND OLIVA, *Text. Res. J.*, 1962, **32**, 497.
[104] FERREL, OLCOTT AND FRANKEL-CONRAT, *J. Amer. Chem. Soc.*, 1948, **70**, 2101.
[105] BIKALES, BLACK AND RAPOPORT, *Text. Res. J.*, 1957, **27**, 80.
[106] MCPHEE AND LIPSON, *Australian J. Chem.*, 1954, **7**, 387.
[107] FARNWORTH, *Biochem. J.*, 1955, **59**, 529.
[108] MOORE, *Text. Res. J.*, 1956, **26**, 936.
[109] JONES AND LUNDGREN, *Text. Res. J.*, 1951, **21**, 620.
[110] ROSE AND LUNDGREN, *ibid.*, 1953, **23**, 930.
[111] FEARNLEY AND SPEAKMAN, *J. Soc. Dy. Col.*, 1952, **68**, 88.

CHAPTER 10

Formation of New Cross-Links

THE presence of disulphide bonds which link main polypeptide chains differentiates wool from other protein fibres. These cross-links are responsible for the complete insolubility of wool in polar solvents, its limited lateral swelling and relatively high wet strength. Regenerated protein fibres, though having a tensile strength comparable to wool when dry, are usually much weaker when wet, because the hydrogen bonds are broken and the dispersive forces weakened by swelling. Wool fibres may be weakened to a similar extent by oxidation, and it was shown (see p. 62) that the tensile strength of the wet fibres is related simply to the number of disulphide bonds remaining.

Conversely the introduction of new cross-links into wool increases the tensile strength; this is not only of scientific interest but offers considerable technical possibilities. Cross-linking can reduce felting considerably, for the ability of wool to felt depends on its ease of migration, which is directly related to its elastic properties (see p. 38). Because the disulphide bond is labile to alkali, it is important to introduce new cross-links before wool can be used in technical processes operating at high alkalinities, e.g. vat dyeing. Moreover, the attack on wool by moth larvae, fungi or bacteria (see p. 296) involves the rupture of disulphide bonds before main-chain enzymatic degradation can occur, so that the resistance to biological damage is greatly increased if cross-links which cannot be broken by enzymes are introduced.

Cross-links of increased chemical stability may be introduced either by opening the disulphide bonds and replacing them by more stable cross-links or by reacting wool with poly-functional compounds which can combine with the reactive amino, imido, carboxyl or phenolic side groups. It is doubtful whether the peptide link is sufficiently reactive for this purpose, although it can be alkylated (see p. 309).

Unfortunately the determination of the number of new cross-links which have been introduced is by no means simple, and none of the usual tests is unambiguous. This is mainly due to the fact that the cross-linking agents are poly-functional and in many cases readily polymerize. Speakman and his colleagues have shown that the internal deposition of polymers increases the strength of wool fibres and generally changes their properties in a similar way to cross linking (see p. 342). Thus the experimental results must be treated with caution and it is particularly important to determine the weight increase of fibres after treatment. If this exceeds 10 per cent. it can be concluded immediately that polymer formation has occurred, in addition to any new cross-links which might also have been formed. It is not always possible to extract the polymer which is formed by reacting wool with a poly-functional compound, since it may be chemically anchored to one of the reactive groups in the wool. This behaviour has been demonstrated by Alexander, Carter and Earland[1] with silicones and by Barr, Capp and Speakman[2] with di-isocyanates (see p. 47).

In the field of synthetic polymers, cross-linking is detected readily by determining changes in solubility, and the introduction of only a few cross-links per macromolecule is sufficient to make rubber and polystyrene insoluble in their usual solvents. This small number of cross-links would not increase the strength of a fibre or sheet made from these materials significantly. In wool research the following methods for estimating the extent of cross-linking have been employed:

(1) *Load–extension.* As the dry tensile strength of wool is independent of cross-linking, the wet tensile strength has always been employed for these measurements and the technique of measuring the work necessary to stretch a fibre by 30 per cent., advanced by Speakman, has usually been used as the criterion. In many cases where cross-linking has undoubtedly occurred (e.g. with formaldehyde under mild conditions) the work required to stretch a fibre in distilled water is unaltered or may even be decreased. This occurs when the "salt links" which contributed originally to the cohesion of the fibre (see p. 213) are replaced by covalent cross-links.[3] This difficulty is overcome by determining the work required to stretch a fibre by 30 per cent. in acid solution, as the negative carboxyl groups are removed by combination with protons and electrical interaction is prevented. The introduction of new cross-links can then be detected by this procedure (e.g. with formaldehyde, see p. 330). In general, however, this method is very insensitive, and it is greatly affected by polymer formation, so that it cannot be considered as a very reliable test for detecting cross-links.

(2) *Supercontraction* (see p. 73). Certain liquids cause wool fibres

to contract in length, and this phenomenon, known as supercontraction, provides the most sensitive method of detecting the presence of cross-links. To illustrate this method, the action of sodium bisulphite[4] which is the oldest and most useful reagent for this purpose, may be considered. By breaking the disulphide bonds, it converts the wool into a rubber-like substance, causing at the same time considerable contraction along the fibre axis. This contraction is prevented by the introduction of new cross-links, provided that the latter are stable to this reagent in boiling solution. Alternatively peracetic acid, which is a highly specific reagent for the disulphide bond (see p. 226) may be used in the place of bisulphite.[5] As this reagent is effective in the cold, it is much milder than boiling bisulphite, and can be used to determine the presence of cross-links which are broken by this latter reagent.

Elöd and Zahn have used phenol extensively as a contracting reagent.[6] The exact mechanism of the contraction in this case is not very clear, but it is believed to depend largely on the breaking of intermicellar hydrogen bonds. It does not depend upon the breaking of disulphide bonds, and when these are replaced by more stable cross-links so that the fibre no longer contracts in bisulphite or peracetic acid, they still contract in phenol. However, the introduction of entirely new cross-links appears to prevent contraction in phenol even when the degree of fibre modification is so small as to be undetectable by contraction in bisulphite.[7] Such mild treatments as boiling for one hour in dilute acid are sufficient to prevent contraction in phenol, and it is doubtful whether the reaction has in fact introduced new cross-links. Consequently this highly sensitive test is difficult to interpret and further information on the supercontraction mechanism is required before it can be used as a diagnostic test for cross-linking.

(3) *Solubility*. As already mentioned, one of the most desirable effects of introducing new cross-links is the increased resistance of the wool towards alkalis. Harris and Smith[8] introduced an alkali solubility test which is widely used for assessing the extent of damage (see p. 292). The treatment consists of immersing the wool substance in 0·10N sodium hydroxide for one hour at 65° C. with uniform agitation and then determining the loss in weight. Under these conditions all the wool will dissolve in about twenty-four hours, and it is obviously very arbitrary to fix a one-hour period. Nevertheless this test is very sensitive both for detecting the damage done to wool and the introduction of new alkali-stable cross-links. Peracetic acid can be

used as an alternative to sodium hydroxide in this kind of test.[5] After treatment with peracetic acid until all the disulphide bonds in wool have been oxidized, 90 per cent. dissolves on subsequent extraction with a weakly alkaline solution ($pH > 7$). If new cross-links which cannot be oxidized by peracetic acid have been introduced the solubility is greatly decreased. A very small number of cross-links is already sufficient to bring about complete insolubility under these conditions, although, unlike the Harris test, this procedure cannot determine the degree of cross-linking.

All these tests are greatly affected if the charged groups in wool are modified. Thus Blackburn and Lindley[9] showed that the contraction in bisulphite is decreased to half when 60 per cent. of the carboxyl groups in wool are esterified. In addition, the solubility of esterified wool fibres in dilute ammonium hydroxide, after oxidation with peracetic acid, is very greatly decreased as there are no negatively charged groups to facilitate dispersion.[7] On the other hand, Speakman[10] has shown that deamination greatly increases the supercontraction in bisulphite.

When examining the cross-linking action of a poly-functional reagent, it is highly desirable to use a similar mono-functional compound as a control, otherwise the change produced by reaction with charged groups without causing cross-linking may be mistaken for a cross-linking reaction. If all these precautions are taken and if a number of the different tests described are used in each case, the presence or absence of cross-linking may be determined with certainty. As such precautions are often neglected and in many cases only one test has been used, it is doubtful whether the ability of a number of reagents to bring about cross-linking has in fact been proved.

Formaldehyde

Formaldehyde has been well known as a cross-linking agent for a whole range of materials, in particular proteins and cellulose, for more than fifty years. The scientific and technical literature concerning its reactions is very extensive, and it is not possible even to attempt to review in this book all the work done and the many patents and technical papers which have been published on the reaction between proteins and formaldehyde. Largely as the result of the work of Fraenkel-Conrat,[11] however, the large number of possible reactions that proteins can undergo with formaldehyde have been clarified and incorporated in a general theory. Although formaldehyde can react with amino, amido, guanidyl, indole and phenolic groups as well as bringing about reduction of the disulphide bond, its ability to form cross-links is not as great as would at first appear. Cross-linking cannot occur between two amino groups, but two guanidyl groups may be joined by one formaldehyde molecule only.[12] Thus Steinhardt, Fugitt and Harris[13] showed that all the amino side chains arising from lysine in wool react almost instantaneously with formaldehyde at room temperature to form methylol groups. On removing the wool from the formaldehyde solution and rinsing with water the reaction is reversed and the formaldehyde lost, the original groups of lysine being reformed. The methylol lysine is a much weaker base and its titration curve is shifted to a lower pH. However, formaldehyde can give stable cross-links between amide and amine groups:

$R-NH_2 + HCHO \rightarrow R.NHCH_2OH$

$R-NH.CH_2OH + H_2N.CO-R' \rightarrow R-NH-CH_2-NH-CO-R'$

The optimum conditions for this reaction (and for proteins in general) is 4 per cent. formaldehyde at 70° C. at pH 3–7.

By means of these methylene condensations, simple amides or guanidines may be introduced into proteins rich in amino groups. In addition, proteins and polypeptides rich in amide groups may be cross-linked between two amide groups by means of an acid solution of formaldehyde and ammonia or primary amines.

Primary amido groups and amino groups may be linked to phenols through formaldehyde by a Mannich type of reaction, but it is not known if two phenolic groups can be bridged in this way.

The original representation of the cross-linking by formaldehyde involving the reaction of two amino side chains is now considered to be incorrect:

$RNH_2 + HCHO + H_2NR' \rightarrow R-NH-CH_2-NH-R'$

Most of the earlier work on the combination of wool with formaldehyde has been invalidated by the work of Fraenkel-Conrat, and only more recent papers will therefore be considered.

Speakman and Peill[14] examined the elastic properties of formaldehyde-treated wool. After 48 hours' reaction at 50° C. with 6 per cent. formaldehyde, the work required to stretch a fibre was increased

by about 5 per cent., if the treatment was performed in the optimum
pH range of 5-7. Unfortunately supercontraction and alkali-solubility
data concerning fibres which have been treated in this way are not
available. Stoves[15] used even more drastic conditions of treatment
(2 per cent. formaldehyde at 100° C.), and showed that some cross-
linking had occurred. In view of the secondary reactions with the
disulphide bond which are known to occur under such conditions
(see p. 251), the interpretation of these results is uncertain.

Middlebrook[16] has extended studies on formaldehyde reactions at
100° C. in acid solution, by applying the paper chromatographic technique to identify reaction products. He concludes that cross-linking
has occurred between the amide group of the combined glutamic acid
and arginine. These cross-links are stable to boiling bisulphite solution when no supercontraction occurs, but are broken by distillation
with phosphoric acid, although formadehyde is not liberated. Fibres
in which the amide groups had been removed were found to supercontract readily after the formaldehyde treatment. This suggests
that acid-labile cross-links are formed between the amide groups of
glutamine and others such as the guanidine groups of arginine.
Middlebrook considers that the formaldehyde that cannot be recovered by acid distillation combines with the asparagine amide
groups to form 6-hydroxytetra-hydropyrimidine-4-carboxylic acid
which is stable to acid hydrolysis. There is no evidence from direct
chemical analysis of the formation of a djenkolic acid cross-link by
the treatment of wool with formaldehyde as postulated by Stoves.[15]
Such cross-links are formed, however, as shown by Brown and
Harris,[17] when wool is treated with formaldehyde after the disulphide
bonds have been reduced with thioglycollic acid. Although formaldehyde reduces cystine, it does not form a new cross-link but always
produces combined thiazolidine-4-carboxylic acid.

Alexander et al.[3] showed that a small quantity of formaldehyde
combines irreversibly with wool under very mild conditions (see
Table 10.1). Direct analysis showed that tyrosine is involved in the
reaction and it was concluded that the cross-link is formed between
the tyrosine and lysine residues.

The new cross-links are stable to alkali but are ruptured by acid
in the manner indicated. This treatment confers great stability to
contraction in bisulphite and greatly reduces the alkali solubility of
the wool. After the wool has been treated with acid and the cross-
link broken, the fibres again supercontract in bisulphite and the alkali

TABLE 10.1

Formaldehyde bound by wool at 60° C. in neutral solution and proportion liberated on distilling with 0·3M-phosphoric acid[3]

Concentration of formaldehyde (M)	Formaldehyde bound by wool (g./g.)	Formaldehyde liberated on distillation (g./g.)	Reversibly bound formaldehyde (percentage of total bound)
0·0033	0·00326	0·00057	15·9
0·0067	0·00629	0·00093	14·8
0·0133	0·00740	0·00129	17·4
0·0330	0·00920	0·00197	21·4

$$\begin{array}{c}
\diagdown\!CH\!-\!CH_2\!-\!\langle\bigcirc\rangle\!-\!OH + HCHO + NH_2(CH_2)_4\!-\!CH\diagup \\
\downarrow \\
\diagdown\!CH\!-\!CH_2\!-\!\langle\bigcirc\rangle\!-\!OH \\
| \\
CH_2\!-\!NH(CH_2)_4\!-\!CH\diagup \\
\downarrow\ \text{acid} \\
\diagdown\!CH\!-\!CH_2\!-\!\langle\bigcirc\rangle\!-\!OH + NH_2(CH_2)_4\!-\!CH\diagup \\
| \\
CH_2OH
\end{array}$$

solubility is increased. Brown et al.[18] reached essentially similar conclusions concerning the alkali solubility of formaldehyde-treated wool. The work required to stretch fibres in acid is increased by about 20 per cent. after this treatment, but the elasticity at pH 7 is unchanged or only slightly reduced. This reinforces the point made on page 325 that cross-linking cannot always be detected by elasticity measurements made in water if the cross-linking agent converts a charged group into one not fully ionized at pH 7. The improvement in all the properties of the wool fibre which can be brought about by this mild treatment with formaldehyde may be of industrial value.

Richards and Speakman[19] have rejected the conclusion of Alexander et al.[3] that tyrosine residues are involved in the reaction of wool with formaldehyde near neutrality. Clearly, if the 3:5 positions in the benzene nucleus of the tyrosine residues are blocked, reaction of wool with iodine (see p. 303) prior to treatment with formaldehyde should inhibit the cross-linking if Alexander et al. are correct.

Although Richards and Speakman have shown that this is so, since the formaldehyde treatment of iodinated fibres is ineffective in reducing supercontraction (see Table 10.2), these workers found that

TABLE 10.2

Influence of treatment with formaldehyde on the supercontraction of untreated and iodinated fibres[19]

Treatment	Supercontraction %	
	Not treated with H_3PO_4	After treatment and heating with H_3PO_4
None	$34\cdot8 \pm 0\cdot3$	$29\cdot7 \pm 0\cdot5$
Iodination	$31\cdot5 \pm 0\cdot6$	$40\cdot0 \pm 0\cdot4$
HCHO, 60° C., pH 6·3 ..	$28\cdot0 \pm 0\cdot3$	$38\cdot4 \pm 0\cdot2$
Iodination and HCHO, 60° C., pH 6·3 ..	$36\cdot8 \pm 0\cdot9$	$29\cdot1 \pm 1\cdot2$
HCHO, 60° C., pH 1·2 ..	$26\cdot7 \pm 0\cdot2$	$37\cdot4 \pm 0\cdot3$
HCHO, reflux, pH 1·2 ..	$1\cdot9 \pm 0\cdot1$	$33\cdot2 \pm 0\cdot4$

formaldehyde treated wool takes up the same amount of iodine as untreated wool, which they consider invalidates the hypothesis of Alexander et al. However, the treated wool was heated with 3M phosphoric acid prior to iodination. Since this would rupture the postulated cross-link to produce a —⟨ ⟩OH residue which
                                                      CH₂OH
can bind two atoms of iodine as does unsubstituted tyrosine[20] the data of Richards and Speakman[19] are not incompatible with the conclusions of Alexander et al.[3]

As pointed out by Speakman and co-workers[21], it is probable that the lack of marked change in mechanical properties of formaldehyde-treated wool, when other tests, such as alkali solubility, suggest that considerable cross-linking has taken place, may be due to reaction occurring principally in amorphous rather than crystalline regions of the fibre.

Mercury salts

Speakman and his collaborators[22] found that mercuric chloride was absorbed strongly by the amino groups, and also that no attack on the disulphide bond occurred in the time of the experiment at room temperature (compare Elöd,[23] who showed that the disulphide bond reacts very slowly). Speakman showed that cross-linking had not occurred as the tensile properties of the fibres were unchanged. After treatment with mercuric acetate, however, the work required to stretch a fibre was increased by 6 per cent. in water and by 15–20 per cent. in acid which indicated cross-linking with the amino groups in this case. Unfortunately there are no data on the supercontraction behaviour of treated fibres, but the resistance to felting is considerably increased. Thus, treatment with mercuric acetate for fifteen minutes in 0·1M mercuric acetate dissolved in 0·1N acetic acid at 25° C. decreased the shrinkage of a sample of wool fabric from 35 per cent. to 5 per cent. After treatment in this way, the weight of the wool is increased by more than 10 per cent. and it is probable that in addition to cross-linking other reactions have also occurred.

Benzoquinone

Benzoquinone is well known to react with amino and sulphydryl groups in the following way:

$$2RSH + 2 \underset{O}{\overset{O}{\bigcirc}} \rightarrow RS-\underset{OH}{\overset{OH}{\bigcirc}}-RS + \underset{OH}{\overset{OH}{\bigcirc}}$$

and was used as early as 1909 by Meunier to strengthen fibres. Subsequently Speakman and Peill[14] studied the reaction between wool and benzoquinone in some detail, and showed that maximum reaction occurred in the pH range of 4–6. After a treatment for 48 hours at 50° C. with a 2 per cent. solution of benzoquinone, the work to stretch a fibre at pH 7 was increased by 20 per cent. and the capacity to felt very markedly reduced. Stoves[24] points out that a treatment of this severity increases the weight of wool by 17 per cent. whereas if cross-linking only had taken place a weight increase of only 4·3 per cent. would have occurred if all the amino groups had reacted. The increase in strength is much greater than that obtained with any other cross-linking agent, which is no doubt due partly to the internal deposition of a resin, as benzoquinone is known to polymerize readily.[25] Unfortunately the relative contributions to the mechanical properties of cross-linking and resin formation cannot be assessed in the absence of supercontraction data. However, Stoves[24] has shown that approximately 50 per cent. of the amino group have reacted with benzoquinone, but that probably only a small proportion of these have cross-linked whereas the remainder are merely functioning to anchor the polymer to the wool, since this cannot be extracted with any solvent.

Poly-functional alkylating agents

Kirst[26] used bi-functional aliphatic diazo compounds and chloromethyl ethers for cross-linking wool in order to decrease its alkali solubility. These reagents, applied from organic solvents, because they readily hydrolyse in water, were found to reduce the alkali solubility by approximately one-half. Elöd and Zahn[27] found that the fibres had lost their tendency to contract in phenol, and their tensile strength was significantly increased. It is not clear which groups participate in the cross-linking reaction, and both carboxyl and amino may react as follows:

$$R-NH_2 + Cl.CH_2.O.CH_2Cl \rightarrow RNH.CH_2.O.CH_2.Cl + HCl$$

$$R'-COOH + Cl.CH_2.O.CH_2NHR \rightarrow$$
$$\rightarrow R'-COO.CH_2.O.CH_2.NHR + HCl$$

The amino link is probably the more important as the ester link is known to saponify readily (see p. 311). Kirst[26] showed, however, that the alkali solubility of wool could be reduced to the same extent by treatment with the mono-functional chloro-methyl ether when, of course, no cross-linking is possible. The effect is undoubtedly due to the introduction of an aliphatic chain into the polypeptide structure which was shown by Harris to increase the strength by increasing the Van der Waals' forces (see p. 340).

Poly-ethylene-imines[28] which have been used widely as cross-linking agents for cellulose, are decomposed only slowly by water and can therefore be applied from aqueous solution, or if insoluble from an aqueous dispersion. Two of the most reactive reagents of this class were found by Alexander et al.[5] to be effective cross-linking agents for wool. Table 10.3 shows the results from a whole series of tests which were employed to determine whether cross-linking had

TABLE 10.3

Cross-linking of wool fibres by bi-functional alkylating agents[5]

Compound used	Contraction after reaction with		Percentage of oxidized fibre insoluble in ammonia
	Bisulphite (%)	Peracetic acid (%)	
None	30·0	12·5	12
(tris-aziridinyl triazine structure)	1·8	2·4	86
CH_2-N(-CH$_2$).CO.NH.(CH$_2$)$_6$.NH.CO.N-(CH_2) bis-aziridine	12·9	6·1	68
diglycidyl ether structure (CH_2.CH—CH$_2$ with O)	0·9	1·5	78
Cl.CH$_2$.CH—CH$_2$ (epoxide)	3·8	2·3	75
CH$_3$SO$_2$O(CH$_2$)$_3$OSO$_2$CH$_3$	6·5	3·5	76
CH$_3$SO$_2$OCH$_2$C≡CCH$_2$OSO$_2$CH$_3$	0·0	1·4	74
1 : 3-difluoro-4 : 6-dinitrobenzene	5·6	9·0	80
1 : 3-dichloro-4 : 6-dinitrobenzene	12·8	11·3	73

taken place. All the compounds employed reduce the alkali solubility considerably, and all the available tests show clearly that cross-linking has occurred. Again reaction is possible both with the carboxyl and amino groups, although as already mentioned the compounds formed with the former would be relatively unstable.

It was found[5] by employing all the tests listed in Table 10.3 that a fibre can still be cross-linked after esterification of the carboxyl groups but not after acetylation of the amino groups and from this it can be concluded that amino groups are involved in this reaction. It is possible that some cross-links may be formed between two carboxyl groups, but were broken under the conditions used in the various tests and therefore not detected.

Simple epoxides were first used by Schlack[29] to increase the affinity of wool for dyes. Later Deuel[30] cross-linked alginic acid and pectin with bi-functional epoxides, which were also used by Speakman successfully on wool. The first compound to be employed[31] was 3 : 4-iso-propylidene-1:2:5:6-dianhydromannitol, but later[32] diepoxybutane (though not diepoxyhexane) was shown to be equally effective. According to Speakman best results are obtained with a neutral aqueous solution, applied at 50° C. for 24 hours. Under these conditions the work to stretch a fibre in water was increased by 5 per cent. and in 0·1N hydrochloric acid by 15 per cent. The fibres also lost their tendency to supercontract and all the tests point to the formation of stable cross-links. The milling shrinkage is only reduced very slightly, e.g. from 47 to 39 per cent. Speakman considers that the cross-links are formed between carboxyl groups, as follows:

$$RCOOH + CH_2\underset{O}{-}CH-(CH_2)_n-CH\underset{O}{-}CH_2 + HOOCR'$$

$$\rightarrow RCOO \cdot CH_2-\underset{OH}{CH}-(CH_2)_n-\underset{OH}{CH}-CH_2-OOCR'$$

Reaction with amino groups was excluded because deaminated fibres can also be strengthened by reaction with the di-epoxides. On the other hand, no evidence for reaction with carboxyl groups could be found from acid-titration data,[5] so that only a few cross-links could have been introduced if indeed the cross-linking does occur with the carboxyl groups. In view of the accuracy with which the acid titration curve can be obtained, it seems more likely that the amino groups are involved, and that the data on de-aminated wool are capable of an alternative explanation, especially in view of the considerable degradation which is known to occur (see p. 316). In support of this conclusion, it is known that esterified but not acetylated fibres can be cross-linked[5] by bis-epoxides.

Epichlorhydrin can also act as a cross-linking agent (see Table 10.3). The second reactive centre in this molecule may be the chlorine atom which is known to react readily with amino groups or alternatively another epoxide group may be formed by elimination of hydrochloric acid (see p. 313).

$$RNH_2 + CH_2\underset{O}{\underset{\diagdown\diagup}{-}}CH-CH_2Cl \rightarrow RNHCH_2\cdot\underset{\underset{OH}{|}}{CH}-CH_2Cl$$
$$\downarrow -HCl$$
$$RNH-CH_2\cdot\underset{\underset{OH}{|}}{CH}-CH_2NHR \leftarrow RNHCH_2\cdot CH\underset{O}{\underset{\diagdown\diagup}{-}}CH_2$$

Glycol esters of methanesulphonic acid (see Table 10.3) were found to be exceptionally effective cross-linking agents. Compounds of this type are alkylating agents which readily react with amino and carboxyl groups but the limited evidence available[5] again suggests that the new cross-links involve the former and not the latter group in the wool fibre. The practical application of these bi-functional alkylating agents should be approached with great caution since some of the compounds are known to be carcinogenic[5] and therefore present a considerable health hazard.

Dinitrofluorobenzene derivatives

It is known that dinitrofluorobenzene reacts with proteins under very mild conditions (see p. 313). Zahn and co-workers have made an extensive investigation of the cross-linking action of bi-functional dinitrofluorobenzene derivatives on silk fibroin, wool and collagen, and since the cross-linked amino acids are stable to acid hydrolysis conditions, it is possible to positively identify them. The general method is to synthesize the suspected cross-linked amino acids and compare their R_F values with any new spots that appear in the paper chromatograms of the hydrolysate of wool after the cross-linking

FORMATION OF NEW CROSS-LINKS

treatment. Zahn and co-workers have made a particular study of the cross-linking action of 1,5-difluoro-2,4-dinitrobenzene and 4,4¹-difluoro-3,3¹-dinitrophenylsulphone on proteins.

After reaction of wool with the former compound, Zahn and Meienhofer[33] have identified the following substances in wool hydrolysates:

showing that two lysine residues, two tyrosine residues or one lysine and one tyrosine residue can cross-link.

An extensive summary of the results of his school on cross-linking reactions, originally incorporated in some twenty publications, has been given by Zahn.[34]

Miscellaneous compounds

The compound formed between urea and glyoxal (Kaurit 140), which is used commercially for cross-linking regenerated cellulose fibres, has been found[35] to decrease the alkali solubility of wool by half, reduce its tendency to felt and increase its tensile strength. It is highly probable therefore that cross-linking takes place in this reaction. Dithioglycollide, obtained by heating thioglycollic acid, can react with the amino groups in proteins as follows:

$$\begin{array}{c} COCH_2 \\ S \diagup \quad \diagdown S + 2RNH_2 \rightarrow 2RNHCOCH_2SH \\ \diagdown CH_2CO \diagup \end{array}$$

$$\rightarrow RNHCOH_2C \cdot SS \cdot CH_2CONHR$$

The new sulphydryl groups formed are readily oxidized to form disulphide cross-links which have been introduced in this way into casein fibres, and also into wool.[36]

Cockburn and Speakman[37] have examined in some detail the action of ninhydrin (1,2,3,-indantrione hydrate) on wool. The increased resistance to extension of treated wool fibres indicates that cross-linking has occurred, but the precise mechanism remains obscure. It has been extablished that the basic groups of wool are involved, but experiments using model substances indicate that ninhydrin reacts monofunctionally rather than bifunctionally. It has been suggested that the reaction product of ninhydrin with basic groups such as arginine:

can undergo powerful hydrogen bonding with hydroxy, phenolic and peptide groups in adjacent chains.

The purple colouration assumed by the wool after reaction must limit the commercial applications of the process. Zahn's compounds[34] suffer from a similar defect.

Di-isocyanates[38] also have been used to cross-link wool, the reactions being basically the same as those described on p. 320.

Rebuilding reduced disulphide bonds

In the course of developing a low-temperature chemical set for wool (see p. 72), Speakman[39] showed in 1936 that the disulphide bond could be reduced with sodium sulphite or bisulphite especially if the latter were dissolved in alcohol. The reduced bond (see p. 246) could be rebuilt with metal salts, in particular the salts of weak acids. Successful rebuilding was also obtained by oxidizing the reduced fibre. Although Speakman published no actual results he envisaged in a patent application[40] the reduction of wool with thiols and the rebuilding of the bonds with alkyl dihalides. This was later shown by Harris and his colleagues[41] to be by far the most useful method of replacing the disulphide bond and is the only method at present used commercially. These workers found[42] that wool which had been modified in this way is much more resistant than untreated wool towards many chemical reagents including alkalis, acids, oxidizing and reducing agents. In addition the treated wool is stained less readily by metals, and is attacked much less readily by biological agents such as moth larvae, carpet beetles and enzymes. The fibres do not supercontract in bisulphite since the new link introduced is not broken by this agent. In a typical treatment the wool is reduced in a 0·02N solution of calcium thioglycollate at 50° C. for two hours, washed, and immediately placed in a phosphate buffer at pH 8 containing an emulsion of ethylene dibromide.

$$R-S-S-R + 2HS-CH_2-COOH \rightarrow 2R-SH + (SCH_2-COOH)_2$$
$$2R-SH + Br(CH_2)_nBr \rightarrow R-S-(CH_2)_n-S-R + 2HBr$$

The properties of fibres cross-linked with different alkylating agents can be seen from Table 10.4.

From a commercial viewpoint the process described above is too slow and expensive, so that more suitable reducing agents were sought. Inorganic reagents such as bisulphite, hydrosulphite and sulphoxylate rapidly reduce wool at elevated temperatures but cause extensive disorientation in the fibre which outweighs the effect of any new cross-links, and the fibres are permanently weakened. Brown and Harris[17] found when the cross-linking agent and reducing agent are applied together very effective and rapid replacement of the disulphide bond may be obtained. Alkylhalides and thiols cannot of course be used together since they react, but this does not apply to inorganic reducing agents such as hydrosulphite or sulphoxylate, which can be used in conjunction with alkyl dihalides or formaldehyde to bring about the following series of reactions:

$$R\text{—}S\text{—}S\text{—}R + Na_2S_2O_4 \rightarrow R.S.Na + R\text{—}S\text{—}S_2O_4Na$$
$$2R.S.Na + (CH_2)_nBr_2 \rightarrow R.S.(CH_2)_n.S.R + 2NaBr$$

It is seen that the alkali solubility is reduced in this one-step process (see Table 10.5) and supercontraction in bisulphite is prevented without appreciable loss of strength, which indicates that perfect rebuilding has occurred. An interesting example of this method of linkage rebuilding[43] is shown by the fact that wool fibres do not supercontract

TABLE 10.4

Rebuilding of reduced wool fibres[42]

Treatment	Reduction in work for 30% extension in water %	Dry breaking strength	Alkali solubility
Untreated	1	1,310	10·5
Reduced	35	1,170	75·0
Reduction + H_2O_2	3	1,310	12·5
,, + CH_2I_2	6	1,380	5·2
,, + $(CH_2)_2Br_2$	8	1,420	6·6
,, + $(CH_2)_3Br_2$	10	1,460	5·3
,, + $(CH_2)_4Br_2$	12	1,380	7·0
,, + CH_3I	33	1,190	33·0
,, + C_2H_5Br	30	1,250	18·3
,, + $C_6H_5CH_2Br$	8	1,310	15·4
,, + $pClC_6H_4CH_2Cl$	5	1,450	12·8
,, + $C_7H_{15}Br$	5	1,380	18·0
,, + $C_{12}H_{25}Br$	4	1,340	18·2
,, + $ClCH_2COOH$	33	1,240	75·0

in bisulphite when this contains ethylene dibromide. This method[17] besides providing a commercially practicable process for setting can also be used for moth-proofing and for stripping dyes without causing damage to the wool. Allyl isothiocyanate[44] was shown to be particularly effective in rebuilding cross-links in reduced wool and can probably be incorporated into the one-step process of Brown and Harris. Benzoquinone also rebuilds reduced disulphide-bonds in wool but is unlikely to be of commercial interest because it discolours the wool.

TABLE 10.5

Properties of wools treated by one-step vs. two-step process[17]

Sample	Condition	30% Index	Alkali solubility (%)
Untreated	—	0·99	9·2
Treated: 1 A	1·6%SFS* 80° C. 1 hr. pH 8	0·41	14·2
1 B	Sample 1A treated with CH_2O	0·63	3·3
1 C	1·6%SFS 0·6% CH_2O 80° C. 1 hr. pH 8	0·92	3·3
2 A	1·6% SFS 95° C. 1 hr. pH 8	0·05	18·3
2 B	Sample 2A treated with $(CH_2)_2Br_2$	0·08	6·1
2 C	1·6% SFS 95° C. 1 hr. pH 8 $(CH_2)_2Br_2$	0·65	4·5

* Sodium formaldehyde-sulphoxylate.

Kirst[26] showed that the bi-functional alkylating agents discussed on page 333 are also capable of replacing the disulphide bonds in wool which have been reduced by thiols. He examined a large range of dihalides and from his data it would appear that 1 : 3-dichlorethyl-4 : 6-dimethylbenzene is more effective for this purpose than the alkyl dihalides. In this case where a large molecule is introduced, additional secondary forces probably contribute to the cohesion of the fibre.

McPhee and Lipson[45] have reacted reduced wool with a number of unsaturated aldehydes, and have obtained evidence for crosslinking. In the case of acraldehyde it is possible that the following reactions occur:

$W-CH_2-SH + CH_2=CH \cdot CHO + HS-CH_2-W \rightarrow$

$\rightarrow WCH_2SCH_2CH_2CH(OH)SCH_2W$

$W-CH_2-NH_2 + CH_2=CH \cdot CHO + HS-CH_2-W \rightarrow$

$\rightarrow WCH_2NHCH(OH)CH_2CH_2SCH_2W$

With α-methacraldehyde, however, the second reaction cannot take place since the lysine content of the wool does not fall on treatment.

N-Ethylmaleimide reacts readily with the thiol groups present in unmodified wool (see p. 317), and it has been shown by Moore and Lundgren[46] that o- and m-phenylenedimaleimide effectively cross-link reduced wool.

The interesting possibility has been put forward by Kessler and Zahn that cystine cross-linkages may be produced in wool by an intra-interchain exchange reaction. It has been shown that mild alkaline treatment of wool fibres results in decreased urea-bisulphite solubility for a virtually unchanged cystine content,[47, 48] and this is held to support the hypothesis that intra-chain disulphide groups are changed to interchenic cross-linkages by the alkali. Swan[49] has criticized this conclusion, although he has confirmed the experimental facts on which it is based. Swan considers the insolubilization to be a denaturation process, which causes many soluble proteins to become insoluble when subjected to quite mild treatments which do not involve the rupture of covalent bonds.

The formation of polymers within the fibre

Speakman showed that by polymerizing monomers from an aqueous or gaseous phase within the fibre, a very great increase in weight with a corresponding increase in strength, decrease in the tendency to felt and improvement in wearing properties of the wool can be obtained without modifying the feel of the fabric seriously. The first method to be developed[50] employed ethylene sulphide (for mode of action see p. 251) which trebled the strength of the fibres after extensive reaction. More rapid deposition of polymer is obtained when the fibres are impregnated with a catalyst before exposure to the vapour of water and monomer. When, for example, a sample of wool is impregnated with a 0·2 per cent. solution of ammonium persulphate and then exposed to the vapours of methyl methacrylate and water at 90° C. for 30 and 60 minutes the resulting gains in weight are 80 and 150 per cent., respectively.[51] Polymerization within the fibre can also be achieved when the monomers are applied from aqueous solution or in the form of a suspension if the fibre is first impregnated with a polymerization catalyst.[52] For example, when using a solution of methacrylic acid containing some hydrogen peroxide the wool should contain impregnated ferrous ions. The hydrogen peroxide reacts with the iron to produce hydroxyl radicals which initiate polymerization. Rapid polymerization occurs within fibres, which have been treated with a reducing agent to give SH groups, on subsequent exposure to a mixture of monomer and hydrogen peroxide (SH groups give OH radicals with H_2O_2).

In a further study of this process, Lipson and Howard[53] have shown that cystine, glutathione, thiourea and thiophenol rapidly cause the polymerization of ethylenic monomers when activated with hydrogen peroxide or ammonium persulphate. With wool, however, it was found that the polymer readily formed in the tip-halves of the fibre after bisulphite treatment, but the root-halves formed little polymer unless previously oxidized. Further, the more reactive A+B fraction of the combined sulphur (see p. 259) was shown to be responsible for the polymer formation. As cysteic acid did not promote the reaction, it was suggested that the active compound is probably a sulphoxide or sulphone (see p. 270). The polymerization then proceeds by a free radical mechanism:

$$\begin{array}{c} | \\ NH \\ | \\ CH-CH_2-S-+CH_2=\overset{R}{\underset{|}{C}}H+CH_2=\overset{R}{\underset{|}{C}}H \\ | \\ CO \\ | \end{array}$$

$$\rightarrow \begin{array}{c} | \\ NH \\ | \\ CH-CH_2-S-CH_2-\overset{R}{\underset{|}{C}}H-CH_2-\overset{R}{\underset{|}{C}}H- \\ | \\ CO \\ | \end{array}$$

The internal deposit can be cross-linked to the wool itself.[52] If, for example, methacrylamide is polymerized within the fibre, the polymer carries amide side chains which can be cross-linked to the amino side chains of the wool with formaldehyde. The cross-linking reaction proceeds most readily at about pH 5 and a remarkable strengthening of the structure is realized by this chemical anchoring of an internal deposit.

Electron-photomicrographs[54] show that the polymer formed within the fibre by the methods described is situated within the macropores of the fibre, between the cuticle and the cortical cells, and probably also between the fibrils. There is no evidence to show that the polymer is deposited between individual molecules, and the suggestion[54] that the polymer fills the 'dead space' between the main peptide chains finds no support. Mercer[54] believes that the increase in strength of wool fibres containing polymer can be explained if the suggestion of Astbury and Woods (see p. 389) is accepted, that in the initial stages of stretching the extra fibrillar phase is responsible for the extension.

REFERENCES

[1] ALEXANDER, CARTER AND EARLAND, *J. Soc. Dy. Col.*, 1949, **65**, 107.
[2] BARR, CAPP AND SPEAKMAN, *ibid.*, 1946, **62**, 338.
[3] ALEXANDER, CARTER AND JOHNSON, *Biochem. J.*, 1951, **48**, 435.
[4] SPEAKMAN, *J. Soc. Chem. Ind.*, 1931, **50**, T1; *Nature*, 1933, **152**, 930.
[5] ALEXANDER, FOX, SMITH AND STACEY, *Biochem. J.*, 1952, **52**, 174.
[6] ELÖD AND ZAHN, *Melliand Textilber.*, 1949, **30**, 17; ZAHN, *Kolloid Z.*, 1944, **108**, 94.
[7] ALEXANDER AND SMITH (unpublished).
[8] HARRIS AND SMITH, *Amer. Dyest. Rep.*, 1936, **25**, 542.
[9] BLACKBURN AND LINDLEY, *J. Soc. Dy. Col.*, 1948, **64**, 305.
[10] SPEAKMAN, *J. Text. Inst.*, 1936, **27**, P231.
[11] FRANKEL-CONRAT, COOPER AND OLCOTT, *J. Amer. Chem. Soc.*, 1945, **67**, 950.
[12] FRANKEL-CONRAT AND OLCOTT, *ibid.*, 1946, **68**, 34; *J. Biol. Chem.* 1948, **174**, 827.
[13] STEINHARDT, FUGITT AND HARRIS, *ibid.*, 1946, **165**, 285.
[14] SPEAKMAN AND PEILL, *J. Text. Inst.*, 1943, **34**, 70T.
[15] STOVES, *Trans. Faraday Soc.*, 1943, **39**, 294.
[16] MIDDLEBROOK, *Biochem. J.*, 1949, **44**, 17.
[17] BROWN AND HARRIS, *Ind. Eng. Chem.*, 1948, **40**, 316.
[18] BROWN, HORNSTEIN AND HARRIS, *Textile Res. J.*, 1951, **21**, 222.
[19] RICHARDS AND SPEAKMAN, *Proc. Int. Wool Text. Res. Conf.*, Australia, 1955, C-308.
[20] ALEXANDER, *ibid.*, C-495.
[21] GHOSH, HOLKER AND SPEAKMAN, *Text. Res. J.*, 1958, **28**, 112.
[22] SPEAKMAN AND COKE, *Trans. Faraday Soc.*, 1939, **35**, 246.
BARR AND SPEAKMAN, *J. Soc. Dy. Col.*, 1944, **60**, 335; MENKART AND SPEAKMAN, *ibid.*, 1947, **63**, 322.
[23] ELÖD, *Kolloid Z.*, 1941, **96**, 284.
[24] STOVES, *Trans. Faraday Soc.*, 1943, **39**, 301.
[25] ERDTMAN, *Proc. Roy. Soc.*, 1933, **143A**, 177.
[26] KIRST, *Melliand Textilber.*, 1947, **28**, 169, 314; 394, 1948, **29**, 236.
[27] ELÖD AND ZAHN, *ibid.*, 1948, **29**, 17, and 269.
[28] BESTIAN, *Ann.*, 1950, **566**, 210.
[29] SCHLACK, British Patent 464,043 (1934).
[30] DEUEL, *Helv. Chim. Acta*, 1947, **30**, 1523.
[31] CAPP AND SPEAKMAN, *J. Soc. Dy. Col.*, 1949, **65**, 402.
[32] FEARNLEY AND SPEAKMAN, *Nature*, 1950, **166**, 743.
[33] ZAHN AND MEIENHOFER, *Melliand Textilber.*, 1956, **37**, 432.
[34] ZAHN, *Proc. Int. Wool Text. Res. Conf.*, Australia, 1955, C-425.
[35] HARTMANN, *Melliand Textilber.*, 1949, **30**, 70.
[36] SCHÖBERL, *Angew. Chem.*, 1948, A60, 7; *Naturw.*, 1949, **36**, 121 and 149.
[37] COCKBURN AND SPEAKMAN, *Proc. Int. Wool Text. Res. Conf. Australia*, C-315.
[38] MOORE AND O'CONNELL, *Text. Res. J.*, 1957, **27**, 783.
[39] SPEAKMAN, *J. Soc. Dy. Col.*, 1936, **52**, 335 and 423.
[40] SPEAKMAN, British Patent 453,701 (1936); see also *Textile Manuf.*, 1941, **67**, 381; 1942, **68**, 129.
[41] PATTERSON, GEIGER, MIZELL AND HARRIS, *J. Res. N.B.S.*, 1941, **27**, 89.
[42] GEIGER, KOBAYASHI AND HARRIS, *ibid.*, 1942, **29**, 381.
[43] BROWN, PENDERGRASS AND HARRIS, *Text. Res., J.* 1950, **20**, 51.
[44] NEISH AND SPEAKMAN, *Nature*, 1949, **164**, 708.

[45] McPhee and Lipson, *Australia J. Chem.*, 1954, **7**, 387.
[46] Moore and Lundgren, *Proc. Int. Wool Text. Res. Conf.*, Australia, 1955, C-355.
[47] Kessler and Zahn, *Text. Res. J.*, 1958, **28**, 357.
[48] Batlow and Kessler, *ibid.*, 1958, **28**, 359.
[49] Swan, *ibid.*, 1959, **29**, 665.
[50] Barr and Speakman, *J. Soc. Dy. Col.*, 1944, **60**, 238.
[51] Speakman and Barr, British Patent 559,787.
[52] Lipson and Speakman, *J. Soc. Dy. Col.*, 1949, **65**, 390.
[53] Lipson and Howard, *Aust. J. Sci. Res.*, 1950, **3**, 324.
[54] Mercer, *J. Text. Inst.*, 1949, **40**, T629.

CHAPTER 11

Chemical Composition

IN addition to the protein material which constitutes the bulk of the fibres, certain fats, sterols and complex lipoids are also present in wool as it is found in nature. Most of these minor components can be extracted by prolonged treatment with alcohol and ether, and are not normally considered to be fundamental constituents of wool. A small quantity of non-extractable mineral matter is also present as shown by the ash content, together with a small percentage (< 0·02 per cent.) of combined phosphorus.

Amino-acid composition

As, however, these non-protein constituents represent a maximum of 1 per cent., wool can be considered for most purposes as made up entirely of protein. Prior to the development by Moore and Stein[1, 2, 3] of the ion-exchange chromatographic technique for the quantitative determination of amino acids, which enables a complete amino acid analysis to be performed on a hydrolysate of a few milligrammes of wool, amino acids had to be determined by tedious separations and colour reactions, which were often non-specific. In Table 11.2 are given the most probable values for the amino acid content of wool according to Astbury[4] and Harris[5] respectively. These estimates are based on analytical figures using the older methods, whilst in Table 11.4 are given three sets of values based on ion-exchange chromatography.

From Tables 11.2 and 11.4 it is seen that the 'best values' suggested by Astbury and Harris were remarkably accurate and their use in the past for representing the composition of wool was well justified. Although all the wool substance can be accounted for in terms of the individual amino acids and their content has been established with some certainty, very little is known about the order in which these acids occur in the peptide chain. It should be noted that two distinct methods are used in the literature for stating the amino acid content of proteins. The amino acid may be expressed as a percentage of the protein. This method gives a total amino-acid composition in excess of 100 per cent. due to the elimination of one molecule of water when two amino acids combine to form a polypeptide link. In the second method the nitrogen content of a given amino acid is expressed as a percentage of the total nitrogen content of the wool, when of course, the total composition is 100 per cent. The value for amide nitrogen

TABLE 11.1

Proportion of polar and non-polar amino acids in wool

Character of amino acid	No. of residues in mol per 10^5 g. wool
acidic	65
basic	82
phenolic	26
hydroxyl	152
amide	81
sulphur-containing	104
remainder	351

represents the number of dicarboxylic acids, i.e. glutamic and aspartic acids, which are present as amides, but the number of molecules of each acid forming an amide bond is unknown as it is impossible to split wool into its component amino acids without first hydrolysing this bond. It is well known that the amino acid composition of wools of different qualities varies (see Table 11.3), and Simmonds[9] and Corfield and Robson[10] consider that the much smaller differences shown when a number of samples of a given quality of wool is analysed are equally real and are not due to deficiencies in the analytical technique. The cystine content also varies widely because of biological degradation, weathering, and exposure to light (see p. 265) and the cystine content is much smaller in the tip portion of the fibres than in the root portion, since the former has been exposed to the degradative influences for a longer period. The root ends of the fibres are therefore used for amino-acid analysis, although this introduces the difficulty of defining the root end of the fibre. In the case of cystine, although the figure obtained probably depends on the extent of the fibre taken to be the root end, it must be borne in mind that it is notoriously difficult to determine this amino acid and widely different results are obtained by different analytica

TABLE 11.2
Probable values for Amino acid composition of wool according to Astbury and Harris

Amino acid	Amount isolated from 100 g. of wool (g.)*	Most probable value (Astbury)	Most probable value (Harris et al.)	No. of residues per 10⁵ g. of wool (Astbury figures)
Alanine	4·4, 4·1	4·13	4·40	46·4
Arginine	10·3, 10·2, 7·8, 6·0, 8·7, 10·4, 9·4, 8·6	10·30	10·40	59·2
Aspartic acid	2·3, 7·27	6·57	7·27	49·4
Cystine	7·3, 7·35, 8·4, 13·1, 9·5, 13·0, 10·0, 13·1, 10·5, 12·72, 11·9, 11·8	11·90	12·20	98·9
Glutamic acid	12·9, 15·27, 16·0	14·10	15·27	95·9
Glycine	6·5	6·50	6·50	87·0
Histidine	0·66, 0·55, 0·7	0·70	0·70	4·5
Hydroxylysine	0·21, 0·10			
Leucine	11·5, 11·3, 11·6, 9·7	11·30	11·30	86·3
Lysine	2·8, 3·30, 2·2, 2·3, 2·5, 3·3	2·65	3·30	18·2
Methionine	0·2, 0·55, 0·7, 0·6, 0·35	0·70	0·71	4·7
Phenylalanine	4·0, 3·8, 1·6	3·75	3·75	22·7
Proline	4·4, 6·7, 7·2	6·80	6·75	59·1
Serine	9·4, 9·5, 9·5	10·30	9·41	98·1
Threonine	6·6	6·40	6·76	53·8
Tryptophan	1·8, 0·7, 1·0	1·80	0·70	8·8
Tyrosine	2·9, 4·8, 4·5, 5·8, 6·1, 5·15, 6·1	4·65	5·80	25·7
Valine	2·8, 4·8, 5·5	4·80	4·72	41·0
Amide nitrogen	1·2, 1·37, 1·2, 1·2, 1·18, 1·4	1·13	1·40	81·0
Total		107·35	109·90	860·0

*These values were obtained by some 40 workers during the past 30 years

CHEMICAL COMPOSITION

TABLE 11.3

Amino Acid composition of Hydrolysates from Wool of Merino 64's, Merino 70's and Corriedale 56's Quality[9]

Amino Acid	Merino 64's		Merino 70's		Corriedale 56's	
	Total Nitrogen (%)	By Wt. (%)	Total Nitrogen (%)	By Wt. (%)	Total Nitrogen (%)	By Wt. (%)
Alanine	3·51	3·7	3·51	3·7	4·37	4·7
Arginine	20·32	10·5	19·35	10·0	18·21	9·5
Aspartic Acid	4·24	6·7	4·68	7·4	4·86	7·8
Amide	7·46	1·4	7·92	1·5	9·27	1·8
Cystine	7·93	11·3	6·50	9·2	6·80	9·8
Glutamic Acid	8·58	15·0	8·54	14·9	9·69	17·1
Glycine	5·80	5·2	6·60	5·9	6·40	5·8
Histidine	1·46	0·9	1·48	0·9	1·59	1·0
*iso*Leucine	1·97	3·1	2·13	3·3	2·38	3·7
Leucine	4·90	7·6	5·37	8·3	5·51	8·7
Lysine	3·25	2·8	3·19	2·8	3·72	3·3
Methionine	0·31	0·6	0·37	0·7	0·37	0·7
Phenylalanine	1·75	3·4	2·28	4·5	2·35	4·7
Proline	5·33	7·3	5·12	7·0	5·52	7·6
Serine	7·25	9·0	8·63	10·7	7·71	9·7
Threonine	4·61	6·6	4·12	5·8	4·84	7·0
Tryptophan	1·73	2·1	1·38	1·7	1·80	2·2
Tyrosine	2·97	6·4	3·09	6·6	3·11	6·8
Valine	3·57	5·0	3·56	4·9	4·50	6·3
Total	96·94	108·6	97·82	109·8	103·00	118·2

methods.[10] In column 5 (Table 11.2) the amino-acid content expressed as the number of gram-molecules of amino acids in 10^5 g. of wool is given for the analytical values chosen by Astbury. The total number of amino-acid residues (860 gram molecules per 10^5 g.) in wool is very similar to that obtained for a large number of proteins, as the number of residues per 10^5 g. of protein generally lies between 800 and 1,000 gram-molecules. In Table 11.1 the number of side groups of different chemical character are shown (e.g. acidic side groups

= glutamic+aspartic−amide). It will be observed that wool contains approximately 65 gram-molecules of acid groups per 10^5 g. of wool so that the total number of acid and basic groups are not equal,

TABLE 11.4

Amino acid analyses of wool by ion-exchange chromatography

Amino acid	N as % total N of wool		
	Simmonds[6]	Corfield and Robson[7]	Graham, Waitkoff and Hier[5]
Alanine	3·51	4·12	—
Amide N	7·46	6·73	—
Arginine	20·30	19·10	21·1
Aspartic acid	4·24	4·38	4·7
Cystine	7·93	7·30	9·9
Glutamic acid	8·58	8·48	9·2
Glycine	5·80	6·29	—
Histidine	1·46	1·91	1·8
isoLeucine	1·97	2·44	3·0
Leucine	4·90	5·85	5·3
Lysine	3·25	3·92	3·9
Methionine	0·39	0·32	0·4
Phenylalanine	1·75	2·12	2·1
Proline	5·33	5·05	6·1
Serine	7·25	8·66	—
Threonine	4·61	5·12	4·8
Tryptophan	1·73	0·82	—
Tyrosine	2·97	2·62	2·7
Valine	3·57	4·16	4·2
Unknown (1)	1·18	—	—
Unknown (2)	0·71	—	—
Total	98·89	99·39	79·2

in disagreement with frequent statements to the contrary. Although significant variations in amino-acid compositions between different keratins such as horn, hair and wool are obtained, it is interesting to note that little variation in the ratios of the amino groups with different polar side chains is found. It has been stressed by Astbury[4] that the similarity between fibrous proteins may lie not so much in an actual identity of amino-acid analysis as in a similar ratio of polar to non-polar groups.

Terminal amino acids

The polypeptide chains constituting proteins are terminated by a free amino group at one end and a free carboxyl group at the other (unless they are present as very large rings, which has been suggested in the case of feather keratin since no end-groups are found[11, 12]). The only definite information on the order of the amino acids in the peptide chains of wool is the nature of the acids occupying these terminal positions. In the case of acids with terminal amino groups it is possible to label them by reaction with dinitrofluorobenzene (see p. 313) and to separate them from a hydrolysate. The proportions of the seven amino acids occupying these positions are given in Table 11.5. From the results[13] with dinitrofluorobenzene it is possible to assess the *average* molecular weight of the different polypeptide chains cross-linked by disulphide bonds which are present in wool. Thus the total number of terminal amino acids which is given

TABLE 11.5

Terminal amino acids in wool and cortical cells from enzymatically degraded wool determined by reaction with dinitrofluorobenzene[13]

Terminal amino acid carrying free amino groups	Mol per 10^6 g. of protein
Aspartic acid	0·63
Glutamic acid	1·25
Serine	1·25
Threonine	4·80
Glycine	5·20
Alanine	1·25
Valine	2·40

approximately as 1·68 mol per 10^5 g. of wool by the results in Table 11.5 leads to a value of 59,000 for the molecular weight. In view of other data which will be considered later, this average value is possibly too high, since wool contains a considerable proportion of short peptide chains in addition to a high molecular weight fraction (see p. 373). The high value of the average molecular weight thus obtained (i.e. the low number of terminal groups found) may be due to one or more of the following reasons. Firstly, a large proportion of the terminal amino groups may not be free, but may form ring compounds or amide links with neighbouring carboxyl groups. Secondly, all the amino groups may not be accessible to dinitrofluorobenzene purely because of steric factors. Thus, ovalbumin has no reactive terminal group and a cyclic structure is therefore possible, although the presence of acetylamino groups[13a] has recently been demonstrated. It is also well known in other systems that some amino groups are sterically inaccessible to dinitrofluorobenzene, and it has been shown that a proportion of the histidine and lysine in several soluble proteins fails to react.[14] For wool it has been found that neither all the lysine[15] nor all the tyrosine[16] had combined with this reagent in three days.

A method[17] for determining the amino acids which provide the terminal carboxyl groups, involving reduction to the alcohol, has been applied to wool[18] and end-groups of glycine, serine, alanine and threonine have been found. Blackburn and Lee[19] have found the same amino acids present as end-groups using the method of hydrazinolysis. The protein is heated with anhydrous hydrazine whereby the amino acid residues are converted into hydrazides with the exception of those terminating the chains and carrying free carboxyl groups, which are liberated in the free state:

$$\underset{R_1}{NH_2.CH.CO}.\underset{R_2}{NH.CH.CO}.\underset{R_3}{NH.CH.COOH} \; N_2H_4 \longrightarrow$$

$$\underset{R_1}{NH_2.CH.CO.NH.NH_2} + \underset{R_2}{NH_2.CH.CO.NH.NH_2} + \underset{R_3}{NH_2.CH.COOH}$$

The hydrazides formed are separated by absorption on a column of a carboxylic ion-exchange resin, while the free amino acids pass through the column and are subsequently identified by paper chromatography. In general, these methods for determining carboxyl end-groups are more drastic than those used for amino groups so that there is correspondingly less certainty about the reliability of the results. An estimate of the total number of terminal carboxyl groups has been deduced from acid-combining data. As shown in Chapter 6, the combination of wool with acid is back titration with carboxyl groups, and although the maximum value is liable to some experimental error, there can be no doubt from the excellent analytical data of Steinhardt and Harris (see p. 183) that 78–82 mol of acid are combined. The most probable analytical figures for the individual amino acids show that the total number of di-carboxylic acids cannot exceed 60–70 mol for 10^5 g. of wool. The difference between the analytical figures and the titration value indicates that at least 10 mol. of terminal carboxyl groups must be present in 10^5 g. of wool. This calculation leads to a *number average* molecular weight for the different polypeptide chains of 10,000. This value is much smaller than that calculated from analysis of the terminal amino groups.

The order of the amino acids along the peptide chain

Although the relative proportion of the different amino acids in many proteins (including wool) is known with considerable accuracy, detailed information concerning the order in which they are incorporated along the chain is only known for the hormone insulin, a protein of relatively low molecular weight. Bergmann and Niemann[20], however, proposed that:

(i) The numbers of molecules of different amino acids occurring in a protein are in simple ratios, such that the numbers can be expressed as $2^m\, 3^n$ where m and n are small integers.

(ii) The total number of amino-acid radicals of all sorts in the smallest analytical unit (minimum mol. wt.) of a protein is of the form $2^m\, 3^n$ and therefore the reciprocal of the fraction of radicals contributed by any one amino acid (i.e. the frequency) is of the form $2^m\, 3^n$.

(iii) The radicals of each sort of amino acid occur at intervals along the peptide chain of which the protein essentially consists, the intervals being of the form $2^m\, 3^n$ places.

Astbury[4] considered that the first and second hypotheses were fundamental to the whole of protein structure but rejected the third. Theoretically, the hypothesis can be tested by its applicability to the analytical data for the amino-acid content of the proteins but, at the time when the theory was advanced, the accuracy of the data available was too low to provide a real test, especially since the probability of expressing any number taken at random from 1–300 in the form $2^m 3^n$ is about 75 per cent. The theory evoked considerable controversy but with the increased accuracy of amino-acid determinations and despite the evidence[4,21] afforded by X-ray data from the highly crystalline keratin porcupine quill, the hypothesis today finds little support.

Structural considerations led Astbury to propose that the polypeptide chains forming an α fold were made up in such a way that amino acids with polar and non-polar side chains alternate. From the nature of the α fold it then follows that all the polar side chains protrude on one side and all the non-polar side chains on the other. In this way aggregation between polar and polar, and non-polar and non-polar side chains is facilitated and a mechanism for the reversible unfolding of the peptide chains is provided. Consden, Gordon and Martin[22, 23] were the first to attempt to determine the order of the amino acids in a protein and this work can be considered as a revolutionary step in protein chemistry. Essentially their method consisted of partially hydrolysing wool fibres with acid and separating the dipeptides with an acidic side chain by ionophoresis on silica gel. Under the influence of an electric field the negatively charged peptides moved to one end of the column, from which they were then cut and analysed by paper chromatography. In this way these workers showed that every amino acid present in wool is combined with aspartic and glutamic acid, and moreover, that the peptide glutamylglutamic acid was present in largest amount. Unfortunately their method is not quantitative and only a very small proportion of the total wool substance can be accounted for in this way. Nevertheless this work for the first time provided some evidence of the mode of linkage between the amino acids which casts considerable doubt on the regular alternating arrangement postulated by Bergmann and Niemann. In addition, analysis of tripeptides in the hydrolysis mixture again confirmed the apparent random distribution of the amino acids within the peptide chain. This work was extended by Consden and Gordon[23] to dipeptides containing cystine, by oxidizing the disulphide bond to a sulphonic acid group by reaction with bromine and handling the cysteic peptides in the partial hydrolyzates in the same way as the aspartic and glutamic peptides. Again, it was found that all the amino acids which could be detected by this analytical procedure were combined to some extent with cystine, as cysteic acid dipeptides containing almost all the amino acids present in wool were isolated. Consden and Gordon[23] were, however, unable to establish the combination of proline and aspartic acid with the cystine because of the experimental technique adopted. Unfortunately the painstaking work of Consden Gordon and Martin is not as informative as might appear at first, since it is probable that the wool fibre is not a homogeneous protein but is of several different components which may all be bound together by disulphide bonds. The order of the amino acids in a homogeneous protein may therefore be more regular than that detected in wool, as the extensive analyses described above were carried out on a mixture of different proteins. At first sight the isolation of dipeptides such as glutamylglutamic acid and other combinations of two polar amino acids as isolated by Martin[22] appears to refute the theory of Astbury[4] postulating alternate polar and non-polar side chains. It is now known however, (see p. 369) that a wool fibre can be separated into two components, one of high molecular weight which is in the α fold and the other of low molecular weight which is incapable of orientation. It is possible therefore that any peptide of two polar amino acids which may be isolated arises from this latter fraction and that the protein in the α fold, to which Astbury's hypothesis applies, may be arranged with alternate polar and non-polar side chains. No conclusion can therefore be drawn as the analytical data neither support nor reject the hypothesis of Astbury[4] or that of Bergmann and Niemann.[20]

The investigations of Consden et al. have been followed by a limited number of studies of the sequential tendencies of the amino acids in wool keratin. Such work has been encouraged by the improvements and modification of the methods originally used by Martin and his colleagues, especially in respect of high-voltage electrophoresis, and also by the success in isolating from wool keratin various fractions that are almost homogeneous (e.g., the kerateines—see p. 363). Thus, the occurrence of cysteylcysteic acid has been demonstrated[24] for α- but not for γ-keratose (in which the cystine groups have been oxidised to cysteic acid). Fell et al.[24a] separated a basic fraction from a tryptic digest of α-keratose and, by means of high-voltage electrophoresis on paper, followed by paper chromatography, isolated the following basic peptides.

(Arg,(Glu–NH$_2$), Val, Leu)–Arg   Ser–Lys
Ala–(Thr, Val, Leu)–Arg   Gly–Arg

Ser–Ser–Arg
Phe–Arg
Gly–Ser–Arg
Ser–Arg
Ala–Lys
Ala–Arg
Lys–Lys

where Arg = Arginine   Leu = Leucine
       Ser = Serine   Glu = Glutamic acid
       Ala = Alanine   Phe = Phenylalanine
       Val = Valine   Thr = Threonine
       Gly = Glycine   Lys = Lysine

The order of amino acid residues within brackets is unknown.

It was noted that free arginine and lysine together with arginine and lysine dipeptides were present in the tryptic digests and Zahn et al. concluded that these two amino acids accumulate in certain parts of the molecule.

Variation in the composition of the various morphological components

Many attempts have been made to determine whether the different morphological components (see Chapter 1) differ in composition, but so far the analyses have not provided a satisfactory answer to this problem, as it is almost impossible to achieve a separation without modifying the fibre chemically. By removing the scales mechanically, Chamberlain[25] found that the sulphur content of descaled fibres is very similar to that of the fibres as a whole. It is difficult, however, to obtain a homogeneous sample of scales mechanically, and as they represent only a very small proportion (perhaps 2–3 per cent.) of the fibre, it is clear that it is most difficult to detect any changes by this technique. For example, if the scales contained twice as much sulphur as the rest of the fibre, the cystine content of the descaled fibre would only be 0·05 per cent. lower than that of the original fibre.

Geiger[26] obtained scale substance by enzymatic degradation of a fibre reduced with thioglycollic acid. The insoluble material obtained in this way gave the analytical figures shown in Table 11.6 which differ considerably from the corresponding analyses of untreated wool. In particular the cystine content of the scales was found to be much greater. It is questionable, however, whether sulphur analyses of a material obtained by a pre-treatment with thioglycollic acid are very significant as sulphur may well be introduced during this reaction, for Schöberl[27] has shown that the anhydride of thioglycollic acid

CHEMICAL COMPOSITION

TABLE 11.6

Composition of untreated wool and of wool scales[26]

Constituent	Untreated wool	Wool scales	
		Found	Corrected value*
Sulphur	3·50	4·83	5·42
Cystine	12·20	18·10†	20·30†
Nitrogen	16·67	13·53	15·17
Arginine	8·6	4·3	4·8
Tyrosine	6·1	3·0	3·3
Serine	9·5	9·9	11·2
Ethyl groups	0·0	4·0	—
Ash	0·2	4·1	—
Lipid	—	2·7	—

* Corrected for the presence of ethyl groups, ash, and bound lipid.
† Calculated from the sulphur content.

readily combines with proteins. Lustig and Kondritzer found that after treatment with concentrated sulphuric acid (see p. 318) the scales could be readily detached from the cortical cells, but found only minor differences in composition. There are no data on the composition of the two components of the scales, the k_1 and k_2 phases, which differ so markedly in their resistance to enzymes, but it seems likely that Geiger[26] separated and analysed only the more resistant component.

The thin outer layer of the scales, the epicuticle, is known to be very resistant to chemical attack and may not be of protein material. It constitutes, however, only about 0·1 per cent of the fibre as a whole (see p. 7). The composition of the cortical cell membranes is also unsettled and a considerable amount of discussion has centred around this component. Specimens isolated by the method of Elöd and Zahn[28] had approximately twice the sulphur content of untreated wool. It is however known, (see p. 11), that these preparations were heavily contaminated with scales; this analysis therefore provides no true indication of the composition of the cortical cell membranes. When isolated from a fibre oxidized with peracetic acid (see p. 368), however, the cortical cell membranes have the composition shown in Table 11.7. It will be observed that the sulphur content is little more than half of that present in the fibre as a whole,[29] and is of the same order as that of the high-molecular component. Subsequent

TABLE 11.7

Composition of wool fractionated by oxidation with peracetic acid

	Intact wool	α keratose	β keratose (cortical cell membranes)	γ keratose
Sulphur %	3·6	2·4	2·2	6·1
Nitrogen %	—	15·9	15·3	—
Carbon %	—	48·7	50·4	45·3
Hydrogen %	—	6·7	7·0	—

workers[30,31] have confirmed the above data. (See also Table 11.12, p. 374). Now it is known that the fibre can be separated into a sulphur-rich fraction and a high molecular weight peptide deficient in sulphur. It is highly probable, therefore, that the preparation of Elöd and Zahn[28] is contaminated with this highly cross-linked cement which may be resistant to breakdown by the enzymes used. The whole question, however, requires further investigation and will probably remain unsettled until the fibre can be separated into its individual components satisfactorily by purely mechanical means. The outstanding feature of the cortical cell membranes is their chemical inertness which enables them to be isolated from the rest of the fibre. Thus, although the whole fibre will in time dissolve in strong sodium hydroxide, the scales and cortical cells dissolve first, leaving the cortical cell membranes behind which then dissolve much more slowly.[32] One of the earliest methods of isolation made use of the fact that the cortical cell membranes resisted enzyme attack even after the wool had been disorientated.[28] Similarly, after oxidation of the disulphide bonds the wool becomes soluble, except for the cortical cell membranes. In spite of the fact that they contain no disulphide cross-links, they are insoluble in all solvents incapable of causing degradation, and dissolve only slowly in strong alkali.[29] When isolated from a fibre which has been oxidized with peracetic acid, the membranes give an X-ray diffraction pattern characteristic of disorientated β keratin and they probably exist in this configuration in the original fibres although this is very difficult to prove. This inference is, however, probably true as the peracetic acid treatment does not destroy the α fold in the remaining 90 per cent. of the fibre. At first sight the insolubility and chemical resistance of the membranes might be attributed to the β configuration, which enables powerful hydrogen bonding between the chains to take place. However, the fibre remains insoluble in the powerful hydrogen-bond-breaking reagents, cupriethylene-diamine and lithium bromide, which readily dissolve typical β proteins (e.g. silk). The facts suggest that a new covalent cross-link other than a disulphide bond is present in the cortical cell membranes and absent in the rest of the wool fibre.[33] There is no evidence in protein chemistry to indicate the existence of such a covalent cross-link in naturally occurring materials although tentative suggestions for the presence of ester and ether groups have frequently been made. It seems possible that an ether cross-link between tyrosine of the following structure may be present:

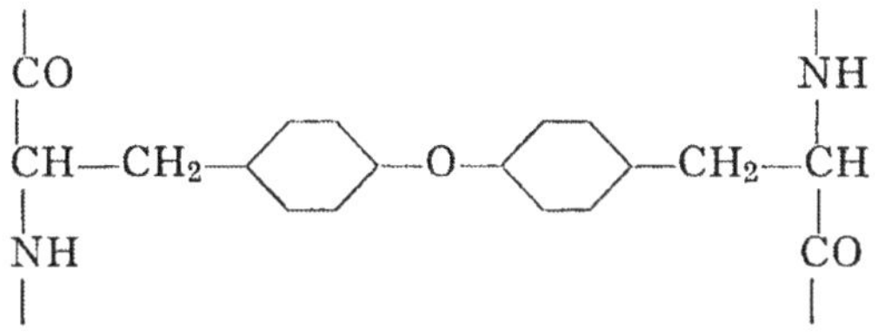

$$
\begin{array}{c}
| \\
CO \\
| \\
CH-CH_2-\langle\rangle-O-\langle\rangle-CH_2-CH \\
| \\
NH \\
|
\end{array}
\qquad
\begin{array}{c}
| \\
NH \\
| \\
\\
| \\
CO \\
|
\end{array}
$$

Thus, reagents which attack tyrosine much more readily than the other amino acids, besides cystine, e.g. chlorine and chlorine dioxide (see p. 304) render the cortical cell membranes soluble in dilute alkali and cupriethylene-diamine. In particular, treatment with chlorine dioxide which confines its attack to the cystine and tyrosine residues in wool enables the cortical cell membranes, in which all the disulphide bonds have already been oxidized, to dissolve in the above solvents. This suggests strongly that the insolubility is due to a cross-link involving tyrosine.[34]

The composition of the medulla of wool fibres is even more obscure since it is absent in finer wools. There is some data[35, 36] to suggest that the medulla is associated with sterols. The sulphur content is, however, somewhat lower than that of the fibre as a whole. For example,[37] in chicken feathers, the cystine content of the medulla was found to be 8·5 per cent. as against 9·5 per cent. in the cortex, and the difference in other keratin fibres is of the same order. Further, Elöd and Zahn[38] showed that the medulla cells are readily digested by pancreatin, unlike cortical cells, which is in agreement with the X-ray data of Rudall that the medulla is not in the α configuration but gives an ill-defined β photograph. The medulla takes up basic dyes preferentially[39] indicating that it is richer in dicarboxylic amino acids than the fibre as a whole. The spindle cells which make up 90 per cent. of the fibres of course determine the analytical figures of the whole fibre, and it has nearly always been found that spindle cells give analyses corresponding to that of the whole fibre when isolated by non-degradative means.

Fractionation into different polypeptides

The insolubility of the keratins in all solvents, which is generally attributed to the large number of disulphide bonds, has greatly hindered the investigation of the macromolecular structure of wool. Probably one of the most important problems in the field of wool chemistry is the discovery of a method of dissolving the wool fibre without causing peptide-bond breakdown. The rôle played by the disulphide bonds in determining the lack of solubility has been deduced from the failure of the wool fibre to dissolve in hydrogen-bond-breaking solvents which readily disperse silk and cellulose. However, Elöd and Zahn[40] showed that part of the wool fibre became soluble in formamide which is capable of breaking strong hydrogen bonds without breakdown of disulphide bonds. As an example of this treatment, 40 per cent. of wool substance dissolves after immersion in formamide at 120° C., and most of this re-precipitates on diluting the formamide with water. The amino-acid analysis of the soluble protein, referred to as formamide keratose, is similar to that of the insoluble residue and shows that the difference in solubility is in no way connected with the cystine content which is very similar in both substances. The soluble product gives an X-ray diffraction pattern typical of disorientated β keratin, and there is evidence to show that it still consists of peptides of fairly high molecular weight.

Attempts to dissolve wool in other solvents without causing severe degradation have led to somewhat ambiguous results as the extent of disulphide bond breakdown was not determined. Thus, it has been shown that wool will dissolve completely in concentrated boiling solutions of calcium or lithium thiocyanate[41] at 180° C. and in phenol and resorcinol at about the same temperature.[42] Although none of these reagents is specific for the disulphide bond they appear to behave as reducing agents leading to the formation of sulphides. In this case, therefore, the action can best be interpreted as a dissolution by alkaline sulphides and not a dispersion by hydrogen-bond-breaking agents.

Following the work of Steinhardt and Fugitt,[43] who showed that cetyl sulphonic acid deamidates wool (see p. 307) without major peptide-bond hydrolysis, Lindley[44] found that wool can be separated into two fractions by treatment with this reagent (0·05N cetyl sulphonic acid at 65° C. for six days followed by extraction with 0·001N sodium hydroxide). Approximately 70 per cent. of the fibre goes into solution and part of this is reprecipitated by acid. The insoluble

TABLE 11.8

Fractionation of wool by treatment with cetyl sulphonic acid[44]
(% *Nitrogen of Total Nitrogen*)

Chemical composition	Untreated wool	Resistant component (i.e. insoluble)	Alkali-soluble portion
Total nitrogen	16·90	14·50	13·40
Total sulphur	3·76	7·51	4·25
Cystine	8·0	20·3	5·2
Phenylalanine	1·9	0·7	1·7
Leucine	6·7	3·6	9·2
Valine	3·6	3·5	4·0
Proline	3·8	7·6	1·0
Tyrosine	2·6	1·3	2·0
Alanine	3·2	3·3	4·0
Arginine	20·0	15·5	—
Hydroxyamino acids	12·8	—	—

(i.e. resistant) material is very rich in cystine (see Table 11.8) and may correspond to the low molecular weight fraction (γ keratose) isolated from wool oxidized by peracetic acid (see p. 374). This fraction, which is very highly cross-linked by disulphide bonds, would be expected to be the most resistant to degradation by acid. It has not been established why this treatment causes the wool to dissolve, but it is probably the result of the introduction of a large number of new positively charged groups resulting from the deamidation, coupled with some peptide-bond hydrolysis.

Solubilization by reduction of the disulphide bond. It has long been known that wool dissolves in dilute alkali after reduction with sodium sulphide and as early as 1925 Bergmann[45] showed that the product thus obtained is soluble in cuprammonium solution from which it can be spun into fibres. It was only comparatively recently however that Happey and Wormell[46] made use of this process to obtain regenerated keratin fibres. These authors dissolved wool in sodium sulphide and precipitated 65 per cent. of the dissolved protein together with a considerable amount of sulphur. After the sulphur has been removed by dissolving the protein in cuprammonium solution and filtering, fibres were spun from this solution by a process similar to that used in the manufacture of viscose rayon. The fibres produced in this way were hardened with formaldehyde and stretched, when they were found to give a fibre photograph typical of orientated β keratin; a fibre in the α configuration could not be obtained. By this process approximately 35 per cent. of the starting material was recovered and converted to fibres, one third being lost in the reduction process and the remainder in the dissolution in cuprammonium and subsequent precipitation in the spinning bath. The reduced keratin in which the disulphide bonds have been severed, only dissolves completely in strongly alkaline solution and the whole process is likely to degrade the peptide chains. Thus, Olofsson and Gralén[47] found that the product obtained by dissolution in sodium sulphide had a molecular weight of only 10,000, although it is now known that a fraction of much higher molecular weight is present in wool. In addition, Goddard amd Michaelis[48] made a very thorough investigation of the dissolution of keratin by alkaline solutions of sodium thioglycollate, and showed that wool did not become appreciably soluble until the pH exceeded 11, although it is now known (see p. 246) that the disulphide bonds are completely reduced at lower pH values. More recent work suggests that the keratin structure is not soluble, even after disulphide-bond breakdown, if no new ionized groups are introduced. At the high pH value needed for dissolution in sodium thioglycollate, ionization of the SH group probably occurs and this determines the solubility. Another serious disadvantage in handling reduced keratin is the continual danger of re-oxidation of the disulphide bonds, which occurs quite rapidly in the atmosphere. Goddard and Michaelis made the very significant observation that after re-oxidation of all the sulphydryl bonds to SS, the reduced keratin (which they call kerateine) assumes the form which is usually known as metakeratin, and remains soluble in solutions of pH less than 11, unlike the original wool. If solutions of metakeratin dissolved in alkali are dialysed, the protein remains in solution but is precipitated by a trace of acid. Also metakeratin is readily digested by enzymes. These various observations lead to the conclusion that the crystalline arrangement of the molecules, and not the presence of the disulphide bonds alone confers on keratin fibres their characteristic resistance to proteolytic enzymes. This conclusion was strikingly confirmed by Elöd and Zahn, who showed that enzymes can digest wool after disorientation of the molecules without causing disulphide-bond breakdown (see p. 360).

The sulphydryl groups of wool (reduced with thioglycollic acid) may be blocked by means of the general reaction:

$$W-SH + RI \rightarrow W-SR + HI$$

Thus with iodoacetic acid, carboxymethylkerateine is formed.

$$\begin{array}{c} | \\ CO \\ | \\ CH-CH_2 \cdot S \cdot CH_2 \cdot COOH \\ | \\ NH \\ | \end{array}$$

Using iodoethanol, hydroxy ethyl kerateine is obtained:

$$\begin{array}{c} | \\ CO \\ | \\ CH-CH_2-S \cdot CH_2 \cdot CH_2 \cdot OH \\ | \\ NH \\ | \end{array}$$

It can be seen from Table 11.9 that the carboxy derivative is soluble in acid and also in dilute alkali such as sodium acetate, whereas the hydroxy derivative is only soluble in relatively strong alkali. The carboxy methyl kerateine dissolved in 0·1M sodium acetate could be separated into two fractions by precipitation with increasing quantities of ammonium sulphate. The first fraction (A) with a high sulphur content (Table 11.9), was precipitated when the solution was 35 per cent. saturated with respect to the salt, and the second fraction (B) at 60 per cent. saturation. Goddard and Michaelis[48] also showed that wool dissolved completely in sodium cyanide, but only after extensive degradation had occurred.

Gillespie and Lennox[49, 50, 51, 52] have extended the work of Goddard and Michaelis and have obtained what is claimed to be an electrophoretically pure wool protein by first extracting the wool with thioglycollate at pH 10·5 and then at 12·3. From the latter solution, a material described as kerateine 2 was precipitated by acidification at pH 4·5. The following S-derivatives of kerateine 2 were prepared by reaction with the corresponding iodides: carboxymethyl, carbamidomethyl, p-carboxyphenylmercuri, methylmercuri and benzyl. In addition, by the use of Raney nickel, it was possible to desulphurize the thiol groups and thus convert the cysteine residues to alanine. As was to be expected, these protein derivatives differed in their solubilities and precipitation points from kerateine 2 itself. The behaviour on electrophoresis of kerateine 2 and those of its derivatives that have been studied is anomolous, and no completely satisfactory explanation has been put forward by Gillespie and Lennox, although a system consisting of monomer in equilibrium with its aggregates seems the most likely.

Lustig, Kondritzer and Moore[53] showed that wool which had been treated with sulphuric acid to introduce additional ionizing groups, e.g. sulphamic acid and sulphonates of aromatic residues (see p. 318), became soluble at a pH between 8 and 9 after reduction with sodium thioglycollate. In this way these workers separated whole fibres into the various fractions given in Table 11.10. They showed in addition that the fractions were electrophoretically homogeneous. This does not necessarily mean however that the molecules in each fraction are identical, but merely that the ratio of charged to uncharged groups is the same at the particular pH of the measurement. The homogeneity is accounted for by the high acidity due to the sulphuric acid groups introduced. Although little main-chain degradation probably occurs during the solution of the proteins, some hydrolysis of the main chain undoubtedly occurs during the sulphonation. It is not unexpected, therefore, that the different fractions have different crystalline structures from that of the original fibre.

Although wool does not dissolve in bisulphite solutions, the addition of a hydrogen-bond-breaking reagent facilitates dispersion. Jones and Mecham[54] showed that 52 per cent. of the wool substance dissolves in a solution of 10 M urea and 0·3 M sodium bisulphite. Mercer[55] found that on subsequent acidification of the urea-bisulphite solution a gummy product was obtained which was soluble only in concentrated urea solutions. In the ultracentrifuge the

TABLE 11.9
Composition of keratin derivatives from wool reduced with thioglycollic acid[48]

Protein	Total S	Total N	Cystine	Cystine-S	Amino-N	Solubility
Native wool	3·35 3·33 3·34	16·29	11·6	3·12	—	Insoluble in M/NaOH or H/Cl.
Kerateine		16·50	12·2	3·24	—	Soluble in 0·1 M/Na₂CO₃ and NH₄OH; insoluble in N/30 HCl.
Metakeratin	3·37	16·50	12·2, 12·1	3·25	0·80	Soluble in 0·1 M/Na₂CO₃ and NH₄OH; insoluble in N/30 HCl.
Carboxymethylkerateine, unfractionated	3·31	16·54	1·42	0·38	—	Soluble in 0·1 M sodium acetate or N/30 HCl.
Carboxymethylkerateine, Fraction A	3·33 3·29	16·32	0·80	0·21	0·80	Soluble in 0·1 M sodium acetate and N/30 HCl.
Carboxymethylkerateine, Fraction B	4·50 4·55	14·95	0·97	0·26	0·50	Soluble in 0·1 M sodium lactate and lactic acid.
Carboxymethylkerateine	2·00	15·90	1·73	0·46	0·78	Soluble in 0·1 M sodium acetate and N/30 HCl.
Carbamylmethylkerateine	2·78 2·84	17·35	2·74 2·40	0·75	—	Insoluble in sodium acetate; soluble in N/30 HCl.
Carboxyethylkerateine	2·12 2·15 2·17	15·84	1·35	0·36	0·76	Soluble in 0·1 M sodium acetate and N/30 HCl.
Hydroxyethylkerateine	2·73 2·76 2·77	15·50 15·30	2·66	0·71	—	Soluble in 0·1 M/Na₂CO₃ or NH₄OH, insoluble in N/30 HCl or 0·1 M sodium acetate.

material was shown to be very heterogeneous, to have an average molecular weight of about 80,000 and to consist of highly asymmetric

TABLE 11.10

Fractionation of hair rendered soluble by sulphonation followed by reduction[53]

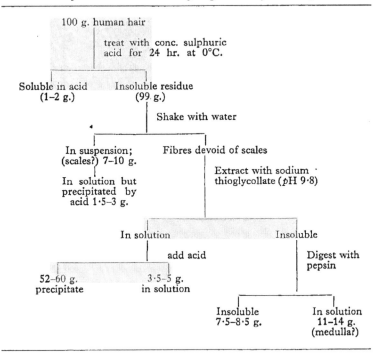

molecules. However, no valid conclusion concerning the macromolecular structure can be inferred since the sulphydryl groups probably re-oxidized during the experiments and a number of molecules may thus have recombined via SS groups giving the heterogeneous solution. The outstanding feature of this preparation was the X-ray pattern which was similar to α keratin. For the first time a substance had been extracted from a wool fibre which still retained the special configuration characteristic of the α fold. Now, however, it has become possible to obtain silk and soluble polypeptides with the same configuration, and it is no longer suggested that a specific biosynthesis is necessary for the formation of the α fold.

On the whole, however, reduction of the disulphide bond followed by dissolution is not a very desirable method for determining the macromolecular structure of keratin fibres, since the conditions are either so severe that main-chain degradation occurs, of if this is avoided, not all of the wool goes into solution, making a complete study of the composition of the fibre impossible.

Solubilization by oxidation of the disulphide bond. In general, the disulphide bond is the most reactive point of the wool structure to oxidation, but with the exception of some of the per-acids (see p. 266) no oxidizing agent confines its attack to this group. Burkhard[56] noted that wool became readily soluble after oxidation with chlorine and nitric acid, but in each case severe degradation occurred, and only a small quantity of high molecular weight product was obtained. Das and Speakman[57] found that wool dissolves in dilute alkali after oxidation with chlorine peroxide and obtained products of varying molecular weight ranging up to 30,000. This reagent, although more specific than those previously examined, was shown by Schmidt and Braunsdorf to oxidize both cystine and tyrosine. The former reaction was not confined to the disulphide bond, since some of the sulphur was eliminated as sulphate (see p. 280). It is therefore probable that in the reaction the protein undergoes main-chain breakdown at the points where cystine and tyrosine side-chains occur. However, the soluble oxidation product obtained may be suitable for the manufacture of regenerated protein fibres. Finally chloroamides and chlorosulphonamides, although probably less specific than chlorine peroxide, are more useful than chlorine for solubilizing wool fibres and keratin in general, because the whole of the wool becomes freely soluble in dilute alkali when all the disulphide bonds have been ruptured.[58] After treatment with chlorine on the other hand, the keratin is not completely soluble[59] (see Fig. 11.1) when all the disulphide bonds have been oxidized, and it has been suggested that this reagent introduces new covalent cross-links in the main chain during its reaction.

Solubilization by oxidation with peracetic acid. As shown on page 266, peracetic acid is as specific as performic acid in its action on proteins, and only oxidizes the side chains containing disulphide bonds and those from the amino acids methionine and tryptophan. These last two amino acids do not occur to any appreciable extent in wool, so that reaction proceeds almost exclusively with the disulphide bond. Moreover, experiments with a large range of di-, tri- and penta-peptides, including those containing cystine and tryptophan, have shown that peracetic acid does not break peptide bonds even when the side chain of one of the constituent amino acids is oxidized.

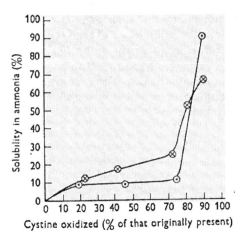

Fig. 11.1.—Relationship between solubility of wool in 3N ammonia and the proportion of cystine oxidized by ⊗ chlorine solution at pH 2 and ⊙ peracetic acid.

This reagent is therefore suitable for studying the macromolecular structure of wool, since the protein becomes freely soluble, except for a resistant fraction of 8–10 per cent. which has been identified as the cortical cell membranes (see p. 11), when more than 90 per cent. of the disulphide bonds have been oxidized.[59] From Fig. 11.1 it is seen that it is necessary to oxidize almost all the disulphide bonds before the wool becomes soluble to any appreciable extent in dilute ammonia, and that the number of disulphide bonds oxidized is not related directly to the proportion of the wool which is rendered soluble.

It has proved possible to separate wool into a number of components by oxidation followed by dissolution in dilute ammonium hydroxide. It is believed that the various fractions, which have been called keratoses, represent different peptide chains which in the original wool were bound together by disulphide bonds. The scheme (see p. 369) shows the method of fractionation.

If the α keratose has been prepared by precipitation with salt all the acid groups are already present in the salt form and the material readily redissolves in distilled water. Preparations obtained in either way are soluble in hydrogen-bond-breaking solvents, such as formic and phosphoric acids, concentrated aqueous solutions of phenol, resorcinol, urea and lithium bromide. Significantly, α keratose gives an X-ray powder diagram typical of proteins in the α configuration, and can be spun from solution into a salt precipitating bath. On stretching, the resulting fibre shows a perfectly orientated α fibre diagram which is almost indistinguishable from that given by a wool fibre[29,60] (see Fig. 11.2). β Keratose gives a powder X-ray diagram typical of a protein in the β configuration, but due to its insolubility cannot be spun into a fibre so that no orientated diagram can be obtained. In Fig.

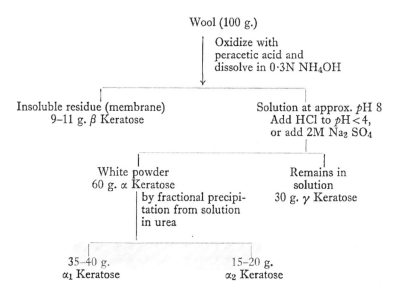

11.3 the powder diagrams of α and β keratose are compared. γ Keratose is highly soluble and cannot be precipitated from solution without difficulty. It can be obtained free from the α form in a relatively pure state by dissolving the oxidized fibre in dilute ammonia, and evaporating the solution to dryness, when a mixture of γ and α is obtained. On extracting the material with a small quantity of 0·001 N. hydrochloric acid, the α fraction is retained and the γ fraction goes into solution.

By evaporating this latter solution down, an almost pure preparation of γ keratose is obtained. The properties of β keratose, which has

370 WOOL: ITS CHEMISTRY AND PHYSICS

been identified with the cortical cell membranes, have been discussed fully on page 359. The α keratose can be obtained from its solution in dilute ammonium hydroxide either by precipitation with acid, when it comes out of solution at a $pH < 4$ or by precipitation with strong electrolytes such as sodium or magnesium sulphate. If the material is obtained by precipitation from acid it is necessary to neutralize the acid groups with dilute alkali, e.g. 0·1M ammonium hydroxide, before it will redissolve in water.

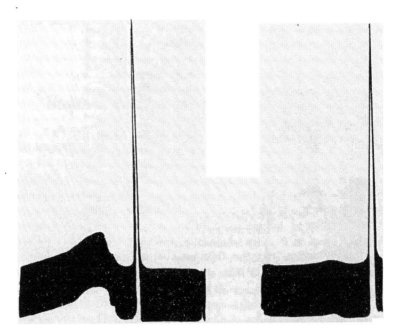

FIG. 11.4.—Electrophoretic pattern of α keratose in solution at pH 8.[64]

FIG. 11.5.—Electrophoretic pattern of α_1 keratose in solution at pH 8.[64]

Electrophoretic examination of the α keratose shows the material to consist of two distinct components[64, 16] (see Fig. 11.4). The components of the mixture, giving rise to the sharp and diffuse bands, can be separated by dissolving the α keratose in a dispersing agent such as concentrated urea solution, and adding acid slowly. The material referred to as α_1, which gives the sharp peak (see Fig. 11.5), precipitates first, followed by the remainder of the protein. The latter gives the diffuse band in the electrophoretic diagram and is referred to as α_2 keratose. Only the α_1 keratose gives an X-ray photograph of a protein

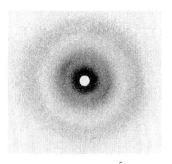

(a) Fibre before stretching.

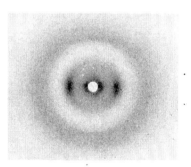

(b) Fibre after stretching.

FIG. 11.2.—X-ray diffraction picture of α keratose spun from aqueous solution into a salt precipitation bath.[60]
(Copper K α radiation; distance 4 cm.)

A B

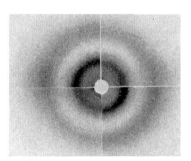

D C

Fig. 11.3.—X-ray diffraction photograph of powdered α keratose (quadrants B & D) and β keratose (quadrants A & C). (Copper Kα radiation; distance 4 cm.)[29]

CHEMICAL COMPOSITION 371

in the α configuration, the α_2 gives an ill-defined powder photograph which suggests that it is in the β form. It must be emphasized that

Fig. 11.6.—Electrophoretic pattern of γ keratose in solution at pH 8.[64]

a_1 keratose S_{20} 2·0

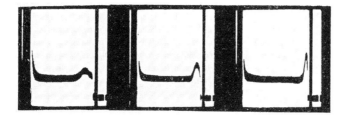

a_2 keratose S_{20} 1·0

Fig. 11.7.—Sedimentation pattern in ultra-centrifuge of α_1 and α_2 keratose (no boundary could be found with γ keratose).[64]

the sharp electrophoretic pattern of the α_1 keratose does not necessarily imply that the material is homogeneous. The sedimentation curves (Fig. 11.7) do not indicate that the material is markedly hetero-dispersed.[64] The sedimentation constant and diffusion constant of α_1, α_2 and γ keratose have been determined,[64, 16] and from these[64, 16] it is possible to calculate the size and shape of the molecules in solution. The values obtained are compared in Table 11.11 with similar values for soluble fractions obtained in other ways. It is seen that α_1 keratose is of high molecular weight and approximately spherical in solution. All the other soluble derivatives of wool obtained either by reduction or oxidation with chlorine peroxide give highly asymmetric molecules, generally of much lower molecular weight. The only material similar to α_1 keratose is the 20–30 per cent. of polypeptide which can be extracted from a solution of urea and bisulphite (see p. 366) and which dissolves only in concentrated urea solutions.

TABLE 11.11

Physical constants of soluble materials obtained from wool in different ways

Method of preparation	Sedimentation constant $\times 10^{13}$	Diffusion constant 10^7	Molecular weight	Shape of molecule
Dissolved in sodium sulphide	0·98	8·70	9,500	asymmetric
Oxidized with chlorine peroxide and dissolved in cuprammonium	0·90	10·00	8,000	,,
Dissolved in urea and bisulphite at pH 8	1·93	1·92	84,000	,,
Dissolved in urea and bisulphite at pH 7	1·45	2·83	42,000	,,
Peracetic acid process α_1 keratose	2·00	3·12	67,000	spherical
Peracetic acid process α_2 keratose	1·04	8·20	13,000	too hetero-dispersed to be measured
Peracetic acid process γ keratose	—	32·50	—	
α_1 Keratose denatured by heating	3·20	3·00	88,000	asymmetric

This material, unlike α_1 keratose, is highly asymmetric, but this may be due to the particular solvent used. As both α_2 keratose and γ keratose are very heterogeneous (see electrophoretic diagrams Fig. 4 and 6), the sedimentation constant and diffusion coefficients lead to only very approximate molecular weights and no significant conclusions as to the shape of these molecules can be made. These fractions may represent the two extremes of one large hetero-dispersed fraction,

Fig. 12.1.—X-ray diagram of α keratin (Lincoln wool) D = 4·0 cm. (*Reproduced from 'Wool Science Review' by permission of the International Wool Secretariat.*)

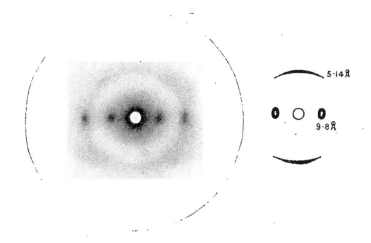

Fig. 12.2.—X-ray diagram of β keratin (Lincoln wool, 70 per cent. extension) D = 4·0 cm. (*Reproduced from 'Wool Science Review' by permission of the International Wool Secretariat.*)

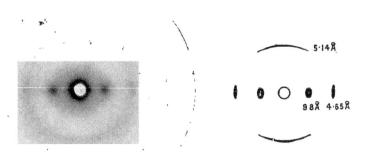

FIG. 12.3.—X-ray diagram of keratin in the course of the α—β transformation (Lincoln wool, extension 35 per cent.) D = 4·0 cm. (*Reproduced from 'Wool Science Review' by permission of the International Wool Secretariat.*)

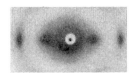

FIG. 12.4.—X-ray diagram of 'set' β keratin. Lincoln wool extended by 100 per cent. and steamed stretched for three hours. D = 4·0 cm. (*Reproduced from 'Wool Science Review' by permission of the International Wool Secretariat.*)

consisting of molecules of all sizes from very small molecular weight to a maximum of about 15,000.

Significant differences in the amino-acid composition have been found by chromatography for all the different fractions obtained from peracetic acid-oxidized wool[29,31] but the N-terminal amino acids of these fractions are the same seven (see p. 351) as those reported for intact wool.[61] The most outstanding difference, however, is the high sulphur content of the γ fraction and the low sulphur content of the α and β keratose (see p. 358). This difference in cystine content and to a lesser extent in other amino acids, illustrates clearly that wool must be considered as made up of a mixture of polypeptides. It is not possible to decide at the moment whether all these different polypeptide chains in the smallest morphological sub-unit (i.e. the fibril) are linked to one another by disulphide bonds. The fact that 30 per cent. of the wool can be extracted with formamide, without disulphide-bond breakdown, would suggest that part of the assembly of polypeptide chains consists of separate phases not necessarily covalently linked to one another. As there is very little information available, however, concerning the nature of the inter-fibrilar material it cannot be decided whether the extraction with formamide removes protein situated between or within the morphological components.

The relationship between the different fractions

As previously mentioned on p. 361, the alkali-insoluble fraction obtained by treating wool with cetyl sulphonic acid[44] is very rich in sulphur and resembles α-keratose. Blackburn[62] has repeated the work of Lindley[44] replacing the cetyl sulphonic acid by dilute hydrochloric acid and his detailed amino acid analyses when compared with those obtained by Corfield et al.[31] show clearly the identity of these two materials. Earland and Wiseman[63] have examined the amino acid analyses of a number of workers and have concluded that their data are consistent with the fact that α-keratose, the cetyl sulphonic acid-soluble fraction and kerateine (thioglycollate-soluble material) are all derived from the same fraction of wool (Fraction 1) and that the cetyl sulphonic acid-insoluble and thioglycollate-insoluble residues and γ-keratose are derived from another fraction (Fraction 2) (see Table 11.12). An examination of 70 analyses covering these fractions shows that not one is inconsistent with the proposed relationship.[63] Evidence has been put forward also which points to the fact that the cystine residues in Fraction 1 are predominantly in the intraform whereas those of Fraction 2 are interchenic.

TABLE 11.12

*The Amino Acid Composition of Hydrolysates of Fractions from 64's Wool**

Amino acid	α-keratose[31]	CSA-[44] soluble	Kerateine[64]	γ Keratose[31]	CSA[44] insoluble
Alanine	4·83	4·0	4·22	2·58	3·3
Amide N	10·25	—	11·56	11·05	—
Arginine	20·8	—	21·12	19·0	15·5
Aspartic acid	6·25	—	5·68	1·79	—
Cystine	3·72	5·20	4·33	14·5	20·30
Glutamic acid	10·9	—	10·82	5·87	—
Glycine	5·16	—	5·27	4·97	—
Histidine	1·24	—	1·42	1·57	—
Isoleucine	2·49	—	2·24	2·14	—
Leucine	7·30	9·2	6·01	2·55	3·6
Lysine	4·60	—	5·03	1·03	—
Phenylalanine	1·94	1·7	1·72	1·15	0·70
Proline	2·69	1·0	3·66	9·85	7·6
Serine	6·70	—	6·47	9·70	—
Threonine	3·45	—	4·44	7·46	—
Tyrosine	2·44	2·0	2·46	1·41	1·3
Valine	3·98	4·0	3·55	4·15	3·5

*% N of total N.

Obviously more thought will have to be given to the relationship the wool fractions isolated by various methods bear to each other and to intact keratin rather than the previous tendency to consider each fraction in isolation.

REFERENCES

[1] MOORE AND STEIN, *Ann. N.Y. Acad. Sci.*, 1948, **49**, 265.
[2] MOORE AND STEIN, *J. Biol. Chem.*, 1948, **176**, 367.
[3] MOORE AND STEIN, *ibid.*, 1951, **192**, 663.
[4] ASTBURY, *J. Chem. Soc.*, 1942, 337.
[5] HARRIS AND BROWN, *Symp. Fibrous Protein*, 1946, 203 (publ. Soc. Dy. Col., Bradford).
[6] SIMMONDS, *Aust. J. Biol. Sci.*, 1954, **7**, 98.
[7] CORFIELD AND ROBSON, *Biochem. J.*, 1955, **59**, 62.
[8] GRAHAM, WAITKOFF AND HIER, *J. Biol. Chem.*, 1949, **177**, 529.
[9] SIMMONDS, *Proc. Int. Wool. Text. Res. Conf.*, Australia, 1955, C-65.
[10] CORFIELD AND ROBSON, *ibid.*, C-75.
[11] WOODIN, *Nature*, 1955, **176**, 1117.
[12] WOODIN, *Biochem. J.* 1956, **63**, 576.

[13] MIDDLEBROOK, *Biochim. Biophys. Acta*, 1951, **4**, 547.
[13a] MARSHALL AND NEUBERGER, *Biochem. J.*, 1961, **78**, 31P.
[14] PORTER, *ibid.*, 1948, **2**, 105.
[15] BLACKBURN, *Biochem. J.*, 1950, **47**, 443.
[16] ALEXANDER, *Kolloid Z.*, 1951, **122**, 8.
[17] FROMAGEOT, MEYER AND PENASSE, *Biochim. Biophys. Acta*, 1950, **6**, 283.
[18] O'CALLAGHAN AND SPEAKMAN, *Proc. Int. Wool Text. Res. Conf.*, Australia, 1955. C-474.
[19] BLACKBURN AND LEE, *ibid.*, C-142; *J. Text. Inst.*, 1954, **45**, T487.
[20] BERGMANN AND NIEMANN, *J. Biol. Chem.*, 1936, **115**, 77; 1937, **118**, 301.
[21] MACARTHUR, *Nature*, 1943, **152**, 38.
[22] MARTIN, *Symp. Fibrous Protein*, 1946, 1 (publ. Soc. Dy. Col., Bradford).
[23] CONSDEN AND GORDON, *Biochem. J.*, 1950, **46**, 8.
[24] CONSDEN, GORDON AND MARTIN, *Biochem. J.*, 1949, **44**, 548: Sanger, Ryle, Smith and Kitai, *Proc. Int. Wool Text. Res. Conf. Aust.*, 1955, C.49.
[24a] FELL, LA FRANCE, AND ZIEGLER, *Sec. Quinquennial Wool Text Res. Conf.* (*J., Text. Inst.*), Harrogate, 1960, T797.
[25] CHAMBERLAIN, *J. Text. Inst.*, 1932, **13T**, 23..
[26] GEIGER, *J. Res. N.B.S.*, 1944, **32**, 127.
[27] SCHÖBERL, *Angew. Chemie*, 1948, A60, 7.
[28] ELÖD AND ZAHN, *Naturwissenschaften*, 1946, **33**, 158.
[29] ALEXANDER AND EARLAND, *Nature*, 1950, **160**, 396.
[30] EARLAND AND KNIGHT, *Biochim. Biophys. Acta*, 1955, **17**, 457.
[31] CORFIELD, ROBSON AND SKINNER, *Biochem. J.*, 1958, **68**, 348.
[32] ZAHN, *Textil-Praxis*, 1948, 3.
[33] ALEXANDER, *Nature*, 1951, **169**, 1081.
[34] ALEXANDER AND SMITH, unpublished.
[35] LLOYD AND MERRIOT, *Biochem. J.*, 1933, **27**, 911.
[36] STOVES, *Nature*, 1946, **157**, 230.
[37] ZAHN, *Das Leder*, 1950, **1**, 222.
[38] ELÖD AND ZAHN, *Melliand Textilber.*, 1944, **25**, 361.
[39] HAUSMAN, *Amer. J. Anat.*, 1920, **27**, 463.
[40] ELÖD AND ZAHN, *Kolloid Z.*, 1944, **108**, 6.
[41] WEIMARN, *ibid.*, 1926, **40**, 120.
[42] HERZOG AND KRAHN, *Hoppe-Seyl. Z.*, 1924, **134**, 290.
[43] STEINHARDT AND FUGITT, *J. Res. N.B.S.*, 1942, **29**, 315.
[44] LINDLEY, *Nature*, 1947, **160**, 190.
[45] BERGMANN, German Patent, 445, 503 (1925).
[46] HAPPEY AND WORMELL, *J. Text. Inst.*, 1949, **40**, T855.
[47] OLOFSSON AND GRALÉN, Proc. 11th Intern. Congr. Pure Appl. Chem., 1947.
[48] GODDARD AND MICHAELIS, *J. Biol. Chem.*, 1934, **106**, 605; 1935, **112**, 361.
[49] GILLESPIE AND LENNOX, *Biochim. Biophys. Acta*, 1953, **12**, 481.
[50] GILLESPIE AND LENNOX, *Aust. J. Biol. Sci.*, 1955, **8**, 97.
[51] GILLESPIE AND LENNOX, *ibid.*, 1955, **8**, 378.
[52] GILLESPIE, *Proc. Int. Wool Text Res. Conf. Australia*, 1955, B-35.
[53] LUSTIG, KONDRITZER AND MOORE, *Arch. Biochem.*, 1945, **8**, 57.
[54] JONES AND MECHAM, *ibid.*, 1943, **3**, 193.
[55] MERCER, *Nature*, 1949, **163**, 18; *J. Polymer Sci.*, 1951, **6**, 671.
[56] BURKHARD, German Patent 432,180.
[57] DAS AND SPEAKMAN, *J. Soc. Dy. Col.*, 1950, **66**, 583.
SPEAKMAN, *J. Text. Inst.*, 1941, **32**, T83.
[58] ALEXANDER, CARTER AND EARLAND, *J. Soc. Dy. Col.*, 1952, **67**, 17.
[59] ALEXANDER, FOX AND HUDSON, *Biochem. J.*, 1951, **49**, 129.
[60] ALEXANDER, CARTER, EARLAND AND MARINER, *Melliand Textilber.*, 1952, **33**, 187.
[61] ALEXANDER AND SMITH, *Proc. Int. Wool Text, Res. Conf. Australia*, 1955, B-56.
[62] BLACKBURN, *Biochem J.*, 1959, **71**, 13P.
[63] EARLAND AND WISEMAN, *Biochim. Biophys. Acta*, 1959, **36**, 273.
[64] ALEXANDER, unpublished.

CHAPTER 12

Stereo-Chemistry and Macromolecular Structure

ALMOST all the information which has been obtained concerning the spatial arrangements of the atoms making up the macromolecules of wool has been derived from studies of X-ray diffraction patterns. It should be stressed at the outset that only highly oriented structures give rise to clearly defined reflections. The proportion of the fibre which is crystalline has not yet been determined and estimates vary widely, figures ranging from 10–50 per cent. having been obtained. It has, however, been shown to be a maximum in the fully stretched and unstretched states.[1] Unfortunately X-ray investigations only provide information about this relatively small proportion of the wool fibre. In the case of fibres of homogeneous composition, the steric arrangement revealed in the crystalline areas may also obtain in the amorphous portions. Wool however is known to consist of several different kinds of macromolecules and the X-ray picture probably provides information concerning one of these only. These considerations limit the conclusions which can be drawn from such experiments concerning the behaviour of a macrostructure such as the wool fibre as a whole, but of course do not detract from the important fundamental information concerning constituent molecules of the keratin class.

Astbury and his colleagues[2-8] carried out a most detailed X-ray investigation of wool and related fibres, and made the most significant discovery that there are two principal X-ray diffraction patterns. The first known as α keratin is given by unstretched wool, and the second known as β keratin is obtained by stretching the fibre; the two forms have been referred to as 'mechanical stereoisomers'. Figures 12.1 and 12.2 show the pictures of α and β keratin, respectively, together with diagrams indicating the repeat distances to which these reflections correspond. Figure 12.3 is an X-ray diagram of a fibre in the course of the α to β transformation when the reflections corresponding to both types of keratin are present.

The two forms can be changed from one to another repeatedly by slow extension in water followed by relaxation. There is some evidence[3] that in a completely relaxed structure the 5·14 Å reflection occurs at 5·06 Å, and is converted to the former value in the initial extension of less than 2 per cent. but as it is very difficult to mount fibres completely without tension the 5·11 Å value is usually obtained for the unstretched fibres.

The β-type pattern has been shown to be present at extensions as low as 5 per cent. with the greatest rate of change taking place at 40 per cent. extension.[1, 9] The intensity changes at extensions less than 20 per cent. are not normally found by photographic techniques but can be detected by using a counter diffractometer.

The X-ray diffraction patterns of both α and β keratin are only affected to a limited extent by water and no difference can be detected in the relatively diffuse spots given by wool between fibres in an atmosphere of 0 to 100 per cent. relative humidity. In the highly orientated α keratin of porcupine quill MacArthur showed[10] that both inter- and intra-micellar changes occur on hydration. The meridian spacings are very slightly increased but an increase of up to 10 per cent. occurs in the equatorial reflection such as that at 9·2 Å which is associated with the side-chain spacings. The non-axial reflections as a whole are poorly defined in a dry specimen of porcupine quill and only become sharp on addition of water. There is some resemblance here to the behaviour of globular proteins which only become truly crystalline and give well-defined X-ray pictures when hydrated. It should be stressed however that no such change occurs in wool, and even in porcupine quill the sharpness of the meridian spacings is unaffected by water.

The effect of steaming a stretched fibre is to give it a permanent set and to change the diffraction pattern of keratin in two ways (see Fig. 12.4). Firstly it gives rise to a better definition of some of the reflections, and secondly it causes other reflections to spread along the hyperbolae, a crystallographic phenomenon associated with a disturbance of the spacings in one direction only. Astbury and Woods[3] have interpreted these changes as indicating a disruption of certain of the side linkages and this deduction provides support for the theory of Speakman that the chemical basis of setting in steam is the disruption of the disulphide cross-links followed by the formation of new cross-links (see p. 69).

From the degree of spread of the spots a very rough estimate of the size of the crystalline areas is possible.[11] The reflections become increasingly diffuse with a decrease in the size of the crystallites and in simple inorganic crystals all definite reflections are lost if the crystallites are less than 50 Å in cross-section. This value seems to impose a lower limit for the crystalline areas in keratin of which the transverse dimension is considered to be of the order of 100 to 200 Å This agrees well with the value calculated by Speakman from swelling experiments (see p. 92).

The structure of β keratin

A detailed examination of the β photograph shows a repeating distance of 3·3 Å along the axis of the fibre and two prominent reflections at right angles to one another at about 9·8 Å and 4·65 Å. By using strips of horn pressed in steam and taking X-ray photographs along the three geometric axes of the strip, Astbury and Sisson[4] demonstrated conclusively that these reflections represent spacings

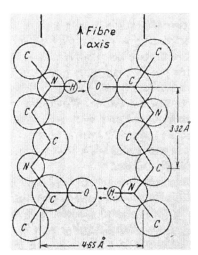

FIG. 12.5.—Arrangement of peptide chains in β keratin.

perpendicular to one another. The β photograph is extremely similar to that of silk[12] and the only major difference is that the 9·8 Å spacing of the β keratin which is thought to represent the separation of the grids due to the side chains is only 4·5–6·1 Å in silk. A higher value would be expected in keratin than in silk for this spacing since the former contains many more bulky side chains. Additional support for identifying the 9–10 Å spacing with the side chains is that this is the only reflection which is unaltered in the transition from α to β.

It can be considered as established[13] that β keratin, like silk and fibres prepared from globular proteins after denaturation, is a zigzag structure of fully extended polypeptide chains, as shown in Fig. 12.5. The chains are bonded to one another in the plane of the paper by hydrogen bonds between neighbouring NH and > CO groups. These grids are separated by side chains which extend perpendicularly to the plane of the paper and this results in a separation of the grids by about 9–10 Å in the case of wool.

The nature of the α fold

There can no longer be any doubt that in α keratin the main chains are folded, and that on stretching, the chains are straightened out. It is now generally accepted that the folded form of the α keratin structure is most likely to be based, with certain modifications, on the 3·7 helix proposed by Pauling and Corey. The earlier structures proposed are still of considerable interest in that they have generally each provided an additional step in formulating the later structures, and although the structures proposed by Astbury (Fig. 12.9) are now generally rejected, it is impossible to overemphasize the influence that this classical work has had in the field of biomolecular structure. For these reasons some of the earlier structures will be discussed in some detail.

The X-ray data by itself can provide very little information. The most important reflection is that of 5·1 Å which gives the length of the repeating units along the main chain. The only other reflection at 9·8 Å was thought by Astbury and Woods[3] to be complex but to be associated with the side-chain distance as in β keratin. The strong reflection at 9·8 Å has been resolved by MacArthur,[10] who used porcupine quill instead of wool, into two reflections at 9·2 Å and 10·5 Å. It is possible, though by no means certain, that the 9·2 Å spacing corresponds to the side-chain spacing while the other of 10·5 Å is attributed to the spacing between pairs of α molecules. The α molecule will presumably run alternatively parallel and antiparallel with respect to the –NH–CO–CHR– units along the chains as they do in β keratin (see Fig. 12.5). Astbury and Bell[5] believe that the stereo-chemical arrangement of the α fold must fulfil the following criteria.

380   WOOL: ITS CHEMISTRY AND PHYSICS

(1) The fold must repeat at a distance of about 5·1 Å.
(2) The chain in the α form must be half as long as in the β form since a wool fibre can undergo a reversible extension under some conditions of up to 100 per cent. It is assumed that the whole of this extension must be derived from the straightening of the secondary fold.
(3) In conformity with experimental observation, the density of the fibre must remain constant on stretching at 1·3 g./cc.
(4) So as to obtain the necessary close-packed structure the side chains must be directed alternately on one side or the other of the plane of the fold.

Consideration of (2) led directly to the conclusion that there must be three amino-acid residues within the fold since the 5·1 Å repeating unit is extended to approximately 10 Å on stretching and in the β fold each residue occupies 3·3 Å (i.e. three residues per 10 Å). Astbury and Bell[5] postulated a fold which appears to be unique in fulfilling all these requirements and figure 12.6 illustrates the change

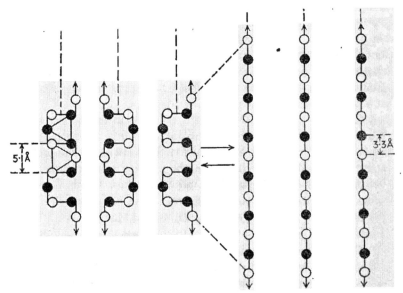

● Represents a side chain projecting UP
○ Represents a side chain projecting DOWN

FIG. 12.6.—Diagrammatic representation of the α to β transformation according to Astbury. (*Reproduced from 'Wool Science Review' by permission of the International Wool Secretariat.*)

occurring in the α–β transformation. It should be emphasized that there is no direct evidence for this particular fold and that the limited X-ray data available could be equally well satisfied by a very large number of different arrangements. It is the assumption that the whole of the reversible extension of 100 per cent. is due to chain folding, i.e. the chain length is doubled in the transformation from α to β, which imposes the suggested fold.

It is, however, by no means evident that the change in the length of the peptide chain must be identical with the macroscopic changes in dimensions of the fibre. In this connexion it should be noted:

(1) that it is impossible to extend a wool fibre reversibly by 100 per cent. without producing permanent damage.

(2) that the β pattern is perfectly formed after an extension of 60–70 per cent. and is unaffected by extension to 100 per cent. Extensions up to 50 per cent. have been found to give changes in the X-ray diffraction photograph that are completely reversible.[15]

(3) that the thermodynamic evidence showing that stretching of a fibre is a process involving unfolding of a molecule, if reliable at all (see p. 67), only applies to the first 30 per cent. of extension, and Bull and Guttmann[16] conclude from elastic data that an elongation of 40 per cent. is involved in the α to β transformation.

(4) that after oxidizing or reducing the disulphide bonds, a fibre can be extended by considerably more than 100 per cent. and that the onset of the α to β transformation only occurs at much higher extensions than in an untreated fibre.[17]

If therefore the 100 per cent. extension is not related to the change in the chain length, the necessity of having the length of the molecule in the α fold half that of the fully extended chain length no longer exists, and a large number of configurations for the α fold is possible. The number of possibilities is reduced if the simplest case of two amino-acid residues per fold is assumed. By rejecting all configurations in which (1) the inter-atomic distances are not those determined in simple molecules; (2) the ordinary valency angles between atoms are not retained; (3) the positions of the side chains undergo a large movement or inversion in the α–β transformation, Zahn[18] proposed a fold for α keratin which is essentially similar to that shown in Fig. 12.7. Several other workers and notably Huggins[19] previously

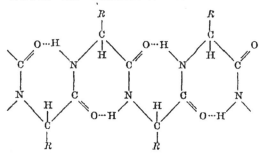

Fig. 12.7.—The folding of the peptide chain into a seven-membered ring.

indicated the possible existence of this type of fold, but since it only provided an increase in length of the peptide chain by 30 per cent. during the α to β extension it was not accepted. It is difficult to see why the typical meridian spacing of this fold should appear at 5·2 Å and not at 2·6 Å since there is a perfect repeat at this value. According to Astbury[21] it is this consideration which led him originally to reject this configuration. Renewed interest in this seven-membered ring structure has been shown as the result of investigations with polarized infra-red light which will be discussed in the next section.

Evidence concerning the α fold from infra-red spectra. Ambrose and Elliott using polarized infra-red on synthetic polypeptides[22] and proteins[23] including mammalian hair showed that in the α form the N—H and C=O stretching vibrations had a dichroism parallel to the chain length whereas in the fully extended or β form the dichroism was perpendicular to the fibre axis. In addition, the N—H deformation frequency showed dichroism perpendicular to the fibre axis in the α configuration and parallel to the fibre axis in the β form. These observations can best be explained by the existence of intramolecular hydrogen bonds between NH and CO groups within the α fold of each individual polypeptide chain. The extent of the dichroism observed in synthetic polypeptides shows that these hydrogen bonds are nearly parallel to the fibre axis. A hydrogen-bonded structure of this kind cannot arise in the type of chain-folding proposed by Astbury for the α fold and Ambrose and Hanby[20] were therefore led to postulate the seven-membered ring structure shown in Fig. 12.7 which is identical with the fold proposed by Zahn[18] and Huggins.[19] It should be stressed that the infra-red evidence demands that the chains must lie approximately straight along the fibre axis and a small-scale folding along the chain of the seven-membered ring as suggested by Huggins is therefore impossible. In this type of fold the α–β transformation can occur without any marked change in the packing of the fold, but merely by rotation of the $\mathrm{\overset{\diagdown}{C}{=}O}{-}\mathrm{\overset{H}{\underset{\diagdown}{N}}}$ group about the asymmetric carbon atom. The absence of a rotation of the side chains about the chain axis during the α to β change was the chief consideration which led Zahn[18] to propose this structure. It should be pointed out, however, that the dichroism observed in wool and even in the highly orientated porcupine quill is very much less than that found for some of the synthetic polypeptides; also the X-ray data cannot be satisfactorily reconciled with the seven-membered ring fold and there is insufficient evidence to warrant its acceptance for the structure of α keratin. Admittedly the spectroscopic data for the polypeptides make it impossible for these to be folded as proposed by Astbury for keratin, but there seems to be no reason why the configuration of the two systems should be the same especially since there are significant differences in the X-ray photographs.[24, 25]

The infra-red evidence for the β configuration fully supports the structure arrived at from X-ray data and provides direct support for the suggestion that in the β fold the hydrogen bonds between N—H and C=O bonds are formed between the molecules and that these are perpendicular to the fibre axis.

A direct consequence of the seven-membered ring model is that fibres in this configuration should be more readily soluble and mechanically weaker than in the β configuration, since all the main-chain hydrogen bonds are taken up in maintaining the intra-molecular fold and only hydrogen bonds from the side chains can contribute to any inter-molecular forces. In the β configuration the hydrogen bonds between NH and CO groups as already shown are all inter-molecular. In direct support of this theory, Bamford, Hanby and Happey[24] showed that synthetic polypeptides in the β form are only soluble in hydrogen-bond-breaking solvents, whereas they dissolve in inert solvents when in the α configuration. Similarly silk dissolves only in potent hydrogen-bond breakers such as solutions of lithium bromide and cupriethylene-diamine, but when converted into the

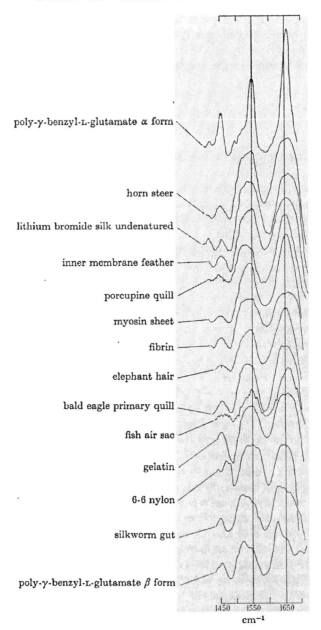

Fig. 12.8.—Comparison of infra-red spectra of several proteins with that of poly-γ-benzyl-L-glutamate in the β-form (lowest curve) and in the α form (top curve).[27] (*Reproduced by permission.*)

α configuration is soluble in water.[26] A direct comparison is impossible in the case of wool since its complete insolubility is determined by the disulphide cross-links. When these are broken a water-soluble material is obtained (α keratose) which is in the α form and on conversion to the β form becomes much less soluble[17] (see p. 388).

At one time it was considered possible to differentiate materials having an α configuration from those in a β configuration by slight differences that occurred in the NH and CO stretching frequencies. Beer *et al.*[27] have, however, cast doubts on these conclusions made by other workers.[28, 29, 30, 31] They have shown (see Fig. 12.8) that there is no definite demarcation between these frequencies in substances known to have a β-configuration and those in an α-configuration.

The Pauling–Corey spiral structure. Pauling and Corey[32] determined with very great precision the inter-atomic distances in a peptide chain and the length of the >NH ... OC< hydrogen bond, and found that neither the Astbury nor the seven-membered ring fold was compatible with their data. They postulated that in a folded protein and synthetic polypeptide all the possible hydrogen bonds are formed and this led them to propose the spiral structure shown in Figs. 12.10 and 12.11. This structure satisfies the infra-red data and is in complete agreement with the X-ray data for certain synthetic polypeptides. It conforms with generally established crystallographic principles and has been widely accepted for those peptides which were thought to have the seven-membered ring fold. Within a few months after the spiral structure was proposed, Perutz[33] found an X-ray spacing corresponding to a repeating unit at 1·5 Å in certain protein crystals and this reflection is now known to be typical of the α fold since it has been found in hair, muscle and related systems. This reflection can be attributed to the periodic structure of the Pauling–Corey helix in which the amino-acid residues occur at this interval along the fibre axis. None of the other suggested structures for the α fold provides a repeat distance at 1·5 Å and this constitutes probably the strongest support at present for the helical structure.[13] Although the X-ray spacings of polypeptides with folded chains correspond closely to those of a spiral structure[25, 34] this is not the case with keratin since the helix does not give the highly characteristic meridional reflection at 5·1 Å. Pauling and Corey[35] believe that they have overcome this objection by postulating that the presence of repeating sequences of amino-acid residues causes the axis of the α spiral to describe a helical course. The presence of a seven-strand cable each composed of six α helices (B-type chain) twisted about a seventh straight helix (A-type chain) (see AB_6 cable, Fig. 12.14) would give the X-ray diagram observed for wool including the 5·1 Å spacing. This theory is, however, difficult to reconcile with the observation[36] that wool fibres can be solubilized and extruded again as fibres giving a 5·1 Å reflection, since in the dissolution process the complex intra-molecular structure postulated would be broken up and unlikely to be reformed.

One other serious objection[37] to the helical structure is that the density of wool is significantly higher than that corresponding to the helix: namely 1·3 instead of the theoretical 1·15. Complete agreement between the theoretical and observed values is not to be expected, but the fact that the theoretical density is too low is difficult to understand. A tentative explanation may be that a small proportion of the fibre only is in the helical form (perhaps only α_1 keratose) and that the remainder is amorphous and of high density.

The present position is that none of the proposed structures is wholly satisfactory, although the Pauling–Corey spiral explains many of the observed facts more adequately than any of the other postulated configurations. The problem of the stereo-chemistry of the α fold remains, for the present therefore, one of unresolved controversy.

The stability of the α fold in wool

Besides the reversible transformation from α to β keratin (see p 376.) the crystalline structure can be irreversibly destroyed by subjecting wool to a number of severe treatments which give rise to a disorientated β structure (or according to Elöd and Zahn to δ-keratin). The X-ray spacings are similar to those found for true β keratin but the reflections are more diffuse and show less, if any, orientation (see Fig. 12.13b). The conversion to a disorientated β structure can be compared to denaturation. All treatments which bring about irreversible supercontraction by the breaking of hydrogen bonds (see p. 75) change the X-ray diffraction pattern to disorientated β. Rupture of all the disulphide bonds by mild reagents (e.g. reduction with sodium thioglycollate or oxidation with peracetic acid) does not destroy the α structure and this confirms the views of Astbury and Woods[3] that these links play no part in determining the crystalline structure of wool. On the whole the α structure is exceedingly stable and wool has to be heated dry at 170° C. or moist at 130° C. or extensively hydrolysed by acids (see p. 300) before conversion to the disorientated β structure takes place. Judged by this criterion wool and other keratins are more difficult to denature than any other protein. Once the disulphide bonds have been severed, however, the structure is readily destroyed, and conversion to the disorientated β structure is brought about by treatments similar to those which lead to denaturation of most other proteins.[17] Thus after reducing or oxidizing all the disulphide bonds the fibre is converted to a disorientated structure by immersion in water at 95° C. for five minutes (see Fig. 12.13). Disorientation by reduction with bisulphite at 100° C.[38] is probably the result of heat denaturation following disulphide-bond breakdown.

The soluble α keratose when precipitated as a powder undergoes the same crystalline changes as the oxidized fibres, and the α fold is disorientated by dry heat at 110° C. and by water at 95° C. in five minutes.[17] α Keratose, which is soluble in hydrogen-bond-breaking solvents, dissolves readily in anhydrous formic acid from which it is precipitated in the α fold on adding water, but if the acid is distilled off under vacuum (40° C.) the keratose left is in the β configuration.[39] Reactions which degrade the molecule by the breaking of peptide bonds bring about conversion into the β configuration. It can therefore be concluded that a soluble globular protein-like molecule can exist in the α fold but in the absence of stabilizing cross-links is converted into the β form by treatments known to produce denaturation. The keratose when converted into the β form by a process which does not degrade the protein, exhibits much greater intermolecular attraction than in the α configuration. Because of this the β form dissolves much less readily, and frequently a sample will not have dissolved completely even after several days in ammonia, unlike the α keratose which only requires a few hours under the same conditions; also the β form swells much less in water and phenol. When the α keratose is converted into the β configuration the molecules are no longer spherical but exhibit a very pronounced asymmetry, and must be considered as rods (see p. 372). Probably the molecule in the β configuration is held in fixed folds by hydrogen bonds and cannot coil up at random as in the α fold. Unlike the conversion to β by stretching, the change to disorientated β is almost invariably irreversible and leads to an opening of the structure. Wool is more readily digested by enzymes and has a decreased anisotropy of swelling (see p. 297) in the disorientated β reversible but not in the β configuration.

The most probable explanation is that in the disorientated form the crystalline areas (or micelles) of wool have been broken up. Since hydrogen-bond-breaking reagents can bring about the destruction of the α structure without breaking any disulphide bonds, hydrogen bonds must be largely responsible for holding the peptide chains in position within the micelles. These bonds however are much more easily broken in the absence of disulphide cross-links which stabilize the micelles possibly by holding them embedded in a disulphide cross-linked cement (see p. 391).

Micellar structure

Physical evidence. Speakman[40] first advanced a theory of the micellar structure of keratin fibres largely as a result of studies of the penetration of hydroxylic liquids into the dry and swollen fibres (see Chapter 4). The penetration was followed by studying the elastic changes, from which it was inferred that molecules with a diameter greater than propyl alcohol cannot enter the fibre. This was taken to mean that the dry fibre contained micropores of this diameter, and as the swelling is almost entirely transverse and results in little change in the X-ray diagram, the liquid is thought to be accommodated in the amorphous regions of the fibre. One of the most important observations made by Speakman was that higher alcohols can penetrate the fibre freely if it is initially swollen with water. From the regain value at which this occurs and the extent of transverse swelling, it is possible to calculate the average pore diameter of the fully swollen fibre and the thickness of micelles.

The value of 40 Å for the former distance agrees closely with more recent values estimated from the internal surface area as given by B.E.T. theory (see p. 101). The thickness of the micelle given by Speakman of approximately 200 Å also agrees favourably with other estimates. Little is known of the shape of the micelle although the anisotropy of the swelling leads to the conclusion that the length is much greater than the width. These, and other, considerations led Speakman to the conclusion that the micelles are in the form of long plates arranged approximately parallel to the fibre axis separated by a distance of 6 Å. The later work of King[41] on the diffusion of neutral molecules into keratin (see p. 128) invalidates the original arguments which gave this value so that the concept of specific pores

FIG. 12.9.—Model of α fold according to Astbury.[14] (*Reproduced from the 'Transactions of the Faraday Society' by permission of the editor.*)

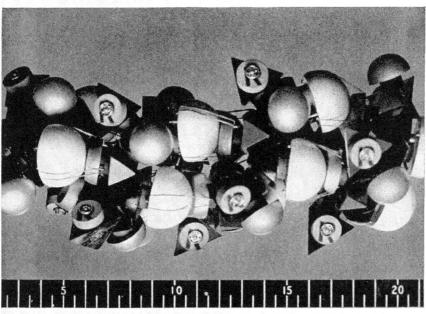

FIG. 12.10.—Model of α fold according to Pauling and Corey.[14] (*Reproduced from the 'Transactions of the Faraday Society' by permission of the editor.*)

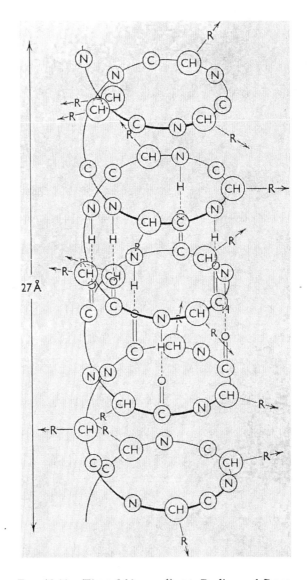

Fig. 12.11.—The α fold according to Pauling and Corey.

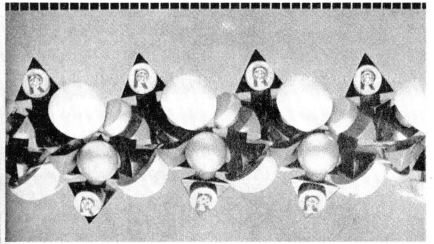

FIG. 12.12.—Model of α fold according to Ambrose and Hanby. (*Reproduced from 'Transactions of the Faraday Society' by permission of the editor.*)

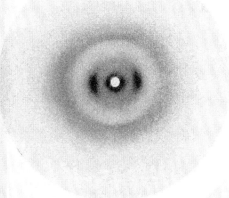

FIG. 12.13.—Stability of the α fold in wool containing no disulphide bonds.

(*) X-ray diagram of wool after oxidation with peracetic acid at room temperature which shows that the α structure is not destroyed by the breaking of disulphide bonds.

(*) X-ray diagram of disorientated keratin obtained by placing the peracetic acid oxydized fibres in water at 95° C.

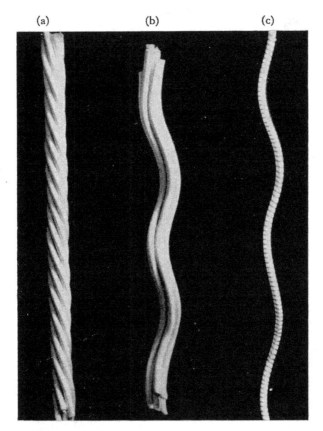

FIG. 12.14.—(a) Pauling and Corey AB_6 cable; (b) Corresponding packing of 7 B-type chains without cabling twist; (c) Single-type chain. (*Reproduced by permission of The Textile Institute*)

within the dry fibre must be rejected. Nevertheless Hermans[42] advances strong evidence showing that dry fibres contain a certain amount of 'free space' due no doubt to the difficulty of packing the main chains with the various side groups closely together.

The diffusion data do not, however, conflict with the molecular model proposed by Astbury and Woods[3] (1933) as a result of a detailed study of the elastic properties of wet and dry fibres. According to this theory, the micelles are embedded in an amorphous material which acts like cement, so that the postulate of a regular network of micropores is not required. The fibres are supposed to be made up of three distinct phases which were originally termed K_1 K_2 and K_3, corresponding to the intermicellar cement, the cell wall and the micellar substance.

Thus on stretching the fibres from 0 to 25 per cent., when the X-ray picture remains virtually unchanged, only the K_1 region is deformed. Further extension, which is only possible in the presence of swelling agents, is thought to take place in the K_2 phase which must be partially crystalline as the X-ray picture changes during the stretching. This is attributed by Astbury to the unfolding of the keratin molecule, and not to inter-molecular slip. It is impossible to extend the fibre by more than 50–60 per cent. in water at ordinary temperatures as it invariably breaks. If however the extension is carried out in hot water or steam, or in alkaline solutions at room temperature, it is possible to extend the fibre to 100 per cent. It is believed that the extension beyond 50 per cent. is controlled by the third phase K_3.

Although chemical evidence strongly supports the division of wool keratin into three phases (see p. 369), it also suggests that the K_2 phase is the material which constitutes the micelles, and that the K_3 phase is a cell wall which surrounds each cortical cell i.e. the cortical cell membrane (see p. 13). More recently Feughelman[43] has attempted to explain the elastic behaviour of wool fibres by a two-phase model. (See also p. 395).

Evidence from supercontraction data. It has been shown in Chapter 3 that supercontraction may be caused either by hydrogen-bond or disulphide-bond breakdown. When caused by reaction with hydrogen-bond breakers, the X-ray picture of the fibres is changed; the typical α fold disappears and in general a disorientated β pattern appears. In some cases, however, all evidence for crystallinity disappears and the fibre gives no diffraction pattern at all.[38] The super-contraction brought about in this way is in many cases reversible, so that the fibre may eventually regain its original crystalline structure. Substitution of the disulphide bond by the more stable lanthionine link does not influence this contraction.[44]

Alternatively, supercontraction may be produced by reagents which break more than 70 per cent. of the disulphide bonds. If this can be achieved at a low temperature, the X-ray picture of a native wool fibre is obtained, and no change in crystallinity occurs (i.e. the α fold is retained).[44] However in the absence of disulphide bonds the α fold is unstable (see p. 388) and the fibre changes into a disorientated β configuration on warming. Consequently if a reagent such as bisulphite is used at elevated temperatures in order to reduce the necessary number of disulphide bonds, a change in structure to the disorientated β configuration is observed. The disappearance of the X-ray picture under these conditions is therefore not due to the supercontraction resulting from disulphide-bond breakdown, but to the denaturing effect which occurs at elevated temperatures. If the disulphide bonds are replaced by lanthionine links, supercontraction is no longer brought about by these disulphide-bond-breaking reagents.

These results are in harmony with the theory of micellar structure developed above. Alexander[44] suggested that the molecules are held together within the micelle by hydrogen bonds, to give rise to crystalline areas in the fibre. The micelles which are supposed to be set in a cystine cross-linked cement, are connected to the latter by disulphide bonds. The disulphide bonds within the micelles which are relatively few in number,[17, 39] are arranged in such a way as to allow relative movement of the molecules within the micelle. Treatment of a fibre with a hydrogen-bond-breaking reagent causes the micelle to buckle, which produces a shortening of the whole fibre as the cement is unaffected and remains rigid, the micelles revert to the original structure with the formation of new hydrogen bonds on the removal of the reagent. This explains why the supercontraction which is brought about by hydrogen-bond-breaking reagents is often reversible.

Chemical evidence for a three-phase composition

Direct chemical support for the micellar model outlined above is provided by the separation of wool into three distinct components on treatment with peracetic acid which oxidizes all the disulphide bonds.[17] It was shown in Chapter 11 that the wool fibre can be divided into a relatively homogeneous protein of high molecular weight, α keratose and a heterogeneous protein of low molecular weight, γ keratose, together with the cortical cell membranes known as β keratose.

The low molecular weight material is found to contain approximately three times as much sulphur as α keratose. This suggests that the micelles which are composed of high molecular weight material are deficient in cystine and are surrounded by a cystine-rich cement which is maintained rigid by three-dimensional disulphide bonding. X-ray photographs of the α keratose show that it is in the α crystalline form, and it is probably this material which is responsible for the observed α keratin photograph of wool. On the other hand, γ keratose gives no reflection typical of the normal X-ray photograph but only some very diffuse rings. This suggests therefore that the cement material is present in the β form, and is completely disorientated and amorphous. Elöd and Zahn[45] have observed some glass-like reflections at approximately 4, 2·1, 1·5 and 1·1 Å which can be identified with amorphous material but these values do not determine whether amorphous material is in the α or β form. These two forms possibly give different infra-red absorption spectra (see p. 386) and since the absorption spectra are not dependent on orientation, an amorphous component may be resolved by this technique. Thus in addition to the ordinary α component, β protein has been detected in longitudinal sections of thick mammalian hair. In view of the foregoing data, this component may be tentatively identified with the cement substance isolated after oxidation as γ keratose.

The cortical cell membranes can probably be identified with the K_3 phase postulated by Astbury and Woods,[3] since they greatly resist extension and break up when fibres are stretched beyond 60 per cent. As irreversible damage always occurs to fibres which have been stretched by more than 50 per cent., it is possible that the third phase of the extension diagram may be due to the rupture of these membranes which then permits the remainder of the fibre to be stressed. This also explains the considerable decrease in strength (c. 30–40 per cent.) which is observed in the fibre which has been allowed to return to its original length after extension beyond the critical point. The chemical work therefore leads to a very similar picture to that derived by Astbury and Woods[3] from purely mechanical studies, and the extension data may be explained equally well by considering the third phase to surround the whole of the cortical cells, as by a membrane which surrounds each individual micelle.

When isolated chemically from oxidized fibres the membrane gives an X-ray diffraction pattern of disorientated β keratin. It is not known definitely whether it is present as β keratin in the native fibre, or whether the change from α to β is promoted by the chemical treatment. Woods et al.[46] claim to have isolated the membrane in the α form from oxidized wool and to have found that it changed very easily to β keratin, e.g. by heating to 40° C. The writers have never succeeded in obtaining the membrane in the α form from wool oxidized with peracetic acid however mild the extraction process,[47] although using some inorganic peracids[48] it is possible to obtain a membrane with some α characteristics. Moreover it has been clearly established that a protein of an α keratin structure can be obtained from the soluble part of the oxidized fibre and that the α fold in this material though not as stable as in native wool (see p. 388) is not changed to the β form under the mild conditions which are reputed to affect the membranes in this way. It is not known for certain therefore whether the membrane is present in the α or β configuration in the native wool, although there is no doubt that it has quite different properties from the remainder of the fibre, and on account of this it can probably be identified with the resistant K_3 phase postulated by Astbury and Woods.[3]

The physical properties of the micelle

Although Speakman[40] established with some certainty that the length of the micelles is much greater than their width, his conclusions on their actual shape must be regarded as tentative. Direct experimental evidence in support of their laminar nature was ultimately produced by Elöd and his colleagues,[45, 49] who isolated the spindle cells from wool by enzyme action and allowed them to settle on a flat surface in the form of a film. The flat parts of the micelles tended to orientate themselves parallel to the surface and X-ray analysis showed the structure to be more orientated in a direction parallel to the film than perpendicular to it, which therefore indicates a laminar structure. In addition it is not known whether the micelles are of uniform shape and although some long-range X-ray spacings have been found in mammalian bristles and in wool it is mere speculation to identify these distances at present with micellar dimensions.

It is impossible at present to establish the number of intramicellar disulphide bonds and their orientation. There can be no doubt, however, that these bonds exert a considerable stabilizing effect on the α configuration, and it seems more probable that some disulphide bonds cross-link the molecules within the micelle-two dimensionally[8]. This type of cross-linking would not prevent disorientation of the micelle by buckling or the break-down of hydrogen bonds which provide the three-dimensional rigidity.

In 1947 Farrant, Rees and Mercer[50] published the results of an electron-microscopic investigation of disintegrated spindle cells which had been gold shadowed so as to increase the resolution. They clearly demonstrated the existence of fine subfibrils (see Chapter 2) but further obtained evidence to show that these are composed of spherical or cylindrical units with a diameter of 100 Å. These results appear to be in direct disagreement with all the previous observations which suggested that the whole fibrous structure is made up of very long and thin molecules lying approximately in the direction of the fibre axis. According to the alternative corpuscular theory strong fibrils are formed by the stepwise joining of the individual keratin molecules via disulphide bonds. Astbury and others had previously suggested that fibrin and actin are made of linear aggregates of symmetrical particles which individually may resemble globular proteins. It is felt that further investigations are necessary before the original laminar structure is rejected, and before these rounded particles are identified with the micelles. There is a strong possibility that the electron photomicrographs of Rees *et al.* are due to artifacts (see p. 5).

Although no accurate estimate of the proportion of the wool fibre which is truly crystalline is available, this is certainly less than 50 per cent. of the total fibre material and may be as low as 10 per cent. It follows that all the long-chain material (i.e. α keratose) cannot be in an ordered configuration in the fibres, so that only part of the micelles is crystalline. It seems reasonable to suppose that the individual micelles contain crystalline areas in much the same way as synthetic fibres contain crystalline regions, one chain may thread its way through both crystalline and amorphous portions giving rise to a structure which is called a fringe micelle. The structure adopted for the wool fibre in this chapter is in effect a synthesis of the early views on micellar structure, and the modern concept of the fringe micelles which had originally been proposed by Gerngross, Herrmann and Abitz[51] and developed later by Meyer[52] to explain the formation and properties of regenerated synthetic fibres. At this stage a complete explanation of the various observations on the accessibility of the fibre to various reagents cannot be given in terms of the proposed macromolecular model. Before attempting to give a partial explanation of most of the facts, it will be convenient to review briefly the most important observations so far recorded, as follows:

(1) Hydrophilic liquids can enter the wool fibre in general, but molecules larger than propyl alcohol only bring about a very limited amount of swelling of the fibres. Similarly the proportion of free carboxyl groups which can be esterified by various alcohols decreases with increase in the molecular weight of the alcohol (see p. 311).
(2) If the fibre is swollen in water, large molecules can enter the fibre readily, but the energy activation of the diffusion processs is almost independent of the size of the molecule (see p. 158).
(3) The X-ray photograph is virtually unchanged if the dry fibre is swollen in water and most other solvents, but all the crystallinity disappears in formic acid[40] and strong solutions of lithium bromide.
(4) Large anions can form salts with all the amino groups in the fibre without changing the arrangement of the molecules in the crystalline areas (see p. 238).

These observations may be explained in the following way, which, however, must be regarded as a considerable over-simplification in the absence of more detailed investigations. It is suggested that water and the simpler hydrophilic solvents penetrate the inter-micellar cement, which is composed entirely of short polypeptide chains, thus causing the transverse swelling typical of these processes. This material then behaves as a homogeneous gel through which relatively large molecules can diffuse by a process of chain displacement requiring an activation energy of 10–14 k. cal./mol. On the other hand the micelles, which are composed of high molecular weight material only, are not swollen by water, although water molecules may penetrate without changing the arrangement. Only molecules with a very high hydrogen-bond energy, e.g. formic acid and lithium bromide, can break down these highly crystalline regions.

The micelles cannot, however, be penetrated by large molecules even from aqueous solutions, which at first sight appears to conflict with the stoichiometric combination of wool with acids of high molecular weight (see p. 224). These conflicting data have not been resolved, in spite of detailed discussion. It may be suggested tentatively that the polypeptide chains constituting the micelle are arranged in such a way that a large proportion of their polar groups (e.g. COOH and NH_2) are directed towards the inter-micellar phase and are therefore equally accessible as the corresponding groups in the amorphous phase to reagents causing esterification and salt formation. The micelles may be laminar or cylindrical with the polypeptide chains parallel to the axis and with the non-polar groups pointing towards the centre. This structure was proposed[44] by analogy with soap micelles in which the polar part of the molecule is in the aqueous phase and the non-polar part within the micelle. The behaviour of the micelle is unlikely to be uniform and differences between the areas of different degrees of order are to be expected.

Small-angle X-ray diffraction patterns

Early X-ray diffraction investigations on wool were limited to what are now referred to as high-angle reflections. These give information regarding repeats of the order 1·5–12 Å within the fibre, such reflections being necessary in postulating main chain configurations (see p. 376 *et seq.*). More recently the development of low-angle spectrometers has enabled much greater spacings to be detected. These spacings correspond to features occurring in the macrostructure of the fibre and give information that is to some extent complementary to that obtained by the electron microscope (see Chapter 1). Since Bear[53] and MacArthur[10] first discovered that small-angle reflections could be obtained from mammalian hairs, further work has largely been confined to investigating intensity variations produced in the reflections by various chemical treatments. Reflections occurring along or near to the meridian of the X-ray diffraction pattern correspond to periodicities occurring along the polypeptide chain, whilst equatorial reflections correspond to lateral regularities.

The low-angle equatorial reflections produced by untreated wool fibres correspond to spacings of about 45, 27 and 16·8 Å which are the 2nd, 3rd and 5th order respectively of an 85Å repeat.[10, 54] Electron micrographs of sections of fibres stained with osmium tetroxide, after partial reduction of the cystine, show cylindrical filaments of approximately 70Å diameter embedded in an osmiophilic ground substance having a richer cystine content than the filaments.[55] The filaments are referred to as microfibrils and have been shown to consist of a number of subfibrils.*[56] The subfibrils appear to be arranged with nine forming a circle with possibly two more in the centre of the circle. Relationships between the various observed arrangements of the microfibrils and the low-angle equatorial reflections have been investigated mathematically.[57, 58]

The meridional reflections are found to be various orders of a 198Å repeat.[10] Reactions with the salts of heavy metals enhance some of the reflections. If each salt reacted specifically with a particular residue, then evidence for a regular periodicity of that residue

TABLE 12.1

Intensification of reflections in keratin produced by heavy metal salts

Reflection intensified as an order of 198 Å.	Heavy metal salt causing intensification	Reflection intensified as an order of 198 Å.	Heavy metal salt causing intensification
3	Mercuric acetate (60) Silver nitrate (60)	8	Silver nitrate (60) Lead acetate (62) Cupric acetate (58, 62) Mercuric acetate (58, 62) Lanthanum acetate (58, 62) Uranyl acetate (58, 62) Silver acetate (58, 62)
4	Iodine in ethanol (61) Mercuric acetate (60) Silver nitrate (60) Silver acetate (58)		
5	Osmium tetroxide (54) Silver nitrate (60)		
		9	Silver nitrate (60)
6	Iodine in ethanol (61) Mercuric acetate (60) Silver nitrate (60)	10	Silver nitrate (60)
		12	Mercuric acetate (54)
7	Mercuric acetate (54) Silver nitrate (60)	16	Mercuric acetate
		19	Mercuric acetate (54)

* The term subfibril can be used in more than one sense. Thus, it is postulated that each of these subfibrils consists of only three α-helices.[56] This is obviously a much smaller order of magnitude than the subfibrils shown in the photomicrograph in Fig. 1.4.

within the fibre would be produced in the form of an intensification of the reflection corresponding to that repeat. Unfortunately most of the salts probably react with more than one residue and consequently intensify more than one reflection.

In the natural fibre, the 3rd order (66Å) spacing is the most intense,[59] suggesting a major discontinuity in electron density at this repeat. The cause of this discontinuity is still a matter of speculation.[56] The increase in intensity of the 6th order reflection by iodine in ethanol is regarded as good evidence for a regular disposition of tyrosine residues at intervals of 33Å.[58] The cystine residues may occur at intervals of 49Å[58] or even 25Å.[60]

REFERENCES

[1] SKERTCHLY AND WOODS, *J. Text. Inst.*, 1960, **51**, T517.
[2] ASTBURY AND STREET, *Phil .Trans. Roy. Soc.*, 1931, **230A**, 75.
[3] ASTBURY AND WOODS, *ibid.*, 1933, **232A**, 333.
[4] ASTBURY AND SISSON, *Proc. Roy. Soc.*, 1935, **150A**, 533.
[5] ASTBURY AND BELL, *Nature*, 1941, **147**, 696.
[6] ASTBURY, *J. Chem. Soc.*, 1942, 337.
[7] ASTBURY, *Advances in Enzymology*, 1943, **3**, 63.
[8] ASTBURY, *Proc. Roy. Soc.*, 1947, **134B**, 303.
[9] BENDIT, *Nature*, 1957, **179**, 535.
[10] MACARTHUR, *Nature*, 1943, **152**, 38.
[11] MACARTHUR, *Symp. Fibrous Proteins*, 1946, 5 (publ. Soc. Dy Col., Bradford).
[12] HERZOG AND JANCKE, *Ber.*, 1920, **53**, 2162.
MEYER AND MARK, *ibid.*, 1928, **61**, 1932.
[13] ASTBURY, *Nature*, 1945, **155**, 501.
[14] ROBINSON AND AMBROSE, *Trans. Faraday Soc.*, 1952, **48**, 854.
[15] WOODS, *J. Text. Inst.*, 1960, **51**, T526.
[16] BULL AND GUTTMANN, *J. Amer. Chem. Soc.*, 1944, **66**, 1253.
[17] ALEXANDER, *Proc. Roy. Soc. Med.*, 1951, **44**, 389; *Kolloid Z.*, 1951, **122**, 8.
[18] ZAHN, *Z. Naturforschung*, 1947, **26**, 104.
[19] HUGGINS, *Chem. Reviews*, 1943, **32**, 135; *Ann. Rev. Biochem.*, 1942, **11**, 27.
[20] AMBROSE AND HANBY, *Nature*, 1949, **163**, 483.
[21] ASTBURY, *Nature*, 1949, **164**, 439.
[22] AMBROSE AND ELLIOTT, *Proc. Roy. Soc.*, 1951, **205A**, 47.
[23] *Ibid.*, 1951, **206A**, 206.
[24] BAMFORD, HANBY AND HAPPEY, *ibid.*, 1951, **205A**, 30.
[25] BAMFORD, BROWN, ELLIOTT, HANBY AND TROTTER, *Nature*, 1952, **169**, 357.
[26] AMBROSE, BAMFORD, ELLIOTT AND HANBY, *ibid.*, 1951, **167**, 264.
[27] BEER, SUTHERLAND, TANNER AND WOOD, *Proc. Roy. Soc. A*, 1958, **249**, 147.
[28] ELLIOTT AND AMBROSE, *Nature*, 1950, **165**, 921.
[29] ELLIOTT, *Proc. Roy. Soc. A*, 1953, **221**, 104.
[30] ELLIOTT, *Proc. Third International Congress of Biochemistry*, New York, Academic Press Inc., 1956.
[31] KEIGHLEY, *M.Sc. Thesis*, University of Liverpool, 1958.
[32] PAULING AND COREY, *Proc. U.S. Nat. Acad. Sci.*, 1951, **37**, 241.
[33] PERUTZ, *Nature*, 1951, **167**, 1053.
[34] COCHRAN AND CRICK, *Nature*, 1952, **169**, 234.
[35] PAULING AND COREY, *Nature*, 1953, **171**, 59.
[36] ALEXANDER, CARTER, EARLAND AND MARINER, *Melliand Textilber.*, 1952, **33**. 187.

[37] ASTBURY, Private communication.
[38] WHEWELL AND WOODS, *Symp. Fibrous Proteins*, 1946, 50. (publ. Soc. Dyers & Cols., Bradford).
[39] ALEXANDER AND EARLAND, *Nature*, 1950, **166**, 396.
[40] SPEAKMAN, *Proc. Roy. Soc.*, 1947, **43**, 552.
[41] KING, *Trans. Faraday Soc.*, 1947, **43**, 552.
[42] HERMANS, *Physics & Chemistry of Cellulose Fibres*, Elsevier, 1949.
[43] FEUGHELMAN, *J. Text. Inst.*, 1960, **51**, T589.
[44] ALEXANDER, *Ann. N.Y. Acad. Sci.*, 1951, **53**, 653.
[45] ELÖD AND ZAHN, *Melliand Textilber*, 1947, **28**, 217.
[46] PEACOCK, SIKORSKI AND WOODS, *Nature*, 1951, **167**, 408.
[47] ALEXANDER, *Nature*, 1951, **168**, 1081.
[48] EARLAND, MACRAE, WESTON AND STATHAM, *Text. Res. J.*, 1955, 25, 963.
[49] NOWOTNY AND ZAHN, *Z. physikal. chem.*, 1943, **192**, 333.
[50] FARRANT, REES AND MERCER, *Nature*, 1947, **159**, 535.
[51] GERNGROSS, HERRMANN AND ABITZ, *Biochem. Z.*, 1930, **228**, 409.
[52] MEYER AND WYK, *Helv. Chim. Acta*, 1937, **20**, 1331.
[53] BEAR, *J. Amer. Chem. Soc.*, 1943, **65**, 1784.
[54] FRASER AND MACRAE, *Nature*, 1957, **179**, 732.
[55] ROGERS, *Am. Acad. Sci.*, N.Y., 1959, **83**, 378.
[56] FRASER, Wool Research Conference, Paris, 1961.
[57] FRASER AND MACRAE, *Biochim. Biophys. Acta*, 1958, **29**, 229.
[58] FRASER, MACRAE AND ROGERS, *J. Text. Inst.*, 1960, **51**, T497.
[59] ONIONS, H. J. WOODS AND P. B. WOODS, *Nature*, 1960, **185**, 157.
[60] SIMPSON AND WOODS, *Nature*, 1960, **185**, 157.
[61] RICHARDS AND SPEAKMAN, *J. Soc. Dyers & Col.*, 1955, **71**, 537.
[62] FRASER AND MACRAE, *Nature*, 1961, **189**, 572.

SUBJECT INDEX

ABSORPTION, effect of epicuticle on rate of, 173
—, heat of, 114
—, of acids (see acids)
—, of phenols, 210
—, rate of, 128, 129, 140, 142
Accessibility, 91, 92, 128, 260, 302
—, determination of by esterification, 311
—, influence of morphological structure, 172, 238
—, interpretation of, 128, 394
Acetylation, 83, 113, 302, 306, 315
Acid, absorption of, 180, 353
— —, effect of hydrolysis, 289
— —, — on mechanical properties, 213
— —, — of salts on, 184
— —, heat of, 186, 223
— —, location of anions, 197, 203
— —, supra-stoichiometry of, 224
— —, theories of, 194, 197, 200, 203
— —, weak acids, 208
Acid orange II, 172
— — —, rate of dyeing with, 162
Actin, 395
Activation energy, 119, 132, 158, 168, 174
— —, of dyeing, 162
Activated reactions, 281
Acrylonitrile, 319
Adsorption, free energy of, 20, 98
—, heat of, 19, 101, 111
—, isotherm, 19
—, multimolecular, 111
—, statistical theory of, 99, 110
—, surface area by, 18
—, Van der Waals', 93
Affinity, see anion affinity
—, see cation affinity
Alanine, 348, 350, 351
Alanylglycine, 95
Alcohols, 88, 91
—, esterification of carboxyl groups by, 311
Aldehyde groups, 252
Aldehydes
—, reaction with wool, 320
—, cross-linking by unsaturated, 341
Alginic acid, 335
Alkali absorption, 210
— —, effect of hydrolysis on, 291
— —, — — formaldehyde on, 213
— —, — — phenol on, 212

Alkali absortion effect of salts on, 213
Alkali absorption, heat of, 187
— —, influence of disulphide bond breakdown, 210
— —, theoretical titration curve, 211
—, effect on disulphide bond, 254
—, effect on wool, 254
—, formation of new crosslinks by, 254
—, — — sulphide by, 254
—, influence on mechanical properties, 295
—, unshrinkable finish by, 257
— solubility, 265, 326, 333
— —, assessment of damage by, 293, 294
— —, influence of disulphide bond breakdown on, 293
— —, — — new crosslinks on breakdown, 293
— —, — — peptide bond breakdown on, 293
— —, — — of temperature on, 293
— —, test, 293
Alkyl dihalides, 64, 339, 340
Alkyl per-acids, 266
Alkylating agents, 249, 336
— —, poly-functional, 333
Allwörden sacs, 7
Allyl isothiocyanate, 340
Amide groups, 112, 288, 307, 329
— —, hydrolysis of, 208
— —, titration of, 182, 224
Amino acid composition, 346, 348, 350
— — —, variation of, 244, 350
Amino acids, acidic, 180, 347
— —, basic, 180, 347
— —, frequency of, 354
— —, order along the peptide chain, 353
— —, pK of, 211
— —, reactivity of, 150, 155
— —, terminal, 351, 352
α-amino acrylic acid, 250, 255, 258, 261
Amino groups, 313, 328, 336, 351
— —, acetylation, 335
— —, combination with benzoquinone, 332
— —, — — mercury, 332
— —, effect of hydrogen peroxide on, 262
— —, elimination of, 315
— —, reaction with alkylating agents, 333, 335

Amino groups, reaction with formaldehyde, 328
— —, — — sulphuric acid, 318
— —, reactivity of, 313
Amorphous regions, 59, 67, 106, 113, 376, 389
n-amyl alcohol, 311
Anhydrocarboxyglycine, 47
Anion affinity, 170, 184, 189, 200
— —, of chloride ion, 223
— —, determination of, 225
— —, of dyes, 226
— —, effect of concentration on, 236
— —, — — hydrogen bond, 230
— —, — — ion size, 229
— —, — — molecular weight, 229
— —, — — structure on, 229
— —, from Gilbert-Rideal theory, 228
— —, from Steinhardt-Harris theory, 226
— —, influence of heat of acid absorption, 189
— —, — — dispersive forces, 230
— —, nature of, 229
Anion exchange, rate of, 168
Anion exchange and dyeing equilibria, 231
Anion exchange equilibria, effect of crosslinking, 232
Anti-chlor, 249
Arginine, 112, 278, 292, 303, 313, 318, 329, 348, 350
Aspartic acid, 348, 350, 351

BASIC dyeing, 240
B.E.T. theory, 19, 92, 99, 112, 146
Benzopurpurine, 171
Benzoquinone, 38, 332
—, polymerization of, 333
Benzyl alcohol, 311
Bergmann-Niemann rule, 353
Biosynthesis, 367
Birefringence, 83
Bisulphite, 249, 250
Bleaching, 249, 262, 294
Bromamines, 280
—, N-bromoacetamide, 280
—, N-brómosuccinimide, 280
Bromate, 73, 281
Bromic acid, 281
Bromine, 272, 276
Butyl alcohol, 257, 311
n-di Butyl ether, of methylol urea, 50

CADMIUM salts, 253
Calcium thiocyanate, 361
Calcium thioglycollate, 339
Capillary condensation, 93, 97
Capillary flow, 16
Carbolan Crimson B.S., 171

Carbonizing, 294, 319
Carboxyl groups, 166, 180, 182
— —, esterification by epoxides, 312
— —, dissociation of, 186, 223
— —, reaction with alkylating agents, 333, 335
Carcinogenic properties of crosslinking agents, 336
Caro's acid, 270, 304
— —, specificity of, 270
Carrotting, 40
Casein, 93, 95, 112, 117, 189
Casein fibres, 338
Cation affinity, 212
— — and basic dyeing, 240
— —, effect of size on, 240
— —, effect of structure on, 241
— —, influence on alkali absorption, 212
Cells, epithelial, 1
Cellulose, 79, 90, 104, 107, 112, 118, 120, 333
—, dyeing of, 165, 172
Cement, intermicellar, 82
Cetyl alcohol, 311
Cetyl iodide, 311
Cetyl sulphonic acid, 361
Chloracetic acid, absorption of, 210
Chloramides, 75, 307
Chloramines, 275, 278, 283
—, mechanism of reaction, 279
—, oxidation of disulphide bond, 279
—, oxidation of tyrosine, 279
Chloramine T, 278
β-chlorethylamine, 308
Chlorination, effect of pH on rate of, 174
—, effect on shrinkage, 45, 47
Chlorine, 75, 272–276
—, effect of pH on reaction, 272
—, formation of Allwörden sacs, 7
—, formation of crosslinks by, 275
—, oxidation of disulphide bond, 276
Chlorine dioxide, 75, 280, 284, 303, 359
— —, oxidation of disulphide bond, 280
— —, — — tyrosine, 280
— —, specificity of, 280, 367
Chloromethyl ethers, 333
Chlorphthalimide, 278
Chlorsulphamic acid, 278, 282
Chlorsulphonic acid, 319
Chlorurea, 278
Chlorzyme process, 31
Chromatography, 272, 281, 321, 329
Chrome dyes, 177
Collagen, 91, 102, 113, 118
Conduction, electrolytic, 119
—, electronic, 119
—, high-frequency, 124
Contact angle, 97
Coomassie Milling Scarlet, 171

SUBJECT INDEX 401

Copper, 265
Corpuscular proteins, 395
Corpuscular structure, 5
Cortex, 4
—, bilateral structure, 5
—, ortho, 5
—, para, 5
—, penetration of dyes into, 173
Cortical cells, 4, 343
— —, effect of ultrasonics on, 5
— —, enzymatic isolation, 4
— —, isolation with acids and alkali, 4
—, membrane, 11, 12, 13, 358, 359, 368, 389, 391, 392
— —, piezoelectricity of, 24
— —, supercontraction of, 74
Cotton, 21, 87, 95, 106
Cramer-Neuberger effect, 303
Creep, 61
—, influence of cuticle, 61
—, — — disulphide bonds, 61, 253
—, interpretation of, 61
—, temperature coefficient of, 62
Crosslinking, 38, 46, 48
— by benzoquinone, 332
— —, by exchange reaction, 342
— — esters of methane sulphonic acid, 334, 336
— — formaldehyde, 327
— — halogeno-dinitrobenzenes, 336, 337
— — mercury salts, 332
—, influence on strength, 63
—, — — supercontraction, 75, 84, 325
Crosslinking agents, carcinogenic properties of, 336
Crosslinks, 249, 275, 282, 324
—, disulphide, 243
—, influence on enzymatic degradation, 324
—, — — felting, 324
—, — — solubility, 324, 325, 326
—, — — tensile strength, 324
—, — — work to stretch, 325
—, introduction of, 324,
Crystalline phase, 238, 249
— —, penetration of dyes, 238
Crystalline regions, 66, 86, 106, 113
Crystalline structure, 83, 254, 295
Cuprammonium hydroxide, 90
—, solubilization, 362
Cupric acetate, 396
Cupriethylene diamine, 386
Cuticle, 5, 24, 343
—, effect on dyeing, 173
—, endo, 6, 7
—, exo, 6, 7
—, sulphur content of, 356
Cysteic acid, 264, 266, 272, 279, 291 343

Cysteic acid, Infra-red spectrum, 269
Cysteine sulphonate, 249, 252
Cystine, 243-285
— in medulla, 3
—, inter-chain, 246
—, intra-chain, 246
—, mode of incorporation, 245
—, state of combination of, 182
—, terminal amino groups, 313
—, variation in reactivity of, 260
Cystine peptides, 261, 271
— —, reduction of, 260, 261

DEAMIDATION, 286, 303, 348
Deamination, 83, 114, 244, 260, 272, 274, 315, 335
—, attack on tyrosine, 316
—, effect on elasticity, 214
—, influence on supercontraction, 327
Degradation, effect of rebuilding disulphide bonds, 339
—, surface, 282
Dehydration, 64, 94
Denaturation, 379, 388
Density, 15, 87, 111, 380, 387
Desorption, 95, 170
—, rate of, 141
Dialanyl cystine anhydride, 261
Diameter of fibre, 87
Dianhydromannitol, 335
Diazo-methane, 302, 308
Dichlor-acetamide, 278
1 : 3 Dichloro 4 : 6-dinitrobenzene, 334
Dichroism, 383, 384
Dielectric constant, 122
— —, effect of formic acid, 122
— —, — — frequency, 122
— —, — — methyl alcohol, 123
Dielectric dispersion, 122, 123
Diepoxides, 335
Diffusion, activation energy, 132, 145, 151, 155, 157, 168, 394
—, across a liquid layer, 152
—, — — membrane, 175
—, boundary, 137, 171
—, effect of agitation, 152
—, — — chemical treatment on rate of, 156
—, — — density on, 164
—, — — Donnan membrane on, 158
—, — — elasticity on, 145
—, — — interruption on rate of, 157, 174
—, — — ion size, 168
—, — — particle size, 151
—, — — polarizability of solute, 145
—, — — swelling on, 130, 132, 137, 143, 146, 169
—, — — variable surface concentration, 134, 161

Diffusion, from aqueous medium, 150
—, — a constant surface concentration, 157
—, gradient, 164, 165
—, influence of morphology on rate of, 173
—, — reaction, 150, 163
—, in a non-steady state, 130
—, in a steady state, 132
—, into rubber, 143
—, layer, 128
Diffusion, mass flow, 146
—, — halomethanes into polystyrene, 134, 146
—, — methyl alcohol into keratin, 144
—, — water in cellulose, 137
—, — water in nylon, 131
—, — water vapour into keratin, 131, 143
—, parabolic rate law, 135, 158, 163, 172
—, through the fibre, 155
—, with adsorption, 135, 165, 172
Diffusion coefficient, 121, 129, 134, 372
— —, effect of concentration on, 130, 136, 140
— — from rate of sorption, 140
— —, intrinsic, 147
— —, mutual, 147, 158
1 : 3 Difluoro 4 : 6 dinitrobenzene, 334
Di-glycine, 113
Diisocyanates
—, cross-linking by, 338
Diketopiperazine, 245
Dimethyl sulphate, 308
Dinitrobenzene derivatives
—, cross linking by, 336
—, 2 : 4 dinitrofluoro benzene, 313
—, 1 : 5 difluoro 2 : 4 dinitrobenzene, 337
—, 4 : 4' difluoro, 3 : 3' dinitrophenylsulphone, 337
Dinitrofluorobenzene, 313, 351
—, effect on set, 70
Dipeptides containing cystine, 355
Dipole moment, influence on sorption, 103
Direct dyes, 172, 225
Directional friction effect, 27, 33, 47, 281
— — —, influence of metallic coatings, 37
— — —, theories of, 33, 36
Disintegration, by enzymatic attack, 296
Disorientation, 83, 293
—, influence of disulphide bond on, 388
Distribution coefficient, effect on diffusion rate, 134, 165
Disulphide bond, 45, 60, 81, 243, 324, 339

Disulphide bond, differences in reactivity, 271, 273, 274, 276
— —, — — — with alkali, 254, 256, 257, 258
— —, — — — — bisulphide, 249, 252, 259
— —, — — — — formaldehyde, 251, 252
— —, — — — — permanganate, 270, 271
— —, — — — — thioglycollic acid, 246, 247, 248
— —, effect of hydrogen peroxide on, 262, 263, 264
— —, — — potassium cyanide, 258, 259
— —, — — ultraviolet light, 265
— —, heterolytic fission, 266
— —, homolytic fission, 266
— —, hydrolysis of, 252
— —, hydrolysis by water, 253
— —, influence on enzymatic digestion, 299
— —, — — extension, 382
— —, — — mechanical properties, 246, 261
— —, — — permanent set, 69
— —, — — strength, 63
— —, inter micellar, 392, 393
— —, oxidation by fluorine, 277
— —, reaction with formaldehyde, 329
— —, rebuilding of, 65, 339, 340, 341
— —, reduction of, 73, 246
— —, — by bisulphite, 250
— —, — by formaldehyde, 252
— —, — followed by crosslinking, 338
— —, — — hydrosulphite, 339
— —, — — sulphoxylate, 339
— —, relation to unshrinkable finish, 281
— —, role in supercontraction, 74
— —, stabilization of micelles by, 389
— —, subfractions of, 260
Disulphide bond breakdown, influence on crystal structure, 388
— — —, leading to disorientation, 387, 388
— interchange reaction, 314
Dithio-glycollide, 338
Djenkolic acid, 252, 261, 329
—, stability, 261
Donnan membrane, 183, 195, 206, 234
— —, effect on diffusion, 158
Drying, 94
—, rate of, 111
Dye acids, 190
Dyeing, 222
— basic, 240
—, effect of aggregation, 239

Dyeing, effect of damage on rate of, 173
—, — of morphological structure, 238
— —, on mechanical properties, 239
—, — on set, 70
Dyeing, exchange mechanism of, 234
—, distribution of dye, 237
—, influence of carboxyl groups on, 308
—, — — electrical double layer, 234
—, neutral, 239
—, polyvalent acid dyes, 238
—, rate of, 160, 170, 233
Dyeing equilibria, effect of salts on, 233
Dyes, stripping of, 340
—, surface active, 231

Egg albumin, 94, 117, 303, 352
— —, titration of, 183
Elasticity, 55, 160
—, at different humidities, 55
—, at different rates of extension, 57
—, detection of chemical changes, 58
—, effect of anions, 214
—, — — deamination, 214
—, — — disorientation, 66
—, — — mercuric acetate on, 160
—, — — pH on, 213
—, — — temperature on, 56, 66
—, — — water on, 96
— equilibrium, 67
—, gel-sol transformation, 57
—, hysteresis, 42
—, influence of chemical structure, 58
—, — on felting, 38
—, long-range, 67
—, rubber-like, 67
—, thermodynamic treatment of, 65
Elasticum, 7
Electrical conductivity, 118
— —, effect of ionic impurities, 120
— —, ionic association, 121
Electrical double layer, 159, 172, 182, 193, 200, 203, 205, 238
— — —, effect of pH on, 194
— — —, — — salts on, 194
— — —, influence on swelling, 219
Electrical properties, 23
— —, effect of swelling on, 118
Electrokinetic potential, 225
Electrophoresis, 182, 370, 371
Electrophoretic mobility, 190, 196
Electron microscope, 8, 343, 393
Emulsion polymerization, 51
Entropy of water sorption, 99, 104, 116, 118
— — — —, contribution to elastic properties, 65
Enzymatic degradation, 324, 357
Enzymatic digestion, difference between α and β forms, 298
— —, influence of disorientation, 297

Enzymatic digestion, disulphide bond, 297
Enzymes, pancreatin, 296
—, proteolytic, 296
Epichlorhydrin, 312, 336
Epicuticle, 6, 173, 357
—, chemical composition, 8, 357
—, damage of, 30
—, detection by staining, 8
—, effect on rate of diffusion, 173
—, influence on friction, 10, 30
—, — — permeability, 9
—, size of, 9
Epoxides, 308, 335
—, formation in wool of, 313
Esterification, 84, 260, 273, 306, 335
—, influence on supercontraction, 327
—, by epoxides, 312
Esters, stability of, 312
Ethyl alcohol, 311
Ethylene glycol, 311
Ethylene-imines, 333
Ethylene sulphide, 251, 342
Evaporation, rate of, 128
Extension, influence on structure, 376
— in steam, 377
—, relation to α–β transformation, 381
—, reversibility of, 57

F-actin, 5
Feather keratin, 1, 385
Felting, 25, 40, 281, 282, 324, 342
—, chemical theory, 38
—, curliness and, 40
—, effect of pH on, 42
—, — — temperature on, 41
—, mechanism of, 38
Fibre axis, 90, 108, 378
— diameter, 1
— diameter, measurement, 15
Fibre roots, influence on felting, 39
— tips, influence on felting, 39
Fibre-twist method, 27, 30
Fibres, of corpuscular protein, 393
—, deformation of, 38, 42
—, descaled, 23, 25
—, migration of, 30, 34, 41, 43
Fibrils, 4
—, micro-fibrils, 4
—, proto-fibrils, 4
—, sub-fibrils, 4, 5
—, tono-fibrils, 6
Fibrin, 5, 395
Fick's law, 130, 139
Fick's second law, 132, 143
Fluorine, 277, 282
a Fold, 379–389
Follicle, 2
Formaldehyde, 249, 252, 327–331
Formamide, 76, 360, 373

Formic acid, 91, 388, 394
Fractionation of wool, 360–372
—— — after deamidation, 361
—— — — oxidation with peracetic acid, 368
—— — — reduction of disulphide bond, 364
—— — — sulphonation, 364
—— — comparison of different products, 372
—— — with cetyl sulphonic acid, 361
—— —, relationship between different fractions, 373, 374
Free energy of absorption, 99, 109, 116
— —, contribution to elastic properties, 65
Free volume, 16, 88
Freeze-drying, influence on sorption, 113
Friction, 25 ff.
—, coefficient of, 26, 31
—, contact area, 29, 33, 37
—, effect of abrasion, 30
—, — — chemical treatment, 30, 32
—, — — enzymes on, 31
—, — — load on, 29
—, — — pH, 28
—, — — soap on, 31
—, — — surface on, 27, 37
—, — — swelling, 28
—, — — temperature, 29
—, inter-fibre, 26
—, lepidometer method, 26
—, measurement of, 25–27
—, relation to chemical reactivity, 33
—, 'slip-stick' method, 26
—, 'violin bow' method, 25, 35
Fungi, 299

GAMMA radiation, 301
Gegen-ions, 159, 189, 225, 237
Gel, 90, 101, 107, 121, 168, 195
Gel, stress free, 109
Gel-sol transformation on stretching, 57
Gelatin, 91, 112, 118, 122
—, titration of, 214
Globular proteins, 377, 379, 388
Glutamic acid, 182, 329, 348, 350, 351
Glutamylglutamic acid, 355
Glutathione, 343
Glycerol, 311
Glycidol, 312
Glycine, 348, 350, 351
Glycollic acid, 67
Glyoxal, 337
Glyoxal-bis-sodium bisulphite, 316
Grease, 2
Growth, effect of nutrition on, 2
— of fibre, 2

HAEMOGLOBIN, 189
Hair, 160
—, mammalian, 1, 160, 383, 385
—, penetration into, 161
Handle, 10, 47, 49, 342
Heat, chemical changes produced by, 300, 301
—, difference between dry and wet, 300
—, dry, 254
—, dry, effect on disulphide bond, 254
—, formation of new crosslinks by, 300
—, supercontraction by, 300
—, of vaporization, 102
—, of wetting, 98, 146, 168
Histidine, 292, 303, 348, 350
—, ionization of, 182
Hofmeister series, 216
Hooke's law, 56, 65
Hydrates, 90, 101, 103, 107, 111
Hydration, 138
Hydrogen bond, 48, 210, 251, 277, 293, 296, 326, 360, 379, 384
— — and anion affinity, 230
— — breaking reagents, 389
— —, effect on set, 70
— —, intra micellar, 390
— —, role in supercontraction, 74
Hydrogen peroxide, 262, 282, 294, 342
— —, adsorption of, 263
— —, catalysed reaction of, 265
— —, rate of reaction with, 263
Hydrolysis, acid, 246, 288, 336
— —, influence of neutral salts, 291
— —, — on disulphide bond, 291
— —, — — dry strength, 290
— —, — — wet strength, 290
Hydrolysis alkaline, 292
— —, destruction of amino acids, 292
Hydrolysis in neutral solutions, 252
—, by water at 100° C., 253
Hydrophilic groups, 97
Hydroquinone, 76
Hydrosulphite, 75, 251, 339
Hydroxy-lysine, 314
Hydroxyl side chains, 347
Hydroxylic liquids, change of mechanical properties by, 389
— —, swelling by, 389
Hysteresis, 57, 95, 111
—, swelling in salt solutions, 217

INDEX, 30 per cent, 58
Infra red, absorption spectra, 383, 384, 393
— —, evidence for structure, 385
— —, polarized, 383
Infra-red spectra of α and β keratins, 385, 386
Insulin, 291, 353
Inter-cellular cement, 5

Inter-facial tension, effect of dyes on, 171
Inter-fibrillar material, 373
Inter micellar cement, 390
— set, 73
Inter-molecular slip, 378, 389
Internal solution, 195, 204
Internal surface, 93
Internal viscosity, 62
Ion Exchange chromatography, 346
— —, amino acids in wool by, 350
Iodine, 277, 303
—, absorption of, 277
Iodoacetic acid, 363
Iodates, 281
Iodo ethanol, 363
Ion exchange, equilibria, 231
— —, rate of, 157
Ionophoresis, 356
di-Isocyanates, 46, 325
Ionizing radiations, 301
Isocyanates, reaction with wool, 320
Isoelectric point, 180, 182, 190, 194, 196, 225
— —, of cotton, 225
— range, 42
— region, 182, 190, 196, 213
Iso-ionic point, 184, 191, 224

JUTE, 22

KEMP wool, 3
Kerateine, 363, 364
—, benzyl, 364
—, carbamidomethyl, 364
—, carboxymethyl, 364
—, p-carboxyphenylmercuri, 364
—, mercuri, 364
—, methyl, 364
α Keratin, 238, 243, 367, 376
— —, Astbury fold, 380, 382
— —, configuration of, 66, 71, 81, 355, 379
— —, helical structure, 386, 387
— —, infra-red frequencies of, 386
— —, seven membered ring fold, 383
— —, seven strand cable structure, 387
— — stability of, 387
β Keratin, 71, 261, 379
— —, disorientated, 79, 82, 290, 387, 389
— —, infra red frequencies of, 386
— —, structure of, 378
meta Keratin, 363
Keratinization, 3
α Keratose, 303, 304, 369, 372, 386, 388, 391
— —, conversion to β form, 388, 389
— —, solubility of, 388
β Keratose, 369, 372, 391, 392

β Keratose, disorientated, 389
— —, shape of molecule, 389
— —, solubility of, 388
γ Keratose, 362, 370, 386, 392
Ketene, 302

LACTOGLOBULIN, hydration of, 113
Langmuir absorption isotherm, 166
Lanthanum acetate, 396
Lanthionine, 75, 84, 210, 243, 254, 256, 276, 282, 294, 297
—, mechanism of formation, 255, 257, 258, 259
—, effect of acetone and dioxan on formation, 257
Larvae, 299
Lead acetate, 396
Lepidometer, 26, 38
Leucine, 348, 350
Levelling, acid dyes, 233, 237
—, rate of, 170
Lipoids, 346
Lithium bromide, 48, 59, 71, 76, 79, 84, 386
— —, mechanism of hydrogen bond breakdown, 79
Localized sites, absorption on, 238
Lysine, 112, 278, 303, 313, 328, 348, 350
—, titration of, 203, 212
—, methylol, 328

MACROMOLECULAR· structure, 361, 367, 377
— —, different phases, 380, 381, 389, 391, 392
— —, evidence from supercontraction, 82
Macropores, 343
Manganese dioxide deposits, 249, 317
Mannich reaction, 328
Maximum acid combining capacity, 181, 185
— alkali combining capacity, 211
Mechanical properties, 91, 124, 208
— —, application of statistical mechanics, 55
— —, influence of alkalis on, 295
— —, — — acid hydrolysis, 288
— —, — — disulphide bond on, 249
— —, — — hydrogen peroxide on, 263
— —, morphology and, 55
— —, relation to dielectric dispersion, 124
Mechano-chemical method, 58
Medulla, 3
—, composition of, 360
—, enzymatic degradation of, 360
Melamine-formaldehyde resins, 48

Membranes, diffusion through, 129, 144
Membrane potentials, 200, 205
Mercuric acetate, 38, 40, 46, 160, 396
Mercury, 253
Mercury salts, 75
Meridian spacing, 377
Methacrylic acid, 342
Methacrylamide, 343
Methane sulphonic acid, esters of, 336
Methanol, 311
Methionine, 243, 266, 348, 350
Methyl bromide, 308
Methyl iodide, 308
Methyl methacrylate, 342
Methyl sulphate, 308
Methylation, 114
Methylene blue, 8
Methylolmelamine, 48
Methylolurea, 50
Methylurea, S-bisalkoxy, 50
Micellar structure, 389, 390
Micellar subdivision, 215
Micelle, 82, 90, 92, 159, 176
Micelle, dimensions of, 92
—, effect on dyeing, 176
—, physical properties of, 393
—, shape of, 389, 393
—, structure of, 239
—, subdivision of, by dyes, 238
Micropores, 389
Microsporum gypseum, 299
Milling, 38, 41, 47, 48
Milling dye, 174, 235
— —, affinity of, 235
Millon's reagent, 292
Mineral matter, 346
Mobility, 170
Mohair, 34
Molecular rearrangement, 60
Molecular weight, 106, 251, 362, 372
— —, average, 352, 367
Mono-ethanolamine bisulphite, 73
Moths, 339
—, mode of attack, 299
—, proofing, 299, 339
—, reduction of disulphide bond by, 299

NEUTRAL dyeing, 239
Neutralization, heat of, 181
Ninhydrin, 338
Nitric acid, 302, 367
Nitrous acid, 315, 316
— —, reaction with tyrosine, 316
Nitrogen, amide, 291, 347, 348, 350
Non-axial-reflections, 377
Nutrition, effect on growth, 2
Nylon, 48, 106, 113, 117
—, absorption of weak acids, 210

Nylon, cold drawing of, 191
—, titration of, 191

OPTICAL interference in swollen polymers, 138
Organosilicon polymers, 46
Osmium tetroxide, 396
Oxidation with H_2O_2, 262
— — peracetic acid, 266
— — potassium permanganate, 270
—, photochemical, 265
—, with chlorine, 164, 272
— of reduced wool, 248
—, residual oxidizing power of wool, 275
Oxidized Wool
—, infra-red spectrum of, 269
Ozone, 265

PADDING, 50
Pancreatin, 294
Papain, 39, 298, 299
Partial hydrolysis, 288
Pauling-Corey helix, 379, 386, 387
Pentane-diol, 311
Pepsin, 296
Peptides, 113
Peptide bond, alkylation of, 309
— —, attack by hydrogen peroxide, 263, 264
— —, influence on strength, 63
Peptide chains, degradation of, 251
Peptide link, 98
Per-acids, 266
Peracetic acid, 75, 81, 266, 274, 282, 326, 368, 390
— —, oxidation of disulphide bond, 267, 268
— —, rate of reaction, 266
— —, specificity of, 266
Perbenzoic acid, 270
Perchromic acid, 270
Performic acid, 266, 267
Permanganate, 270, 271, 282, 284, 317
—, degradation by, 270
—, oxidation of disulphide bond, 270
Permanent set, 68, 72, 253, 261, 377
— —, chemical mechanism, 69
— —, disruption of side chains, 377
— —, effect of deamination, 69
— —, influence of acidity, 70
— —, — — disulphide bonds, 70
— —, — — dyes, 70
— —, — — hydrogen bonds, 70
— —, — — pH on, 71
— —, — — X-ray pattern, 377
— —, production at low temperatures, 73
— —, reversing of, 71
Permanent waving, 73, 248

SUBJECT INDEX

Permeability, 16
Permono-sulphuric acid, see Caros Acid
Perphosphomolybdic acid, 269
Perphosphotungstic acid, 269
Persulphate, 73
Pervanadic acid, 270
Phenol, 76, 84, 326, 361, 389
—, absorption of, 210
—, influence on strength, 64
—, swelling by, 64
Phenylalanine, 318, 348, 350
Phenylenedimaleimides, 342
Phosphoric acid, 319
Phosphorus pentoxide, 319
Photochemical oxidation, 265
Piezoelectricity, 24
Plastic flow, 59, 67
— —, rearrangement of hydrogen bonds, 59
— —, role of amorphous regions, 59
Poisson ratio, 109
Polar groups, accessibility, 394
Polar Yellow, 171
Polyamides, cross-linked with sulphur, 256, 259
Polyglycine, 52, 117
Polyglycine dl. alanine, 113
Polymer, chemical combination with, 343
—, deposition of, 46, 325
—, ethylene sulphide, 46
—, mol. wt. of, 48
—, thermal setting, 48
Polymer deposition, changes in chemical properties, 342
— —, effect of plasticizer, 51
Polymer formation, 251
— — interfacial, 48
— — within the fibre, 342
Polymethyl methacrylate, 33, 51
Polypeptides, 113, 328, 367, 384, 383
—, synthetic, 113, 383, 386
Polypeptide chains, 82, 113, 122, 123
— —, rotation of, 124
Polystyrene, 33, 51
Polythene, 33, 131
Polyvinyl alcohol, 102
Porcupine quill, 34, 354, 377, 380, 385
Pores, 17, 22, 86, 88, 90, 92, 95, 118, 128
—, condensation in, 19
—, diameter, 389
—, size, 168
Porosity, 16
Potassium cyanide, 258
Potassium nitroso disulphonate, 304
Proline, 112, 348, 350, 356
Propanol, 311, 384
Propiolactone, 320
Propylene oxide, 312

Protein film, 101, 190
Proteins, pK of, 192, 195
Pyro-electricity, 24

QUATERNARY ammonium complex, 278
Quality, 1, 15
Quinones, 46, 303

RADIUS, mean hydraulic, 16
Ratchet theory of friction, 33
Recovery from deformation, influence of humidity, 60
— — —, — — temperature, 60
Reduced wool, formation of, 246
— —, new crosslinks in, 248
— —, reoxidation of, 248
— —, solubility of, 246
Reducing agents, effect on disulphide bonds, 246
Refractive index, 122
Regain, 87, 89, 94, 114, 118, 128, 132
Regenerated keratin fibres, 362
— protein, 324
— —, fibres, 367
Relaxation, effect on elasticity, 58
Resorcinol, 361
Rigidity, 96, 98, 107
Ring dyeing, 171, 173
Rubber, 326
—, sorption of solvents by, 104

SALT links, 180, 183, 210, 213, 325
—, between sulphonic acid and amino groups, 269
— —, hydrogen bond structure of, 213
Salt solutions, effect on elastic properties, 64
Saponification, 310, 311
Scale structure, 22
— —, effect on friction and shrinkage, 34
Scales, 5
—, chemical reactivity of, 6
—, composition of, 357
—, different layers of, 6
—, dyeing of, 173
—, influence on felting, 43, 44
—, isolation of, 5
—, masking of by polymers, 46, 48
—, piezoelectricity, 24, 36
—, profile of, 36
—, removal of, 43
—, triboelectricity of, 36
Scaliness, 25
Secondary forces in crosslinking, 65
Scouring, 254
Sedimentation constant, 371
Sedimentation of soluble wool products, 370
Serine, 112, 288, 318, 348, 350

Set, 68–73
—, relaxation of, 68
—, chemical production of, 248
Shearing stresses, 107
Shrinking (see felting)
Side chains, amide, 347
— —, acidic, 307, 347
— —, basic, 313, 347
— —, phenolic, 297, 302, 347
— —, polar, 87, 112, 259, 307, 347, 352
— —, polar and non-polar, alternation of, 353, 358
Silicones, 47
Silicon tetrachloride, 47
Silk, 3, 19, 70, 72, 79, 101, 102, 106, 114, 117, 122, 257, 264, 303, 306, 367, 379
—, absorption of weak acids, 210
—, acid combination of, 205
—, titration of, 182
Silver acetate, 396
Silver nitrate, 396
Skin, 1
Sky Blue FF, 171
Sodium bisulphite, 75, 250, 326, 339
Sodium chlorate, 281
Sodium chlorite, 281
Sodium cyanide, 75, 364
Sodium hydrosulphite, 294
Sodium hypochlorite, 305
Sodium sulphide, 75, 251, 254, 362
Sodium sulphite, 339
Sodium sulphoxylate, 251
Sodium thioglycollate, 362, 363, 388
Softening point, 77, 79
— —, influence of disulphide bonds, 79
Soluble proteins, acid titration of, 188
— —, dissociation constant of, 193
— —, effect of salts on titration, 194
— —, electrophoretic mobility, 190
— —, pK of, 195
— —, titration of, 182, 190
Solubility, 325, 326
—, difference between α and β keratin, 384, 385
Solubilization, 360–373
— by bisulphite-urea, 366, 367
— — calcium thiocyanate, 360
— — chlorine, 275
— — chlorine peroxide, 280, 367
— — chloramides, 367
— — chlorsulphamic acid, 278
— — chlorurea, 278
— — cyanide, 363
— — formamide, 360
— — hydrogen peroxide, 264
— — lithium thiocyanate, 360
— — phenol, 360
— — reduction of the disulphide bond, 361

Solubilization by resorcinol, 360
— — sodium sulphide, 250
— — oxidation of the disulphide bond, 367–373
— — — with peracetic acid, 367–373
Solway Blue, S.E.N., 177
Sorption, differential heat of, 114
—, entropy of, 104
—, free energy of, 104
—, heat of, 110, 118
—, isotherms, 93, 114
— — for acetone, 92
— — — alcohols, 87, 91, 92
— — — formic acid, 91, 111
— — — pyridine, 91
— — — water, 86
—, multimolecular, 98
—, rate of, 128
—, solution theory of, 102, 103
Specific conductivity, 118
Specific volume, 89, 106, 147
"Spot welding", 47, 49, 51
Statistical mechanics, application to elasticity, 55
Stereochemistry, 376
Stereoisomers, mechanical, 376
Sterols, 346, 360
Strength dry, breaking, 57, 92
— —, effect on acid hydrolysis, 290
— —, — of disulphide bond hydrolysis on, 254
— of fibre, influence of different bonds on, 62
— — —, — — disulphide bonds, 62
— — —, — — hydrogen bonds, 63
— — —, — — new crosslinks, 63
— — —, — — peptide bond breakdown, 63
— wet, 249
— —, effect on acid hydrolysis, 290
— —, — of bisulphite on, 251
— —, — of reducing agents, 249
'Stress-free' isotherms, 93, 109
Stretching, influence on reactivity, 261
— kinetics of, 61
Stripping, 294
Sulphenic acid, 253, 255, 265
Sulphinic acid, 270
Sulphonation, 318
—, influence on solubility of, 364
Sulphone, 270, 343
Sulphoxide, 275, 343
Sulphoxylate, 341
Sulphur
—, content of wool, 243, 244
—, methods for analysis, 243, 244
—, non-disulphide, 243
Sulphur content, distribution of, 243
Sulphuric acid, 262, 318, 364
Sulphuryl chloride, 282

SUBJECT INDEX

Sulphydryl groups, 338
— —, crosslinking of by alkyl dihalides, 339
Supercontraction, 73, 254, 261, 275, 279, 288, 325, 311
— by disulphide bond breakdown, 74
— — hydrogen bond breakdown, 74
— — — steaming, 73
—, disoriented β structure, 388
—, effect of new crosslinks, 75
—, evidence of two phase structure, 82
—, influence of side chains on, 83
—, mechanism of, 81
—, relation to swelling, 84
—, reversible, 79
—, α structure, 388
Supra-stoichiometry of acid absorption, 185, 224
Surface area, 15, 86, 92
— —, geometrical, 21
— —, specific, 16, 21
Surface activity of dyes, 171
Surface degradation, 282
Surface internal, 22
Surface potential, 191, 200
Swelling, 20, 33, 44, 86, 107, 130, 168, 180, 208, 217, 389
— anisotropy, 83, 87, 90, 92, 290
— crystals, 92
—, differential, 108
—, Donnan theory of, 217
Swelling, effect of acid on, 197, 204
—, — — acid on pretreated fibres, 218
—, — — alkali on, 214
—, — — pH on, 215
—, — — salts on, 216
—, free energy of, 109
—, heat of, 109
—, influence of salts on, 205
—, — on supercontraction, 84
—, ion hydration and, 217
—, osmotic pressure and, 205, 217
— by phenol, 84
— pressure, 108
— temperature, minimum, 41
Swelling in acid, effect of disulphide bond breakdown, 218
Swelling and elasticity, 216

TEFLON, 33
Temporary set, 68, 72
Tensile strength, 46, 51, 91
Tensile strength, wet, 324
Terminal amino groups, 351
Terminal carboxyl groups, 352
Thiazolidine 4-carboxylic acid
—, stability, 216
Thioglycollic acid, 252, 338, 357, 363
— —, influence of pH on reaction, 247, 248

Thioglycollic acid, influence of, temperature on reaction, 247, 249
— —, rate of reaction with, 247, 248
Thiol Groups
—, content of wool, 317
—, reaction with N-ethylmaleimide, 317
—, organomercuri chlorides, 317
Thiols, 249
Thiophenol, 343
Thiosulphate, 250
Thiourea, 343
Thiazolidine–4 carboxylic acid, 252, 329
Three-phase composition, 389, 390
Threonine, 348, 350, 351
Tetraglycine, 113
Titration of alkali, mechanism of, 189
Titration curves, 167, 180, 314, 328
— —, data, 335
— —, effect of chemical treatment on, 185
— —, — — salts on, 184
— —, — — temperature on, 181, 186
— —, form of, 183
— —, heat of, 181
— —, influence of salts on, 196
— —, — — swelling, 185
$\alpha \rightarrow \beta$ Transformation, 62, 376, 377, 381
— —, influence on elasticity, 67
— —, in setting, 72
Triboelectricity, 23
Tryptophan, 266, 288, 292, 348, 350
Two-phase model, 82, 84
Tyrosine, 46, 250, 279, 292, 329, 348
—, accessibility of, 302
—, acetylation, 302
—, apparent pK of, 302
—, attack by nitrous acid, 316
—, colouration of wool by, 304
—, crosslinking by, 359
—, dinitrobenzene ether, 302
—, iodination of, 303, 313
—, ionization of, 211
—, nitration of, 302
—, oxidation of, 303
— peptides, 305
—, reaction with formaldehyde, 329
—, — — mineral acids, 316, 317
—, reactivity of, 302–307
—, relation to production of unshrinkable finish, 283, 284
—, resistance to peracetic acid, 305
—, titration of, 188
—, variation in reactivity, 305

UREA, 72, 337, 366
Urea-formaldehyde resins, 48, 50
Ultrasonics, disintegration by, 5, 11
Ultra-violet light, effect of, 265, 293
Unshrinkable finish, 249, 251, 257, 265, 270, 271, 278, 282, 283

Unshrinkable finish, by enzymatic treatment, 299
— —, effect of new crosslinks, 282
— —. — — pH of chlorine solution, 282
— —, importance of disulphide bond breakdown, 282
— — and reaction of tyrosine, 283, 284
—, with peracetic acid and hypochlorite, 284
Uranyl acetate, 396

VALINE, 348, 350
Van der Waals' forces, 231, 240, 333
Variable diffusion coefficient, measurement of, 135
Vinyl chloride, 51
Vinyl polymers, 46, 51, 93, 113
Viscose rayon, 362
Viscosity, internal, 67

WATER, effect on disulphide bond, 254
—, — — wool at 100° C, 254
Water repellency, 86
Water, sorption of, 84, 128
Water content, effect on elasticity, 56
Weak acids, anion affinity of, 208
— —, distribution coefficient of, 209
— —, effect on mechanical properties, 208
— —, — — X-ray pattern, 210

Weak acids, sorption of, 208
Weakening of fibre, by extension, 60
Weathering, atmospheric, 243, 266, 347
Wetting, heat of, 20, 110, 114, 117
Wood, sorption of water by, 96
—, swelling of, 107
Wool gelatin, 289
—, amino acid content, 289

X-RAY analysis, 90, 107
X-ray analysis of dyes, 177
— — of dyed fibres, 239
— — data, 354
— — diagrams, 370
— — diffraction, 377
— — diffraction patterns, small angle, 394
— — photograph, influence of formic acid, 394
X-ray photograph, influence of lithium bromide on, 394
— —, — — swelling on, 394

YARN, 40, 118
—, shrinkage of, 41
Young's modulus, 109

ZEIN, 114
Zinc chloride, effect on supercontraction, 290
Zwitterions, 180

AUTHOR INDEX

Abitz, 393
Adams, 180
Adamson, 168, 232, 237
Agar, 152
Allwörden, 7, 8
Ambrose, 383, 384, 386
Anderson, 15, 17
Andrews, 265
Anson, 303
Ashpole, 125
Asquith, 69, 70, 255
Astbury, 24, 56, 57, 59, 66, 67, 68, 70, 72, 73, 74, 90, 106, 110, 176, 182, 238, 239, 244, 295, 346, 351, 353, 354, 356, 376, 378, 379, 381, 383, 387, 388, 389, 391, 392.
Atkin, 214
Auber, 5
Aulabaugh, 49

Badenhuizen, 111
Bagnall, 145, 146, 155
Bailey, 48, 243
Baker, 103, 104, 114, 123
Bakhmeteff, 15
Baldwin, 47
Bamford, 384, 386, 387
Barkas, 95, 96, 107, 111
Barker, 6, 15
Barr, 46, 251, 325, 332, 342
Barrer, 104, 105, 133, 134, 135, 144, 156, 238
Bath, 113
Baudisch, 262
Baxter, 119, 122, 128, 200
Bear, 395
Beaven, 303
Beek, 224
Beer, 385, 386
Bekker, 3
Bell, 50, 376, 379, 381
Bendit, 377
Berg, 38, 40
Bergmann, 255, 261, 353, 356, 362
Bestian, 333
Bidder, 182
Bikales, 319
Bjerrum, 180
Black, 319
Blackburn, 3, 12, 68, 69, 71, 73, 83, 113, 251, 261, 267, 277, 278, 299, 303, 308, 309, 313, 327, 352, 373
Blaine, 20, 21, 22, 86, 92, 93, 102

Block, 296
Blow, 17
Bogaty, 217
Bohm, 27, 38
Boltzmann, 136, 138, 139
Booth, 163
Bowden, 26, 33
Boxser, 42
Boyd, 168, 237
Bradley, 102, 103
Braundsdorf, 280, 284
Bravermann, 152
Breine, 262
Briggs, 112
Brown, 61, 62, 75, 249, 329, 330, 339, 340, 341, 346, 384, 387
Brunauer, 18, 19, 21, 93, 94, 98, 111, 146
Brunner, 152
Buckingham, 103, 104
Bull, 19, 57, 65, 94, 95, 101, 104, 111, 115, 116, 288, 382
Burley, 317
Burgess, 6, 296
Burkhard, 367
Burton, 123
Butler, 145, 146, 155

Cady, 129
Cannan, 191, 193, 194, 292
Capp, 46, 325, 335
Carman, 16, 17
Carter, 45, 46, 47, 48, 50, 60, 70, 74, 150, 156, 249, 262, 266, 267, 270, 278, 279, 282, 283, 284, 303, 304, 306, 307, 308, 309, 310, 311, 312, 313, 325, 329, 330, 367, 369, 387
Cassie, 17, 96, 97, 98, 100, 105, 107, 108, 110, 116, 118, 128, 145
Chambard, 296
Chamberlain, 26, 28, 356
Chang, 27, 38, 41, 42, 44, 116
Charlesworth, 282
Charman, 171, 230
Chen, 111
Chibnall, 314
Chiltern, 152
Clegg, 176, 229
Cochran, 387
Cockburn, 255, 316, 338
Cohn, 181, 186, 211, 212
Coke, 332
Colburn, 152

Comte, 296
Consden, 264, 266, 272, 277, 355, 356
Cook, 95, 102, 113
Cooper, 111
Corey, 386, 387
Corfield, 244, 292, 347, 350, 358, 373
Cramer, 212, 303
Crank, 130, 134, 135, 136, 137, 138, 139, 140, 141, 142, 144, 146, 147, 148, 149, 166, 171, 172, 174
Crawshaw, 266
Crick, 387
Crisp, 238
Crowell, 152
Crowder, 252
Cullum, 314
Curtis, 123
Cuthbertson, 243, 255, 256, 258

DALTON, 187
Dankwerts, 163
Das, 74, 280, 303, 304, 367
Davies, 103, 104
Davis, 116, 118, 288, 319
Davidson, 284
Dawson, 70, 176, 238, 239
Daynes, 142
Dean, 314
De Boer, 102
Denis, 303
Desnuelle, 302
Deuel, 335
Diehl, 285
Dole, 94, 101, 114, 116, 118
Donnan, 203, 206
Douglas, 214
Drake, 303
Drucker, 255, 316
Duncan, 232
Duspiva, 299

EARP, 118
Edsall, 211, 212
Eilers, 114
Ellis, 113
Elliott, 69, 70, 214, 215, 238, 270, 271, 282, 284, 315, 316, 383, 384, 386, 387
Elöd, 3, 6, 11, 66, 74, 75, 77, 82, 84, 205, 208, 210, 233, 234, 245, 246, 249, 251, 253, 264, 289, 290, 292, 293, 295, 296, 297, 298, 299, 300, 326, 332, 333, 358, 359, 360, 391, 392
Elsworth, 249
Emersleben, 17
Emmett, 18, 19, 21, 98, 111, 146
Erdas, 190, 191
Erdtman, 333
Errera, 122
Euken, 152
Eyres, 140

Eyring, 59, 108, 213

FAGE, 152
FALLER, 114, 118
FARNWORTH, 69, 73, 282, 283, 320
Farrant, 5, 393
Fearnley, 321, 335
Feodroff, 15
Ferrel, 319
Feughelman, 389
Fiddel, 170, 172
Fidler, 314
Fikentsher, 208
Finsh, 122, 145
Fisher, 111, 128
Flory, 94, 104
Ford, 308, 309, 310, 311, 312, 313
Fourt, 57, 248, 249
Fowler, 17, 18, 200
Fox, 61, 62, 122, 145, 261, 266, 267, 268, 270, 271, 272, 273, 274, 275, 276, 282, 305, 306, 312, 316, 326, 327, 333, 334, 335, 368
Fraenkel-Conrat, 309, 312, 319, 327, 328
Fraser, 5, 395, 396
Freney, 30, 38, 257
Freundlich, 239
Fricke, 123
Frishman, 26, 27, 157
Fritze, 302
Frohlich, 208, 210
Fromageot, 296, 352
Fugitt, 166, 170, 181, 182, 184, 185, 187, 188, 189, 197, 208, 209, 213, 222, 223, 224, 226, 227, 229, 231, 232, 233, 291, 308, 328, 361
Fuller, 114
Fuoss, 121, 122

GAGLIARDI, 86, 97
Garvie, 171
Gee, 105
Geiger, 65, 244, 247, 248, 249, 297, 299, 339, 340, 357
Gerngross, 394
Ghosh, 331
Gilbert, 20, 98, 99, 200, 202, 203, 229, 230, 235
Gillespie, 364
Glueckauf, 232
Glynn, 243
Goddard, 246, 362, 365
Godson, 174
Golden, 5
Gonon, 61, 67
Good, 103, 104
Goodall, 224, 238
Goodings, 31, 116
Gordon, 129, 264, 266, 272, 277, 291, 355, 356

Gorter, 101
Gortner, 112
Gough, 10, 150, 152, 153, 154, 155, 156, 163, 164, 174, 175, 272, 275, 283, 284, 302, 304, 305, 306
Gouy, 22
Graham, 350
Gralén, 7, 8, 9, 10, 12, 26, 27, 28, 29, 30, 31, 32, 362
Green, 94, 112
Gregor, 232
Groves, 113
Guggenheim, 200, 203, 206
Gurney, 163
Gutmann, 57, 382
Gutman, 152

HAILWOOD, 104
Hale, 168, 169, 231
Haller, 277
Halsey, 59, 95, 102, 113, 239
Hanby, 384, 386, 387
Hanford, 33
Hamer, 187
Hamm, 258
Happey, 362, 384, 386
Harkins, 19
Harned, 181, 186, 187
Harnish, 255
Harris, 3, 4, 6, 9, 26, 27, 40, 57, 58, 61, 62, 65, 75, 157, 166, 170, 181, 182, 183, 184, 185, 186, 187, 188, 189, 191, 192, 195, 197, 199, 205, 208, 209, 210, 213, 215, 216, 217, 222, 223, 224, 225, 226, 227, 229, 231, 232, 233, 244, 247, 248, 249, 252, 254, 262, 263, 264, 265, 292, 293, 294, 297, 299, 300, 302, 308, 316, 317, 326, 328, 329, 330, 339, 340, 341, 346
Harrison, 11, 55, 290
Hartley, 130, 137, 138, 146, 147, 172
Hartmann, 337
Hartree, 140
Haselbach, 61, 67
Haselmann, 6, 7, 12, 13
Hauser, 93
Hausman, 360
Hawn, 5
Hedges, 86, 96, 106, 114, 115, 116, 128, 168
Henry, 136, 137, 140, 141
Hermans, 87, 88, 89, 95, 103, 106, 107, 111, 137, 138, 163, 168, 389
Herriot, 302
Herrmann, 394
Hertel, 15, 17, 18
Herzog, 361, 378
Hier, 350
High, 301

Hildebrand, 257, 259
Hill, 98, 99, 163
Hille, 319
Hiller, 314
Hind, 71
Hipp, 113
Hirst, 42, 180, 182, 205, 213, 216, 222
Hixson, 152
Hobday, 238
Hober, 112
Hock, 3, 4, 6, 9
Hofmeister, 216
Holiday, 303
Holker, 331
Homiller, 266
Hoover, 94, 95, 96, 102, 106, 113, 114, 121
Hopkins, 140
Horden, 317
Horio, 5
Horn, 254
Hornstein, 330
Horrabin, 104
Hove, 277
Howard, 26, 35, 343
Huggins, 104, 383, 384
Human, 6
Humoller, 291

INGHAM, 140
Ishida, 33
Isings, 9

JABLCZYNSKI, 152
Jackson, 48, 140, 268
Jacobsen, 219
Jagger, 60
Jancke, 378
Jeffries, 5
Johnson, 13, 281, 325, 329, 330
Jones, 112, 254, 320, 366
Jope, 303
Jura, 19
Justin-Mueller, 33, 38

KAMMERER, 270
Kanagy, 316
Karger, 55
Katchman, 95, 113
Katz, 59, 94, 103, 107
Keighley, 386
Kersten, 315
Kessler, 342
Kilbrick, 191, 194
King, 3, 15, 25, 26, 27, 28, 29, 86, 87, 88, 89, 90, 91, 106, 119, 120, 121, 122, 123, 124, 128, 130, 131, 140, 142, 143, 144, 145, 149, 152, 166, 168, 170, 389
Kirk, 187

Kirst, 333, 341
Kitchen, 282
Kitchener, 151, 159, 166, 168, 194, 195, 196, 197, 208, 224, 229, 232, 235, 239
Knight, 268, 269, 289, 358
Knudsen, 145
Kobayashi, 249, 299, 339, 340
Kohler, 302
Kondo, 5
Kondritzer, 318, 364, 366
Köpke, 173
Korn, 94, 95, 96, 106, 113, 114, 121
Kozeny, 17
Krahn, 361
Kratky, 295
Kraus, 122
Kressman, 151, 168, 229, 232
Kunin, 168
Kunitz, 257, 259

LABOUT, 114
Lagermalm, 9, 12
Landolt, 49
Langheld, 303
Langmuir, 19
Lavine, 270, 275, 281
Leach, 288, 317
Leden, 26
Lee, 261, 352
Lehmann, 11, 33, 38
Lemin, 173, 185, 191, 229, 235, 236, 319
Lennard-Jones, 145
Lennox, 300, 364
Leveau, 6
Levich, 152
Lewis, 244
Lindberg, 7, 8, 11, 26, 28, 29, 30, 31, 32, 165, 173
Linderstrom-Lang, 219, 299
Lindley, 5, 10, 60, 68, 69, 70, 71, 73, 83, 255, 309, 315, 316, 327, 361, 373
Lindquist, 301
Lipson, 26, 35, 48, 257, 320, 341, 342, 343
Lister, 232
Liu, 41, 42
Lloyd, 103, 112, 182, 360
Loeb, 205
Lowther, 267
Lundgren, 301, 302, 320, 342
Lustig, 318, 364, 366

MACARTHUR, 354, 377, 379, 395
Macauley, 232
MacFayden, 314
Maclaren, 317
MacRae, 270, 392, 395, 396
Makinson, 26, 27, 29, 37

Magne, 113
Mandels, 299
Manogue, 13
Maresh, 49, 176, 239
Mariner, 369, 387
Mark, 378
Marsh, 118
Marshall, 49, 240, 352
Marston, 3
Martin, 23, 24, 25, 26, 28, 30, 33, 34, 38, 39, 40, 238, 266, 272, 277, 291, 355, 356
Mauthner, 258
Mayer, 99
Mazinque, 281, 301
McAllan, 314
McBain, 103, 104
McCleary, 49, 301
McLaren, 86, 93, 94, 95, 111, 113, 116, 117, 118
McLelland, 268
McMeekin, 113
McMurdy, 6
McPhee, 284, 285, 320, 341
Mease, 317
McQue, 299
Medley, 23, 29, 119, 120, 121, 122
Mecham, 300, 366
Meek, 124
Meienhofer, 289, 337
Mellon, 94, 95, 96, 102, 106, 113, 114, 121
Menkart, 26, 28, 34, 38, 39, 41, 42, 160, 332
Mercer, 3, 4, 5, 6, 7, 11, 13, 26, 27, 28, 30, 31, 34, 36, 37, 41, 42, 44, 59, 343, 344, 366, 393
Merriot, 360
Meunier, 296
Meyer, 61, 67, 111, 208, 222, 352, 378, 394
Michaelis, 246, 362, 365
Middlebrook, 60, 70, 251, 252, 298, 308, 313, 315, 329, 351
Miller, 291
Millson, 6, 7, 8, 10, 36, 170, 172, 173, 176
Mirsky, 303
Misch, 111
Mittelmann, 26, 30
Mizell, 57, 65, 244, 247, 248, 249, 297, 319, 339
Monge, 25
Moore, 320, 338, 342, 346, 364, 366
Moran, 103, 112
Morrow, 26
Morton, 165, 170, 205
Moss, 13
Mott, 163
Moxton, 17

Müller, 7, 8, 9
Murphy, 118
Myers, 168, 244

NEALE, 165, 171, 172, 234
Neish, 47, 282, 340
Nernst, 152
Neuberger, 212, 303, 352
Neurath, 101
Nicolet, 255
Nieman, 353, 356
Nilssen, 173, 270, 271, 282, 284
Northrup, 302
Nowotny, 74, 246, 253, 264, 289, 290, 292, 293, 295, 300, 392
Noyes, 152

O'CALLAGHAN, 352
O'Connell, 301, 338
O'Donnell, 269
Olcott, 300, 309, 312, 319, 328
Oliva, 319
Olofsson, 27, 362
Onions, 396
Orr, 145, 146, 155
O'Sullivan, 120
Osterloh, 257
Ottesen, 94, 113
Overbeeck, 193
Owen, 186
Overbèke, 281, 301

PALMER, 15, 191, 194
PARK, 134, 141, 142, 144, 145, 148
Parker, 123
Partington, 145, 146, 155
Partridge, 288
Patel, 171
Paterson, 135
Patterson, 65, 135, 244, 247, 248, 249, 292, 297, 302, 308, 339
Pauling, 112, 113, 386, 387
Peacock, 392
Peill, 328, 332
Peirce, 96, 97, 105
Pendergrass, 75, 340,
Penasse, 352
Pepper, 232
Perry, 5
Perutz, 113, 386
Peters, 167, 205, 206, 207, 240, 291
Pfeiffer, 230
Philip, 7, 8, 11, 12
Phillips, 44, 60, 70, 103, 112, 113, 243, 249, 251, 252, 254, 255, 256, 258, 259, 260, 262, 277, 278, 291, 303, 308, 315
Poiseuille, 98
Porai-Koschitz, 238
Porchard, 296

Portas, 113
Porter, 5, 352
Preston, 111, 284
Proctor, 204, 208, 216

RAMSAY, 3, 4, 6, 9
Rao, 95
Raoult, 94, 101
Rau, 304
Rapoport, 319
Raven, 256, 259, 274, 280
Raynes, 283
Reed, 5
Rees, 4, 6, 6, 7, 36, 393
Reichenberg, 168, 231, 232
Reitz, 319
Reumuth, 6, 11
Richards, 303, 330, 331, 396
Rideal, 200, 202, 203
Rigelhaupt, 28
Ringel, 254
Ripa, 60, 61
Robinson, 137, 138, 139, 144, 148, 149
Robson, 244, 292, 347, 350, 358, 373
Rogers, 5, 395, 396
Rose, 320
Rouette, 300
Rouse, 131
Rovery, 302
Rowen, 20, 21, 22, 86, 92, 93, 97, 101, 102, 103, 104, 105, 111, 117
Royce, 33
Royer, 49, 170, 172, 173, 176, 239, 301
Rudall, 33, 34, 35, 70
Rust, 51
Rutherford, 210, 264, 292, 300, 302, 308, 316, 317

SACK, 122
Saidel, 255
Salley, 47, 49
Sanford, 291
Sanger, 266, 291, 313, 314
Sansiva, 95
Sarjant, 140
Satlow, 342
Schlack, 308, 312, 335
Schmeidler, 156, 157, 158, 167
Schmid, 55
Schmidt, 4, 6, 187, 280, 284, 300
Schoberl, 252, 254, 257, 258, 261, 298, 338, 357
Schofield, 22
Schrim, 230
Schubert, 237
Schuckmann, 296
Schuringa, 9
Sekora, 295
Selim, 28
Shack, 152

Shah, 59
Shaw, 23, 122
Shinn, 255
Shooter, 33
Shorter, 39, 55, 57, 114, 128
Sikorski, 57, 392
Silva, 205
Simha, 101, 103, 104, 105
Simmonds, 243, 347, 349, 350
Simms, 198
Simpson, 396
Sisson, 376, 378
Siu, 299
Skertchly, 376, 377
Skinner, 224, 358, 373
Smith, 26, 27, 96, 110, 159, 161, 172, 176, 225, 262, 263, 264, 265, 292, 293, 294, 312, 316, 326, 327, 333, 334, 335, 336, 360, 373
Smyth, 303
Snellman, 190, 191
Sookne, 58, 182, 183, 191, 195, 216, 217, 225
Speakman, 6, 10, 25, 26, 27, 28, 30, 31, 34, 37, 38, 39, 41, 42, 44, 45, 46, 47, 55, 56, 58, 59, 60, 61, 69, 70, 71, 74, 75, 86, 90, 91, 92, 94, 95, 96, 97, 111, 113, 116, 128, 159, 160, 161, 167, 168, 172, 176, 180, 182, 186, 205, 206, 208, 213, 214, 215, 216, 217, 218, 222, 229, 238, 251, 253, 254, 261, 270, 271, 280, 281, 282, 284, 291, 303, 304, 315, 316, 318, 321, 325, 326, 327, 328, 330, 331, 332, 335, 338, 339, 340, 342, 352, 367, 389, 394, 396
Sponsler, 113
Stacey, 172, 312, 326, 327, 333, 334, 335, 336
Stahl, 299
Stakheyewa, 4
Stam, 108
Stamm, 118
Standing, 165, 172, 240
Statham, 270, 281, 392
Stather, 255, 261
Stearn, 213
Stein, 346
Steinhardt, 166, 170, 181, 182, 183, 184, 185, 187, 188, 189, 192, 197, 199, 205, 208, 209, 212, 213, 222, 223, 224, 226, 227, 229, 231, 232, 233, 240, 241, 291, 308, 328, 361
Stell, 304
Stern, 225
Stevenson, 283
Stirm, 300
Stock, 47, 49
Stott, 25, 27, 38, 41, 42, 44, 116,1 80, 182, 186, 208, 213, 214, 215, 216, 238

Stoves, 250, 252, 254, 265, 329, 332, 333, 360
Street, 56, 57, 376
Stringfellow, 171
Subba, 95
Sullivan, 16, 17, 18
Sutherland, 385, 386
Swan, 250, 255, 257, 259, 317, 342
Synge, 291

TABOR, 33
Talibuddin, 22
Tanner, 385, 386
Tarbell, 255
Taylor, 239
Teller, 18, 19, 98, 111, 146
Teuber, 304
Thomas, 33
Thomson, 37
Thompson, 269
Thovert, 138
Tiler, 244
Tobolsky, 59
Toennies, 266, 270
Tomlinson, 36
Traumann, 317
Treloar, 107
Trotman, 299
Trotter, 384, 387
Trouton, 128
Turl, 6, 7, 8, 10, 36, 173
Turnbull, 123

UBBELOHDE, 105
Ultee, 9
Urquhart, 96, 118, 165, 240

VALKO, 165
Van Name, 152
Van Slyke, 314, 316
Van der Waal, 231
Vermas, 87, 137, 168
Verwey, 193
Du Vigneaud, 291
Vickerstaff, 172, 173, 185, 191, 224, 228, 230, 235, 236, 240, 319
Vickery, 296
Von Muralt, 198
Vreedenberg, 163

WAGNER, 163, 257, 270
Waitkoff, 350
Wakeham, 113
Waldschmitz, 296
Walden, 301
Walker, 118
Warburton, 17, 109
Ward, 301
Warner, 292
Warren, 268

Warwicker, 165, 172, 240
Waterhouse, 299
Watkins, 173
Watson, 135
Weber, 198
Weidinger, 111
Weimarn, 361
Weston, 268, 269, 270, 392
Whewell, 8, 10, 28, 30, 75, 79, 80, 217, 254, 282, 315, 388, 389
White, 108
Whitney, 152
Wiener, 138
Wilderman, 152
Willavoys, 103, 104
Willis, 165, 172, 240
Williams, 118, 129, 281
Williamson, 316
Wilson, 134, 135, 165, 170, 172, 204, 208, 216
Windle, 122
Winnek, 187
Wiseman, 304, 373, 374
Witt, 33, 38

Wolcznk, 152
Wood, 257, 385, 386
Woodin, 269, 351
Woods, 4, 8, 10, 30, 57, 59, 66, 67, 68, 72, 73, 74, 75, 79, 80, 90, 106, 244, 281, 295, 376, 377, 379, 382, 388, 389, 391, 392, 396
Wormell, 362
Wyk, 61, 67
Wyman, 186, 187

YAGER, 123

ZAHN, 3, 6, 7, 9, 11, 12, 13, 63, 66, 74, 75, 76, 77, 82, 84, 246, 251, 253, 257, 259, 264, 285, 289, 290, 292, 293, 295, 296, 297, 298, 299, 300, 301, 302, 315, 317, 319, 326, 333, 337, 338, 342, 358, 359, 360, 383, 384, 391, 392
Zaiser, 212, 213, 240, 241
Zuber, 317
Zwicker, 102

Date Due

MAY 19 1976	APR 18 1976	
NOV 02 1992		
	NOV 17 1998	
NOV 26 1992		
MAR 09 1998		
FEB 27 1998		
JAN 15 2000		

CAT. NO. 23 233   PRINTED IN U.S.A.

TS1547 .A64 1963

Alexander, Peter

Wool.

53860

DATE	ISSUED TO

53860

TS   Alexander, Peter
1547   Wool. 2d ed.
A64
1963

Trent
University

CPSIA information can be obtained
at www.ICGtesting.com
Printed in the USA
LVHW081749030922
727556LV00010B/959